DIAMINO AMINO ACIDS

Diamino Amino Acids

Hanna A. Schäfer and Lilli M. Wohlbier
Editors

Nova Biomedical Books
New York

For permission to use material from this book please contact us:
Telephone 631-231-7269; Fax 631-231-8175
Web Site: http://www.novapublishers.com

LIBRARY OF CONGRESS CATALOGING-IN-PUBLICATION DATA

Diamino amino acids / [edited by] Hanna A. Schäfer and Lilli M. Wohlbier.
p. ; cm.
Includes bibiliographical references and index.
ISBN 978-1-60456-892-9 (hardcover)
1. Diamino amino acids. I. Schäfer, Hanna A. II. Wohlbier, Lilli M.
[DNLM: 1. Amino Acids, Diamino. QU 60 D5367 2008]
QP562.D53D53 2008
572'.65--dc22

2008026975

Published by Nova Science Publishers, Inc. ✢ New York

CONTENTS

Preface		**vii**
Research and Review Articles		
Chapter I	Asymmetric Dimethyl L-Arginine (ADMA) and Delayed Cerebral Vasospasm after Aneurysmal Subarachnoid Hemorrhage (aSAH) *Carla S. Jung and Ryszard M. Pluta*	**1**
Chapter II	Neurotropic and Immunotropic Effects of Amino Acid L-Lysine *Liudmila Severyanova and Maxim Dolgintsev*	**31**
Chapter III	D-Serine and Schizophrenia *Kenji Hashimoto*	**103**
Chapter IV	Analysis of Methylated Arginines by High-Performance Liquid Chromatography-Fluorescence Detection: A Review *Makoto Tsunoda*	**123**
Chapter V	Yield of Compatible Solute Hydroxyectoine through Conversion from Ectoine by Recombinant Escherichia Coli *Shinichi Nagata, Daochen Zhu, Shunyan Cui and Hideaki Sasaki*	**139**
Chapter VI	Significant Yield of Ectoine by Halomonas sp. EG6 During Osmotic Downshock Treatment *Omneya Osman, Yaoqiang Wang, Daochen Zhu, Shoko Tanabe-Hosoi and Shinichi Nagata*	**155**

Chapter VII Increase in Amino Acids and Ammonia During Liquefaction of Semen **171**
Thomas W. Stief

Chapter VIII Selected Diamino Acids in Soils of Differently Managed Mountain Meadow and Forest Ecosystems: Assessment of Their Role in Ecosystem Nutrition **183**
Pavel Formanek, Klement Rejsek, Valerie Vranova and Borivoj Klejdus

Short Communication

Communication A Arginine Conserves the Hemostasis Activation State of Plasma Even Against Freezing/Thawing **219**
Thomas W. Stief

Index **235**

PREFACE

Diamino Amino Acids are one of the 12 groups of amino acids.Amino Acids are the chemical units or "building blocks" of the body that make up proteins. Protein substances make up the muscles, tendons, organs, glands, nails, and hair. Growth, repair and maintenance of all cells are dependent upon them. Next to water, protein makes up the greatest portion of our body weight. Amino Acids that must be obtained from the diet are called "Essential Amino Acids" other Amino Acids that the body can manufacture from other sources are called "NonEssential Amino Acids."

Chapter I - L-arginine is a semi-essential alpha diamino acid in which the second amino group is part of the guanidinium group at a distal carbon at the delta position. Because of the presence of this second amino group, L-arginine is a source of nitric oxide (NO) that is cleaved from the terminal guanidino nitrogen atom by nitric oxide synthase (NOS). NO, described by Robert F. Furchgott as an endothelium derived relaxing factor (EDRF) evokes, because of its free radical properties and affinity to heme, ferrous iron and cysteine, a wide spectrum of physiological and pathophysiological effects. For many years, different exogenous NOS inhibitors were used to elucidate the role of NOS and NO in health and disease.

In 1992, Patrick Vallance and colleagues described elevated levels of the endogenous NOS inhibitor, asymmetric dimethylarginine (ADMA) and associated its presence with kidney failure and essential hypertension. Recently, ADMA was also linked to cardiovascular diseases, erectile dysfunction, preeclampsia, hypercholesterolemia, and diabetes mellitus.

Di-methylarginines are produced by post-translational double methylation of L-arginine which is catalyzed by a family of protein N-methyltransferases (PRMT), using S-adenosylmethionine as a methyl group donor. Symmetric (SDMA) and asymmetric (ADMA) forms of di-methylarginine are derived during this process. SDMA has no known physiological effect, whereas ADMA acts as an endogenous inhibitor of NOS. ADMA is excreted in urine or eliminated by dimethylarginine dimethylaminohydrolase (DDAH I or II). One of the products of ADMA hydrolysis is L-citrulline, also a by-product and a marker of NO production by NOS, which in turn inhibits DDAH. ADMA hydrolysis increases NOS activity and NO production. ADMA-related endogenous inhibition of NO production by NOS can be modified by: 1) changing the availability of L-arginine; 2) changing the availability of L-citrulline; 3) stimulating or inhibiting DDAH activity; 4) modifying methylation via regulating availability of adenosylmethionine; or 5) modifying PRMT activity.

Since the pathophysiological and pharmacological effects of NO include cardiovascular, neurological, immunological, infectious, and tumorigenic processes, there is rapidly growing interest in the endogenous and exogenous modification of NO-controlled processes. This review summarizes research elucidating the role of NO, NOS and ADMA in regulating cerebral blood flow in health and in brain pathologies, focusing on development and prevention of delayed cerebral vasospasm after subarachnoid hemorrhage (SAH) evoked by a ruptured intracranial aneurysm.

Chapter II - *The aim* of the study is to investigate L-lysine effects on the integrative brain activity and on the immunity as well as the interaction between the nervous and immune regulatory systems in the painful-emotional stress condition.

Methods: To assay the neurotropic amino acid effects on the following behavioral reactions of different biological significance were studied.

- Footshock aggression, tail-flick and hot plate tests (in Wistar rats) were used to compare peculiarities of L-lysine action on pain sensitivity and pain-induced behavior due to the stimulation of different nociceptor kinds.
- The learning in T-maze on the base of food reinforcement and exploratory activity of rats were also investigated.

To estimate the possible L-lysine influence on immunity the humoral immune response and delayed hypersensitivity phenomenon were induced by means of the corresponding ram erythrocytes injection to Wistar rats and CBA mice. Such congenital defensive mechanisms as bactericidal and phagocytic neutrophil activities were studied.

The intraperitoneal administration of L-lysine in the doses arranged from 0.15 up to 5000 mcg/kg was performed 15 min before experimental session. All control animals were injected with saline.

Results: L-lysine effects seemed to be algesic or analgesic in depending on the utilized pain model and administered dose. It should be stressed that the predominant role in the determination of pain-induced behavior features is played not by the painful stimulus intensity but by the biological significance of pain stressful situation and consequently by the supraspinal integrating mechanisms.

L-lysine administration promoted the formation and consolidation of the conditioned nutritional behavior in T-maze, but decreased the exploratory activity of rats.

It was shown that systemic L-lysine administration during the productive phase increases the humoral immune response value in rats but decreases the development of the delayed hypersensitivity reaction in mice.

In pain stress, provoked by electroshocks lasting 2 min per every day during productive phase immunosupression was revealed. Combination of pain stress and L-lysine administration during this period of immune reaction development is followed by the decrease in immune response.

The congenital defensive phenomena seemed to be less influenced by L-lysine. Thus, oxygen-dependent bactericidal neutrophil activity was enhanced only at the higher amino acid doses. However the phagocytic neutrophil activity was inhibited in the dose-dependent manner.

Conclusion: On the whole the obtained findings is the original knowledge about the amino acid incorporated into the active proteins the content of which is high in the upper brain structures particularly in the cerebral cortex.

According to these result the treatment strategies of L-lysine administration may be proved including the composition of L-lysine-containing diets and their using to modulate behavioral reactivity, to decrease emotional overstrain as well to correct immunodeficiency conditions.

Chapter III - Several lines of evidence suggest that a dysfunction in glutamatergic neurotransmission via N-methyl-D-aspartate (NMDA) receptors is involved in the pathophysiology of schizophrenia, and that D-serine, an endogenous agonist at the glycine sites of NMDA receptors, plays a role in neurotransmission via these receptors. Previously, the authors reported that serum levels of D-serine were significantly decreased in patients with schizophrenia, and that the ratio of D-serine to total serine in the cerebrospinal fluid (CSF) of first-episode and drug-naive patients with schizophrenia was also decreased. D-Serine is synthesized from L-serine by the enzyme serine racemase and is degraded by D-amino acid oxidase (DAO). Extracellular D-serine is removed by alanine-serine-cysteine transporter-1 (Asc-1) in neurons. These findings suggest that synthetic and/or metabolic pathways for D-serine may be altered in the brain of schizophrenia patients. Furthermore, D-serine has been shown to be effective for treating several symptoms of this disorder. Accumulating evidence implicates DAO and its interacting gene G72 in the pathogenesis of schizophrenia. In this chapter, the author reviews the role of D-serine, an endogenous co-agonist of glycine modulatory sites on NMDA receptors, in the pathophysiology of schizophrenia. The author also discusses the glycine modulatory sites on NMDA receptors as potential therapeutic targets for schizophrenia.

Chapter IV - N^G-Monomethyl-L-arginine (L-NMMA), N^G,N^G-dimethyl-L-arginine (ADMA), and N^G,$N^{G'}$-dimethyl-L-arginine (SDMA) are emerging cardiovascular risk factors that may impair endothelial function by interfering with nitric oxide synthesis. Especially, ADMA was found to be correlated with several diseases such as hypercholesterolemia, hypertension, type 2 diabetes and hyperhomocysteinemia. Hence, accurate measurement of these methylated arginines to clarify their physiological roles *in vivo* must be established. In recent years, various analytical methods have been developed for the analysis of methylated arginines in bio samples. The most widely used assays are based on solid-phase extraction of basic components in bio samples, followed by derivatization and chromatographic analysis by high-performance liquid chromatography (HPLC) with fluorescence detection. As a fluorescent derivatization reagent, *o*-phthaldialdehyde (OPA), 6-aminoquinolyl-*N*-hydroxysuccinimidyl carbamate (AQC) and 4-fluoro-7-nitro-2,1,3-benzoxadiazole (NBD-F) have been used. This chapter reviews the various HPLC-fluorescence methods described so far for the analysis of methylated arginine in bio samples such as plasma, serum or urine. It discusses the main experimental conditions used for sample pretreatment and for analysis including fluorescence derivatization and chromatographic conditions.

Chapter V - *Halomonas ventosae* DL7 synthesized ectoine and hydroxyectoine as major and minor compatible solutes, respectively, when grown in the presence of 3 M NaCl. The amount of hydroxyectoine synthesized increased with an increase of NaCl concentrations and incubation temperatures. To achieve high amount of hydroxyectoine, an attempt was made to examine the conversion of ectoine accumulated in the cells to hydroxyectoine using the recombinant *Escherichia coli* which harbors the ectoine hydroxylase gene (*thpD*) from *Streptomyces chrysomallus*. The presence of carbon source and NaCl was indispensable for the sufficient uptake of ectoine by recombinant *E. coli* BL21-thpD strain, and thus the authors attempted to examine what kind of carbon sources are required for activity expression of uptake and conversion in M63 medium. As a result, both activities on uptake as well as conversion were maximal under the co-existence of 10 mM glucose and 50 mM xylose. Other factors to affect both activities such as incubation temperatures, external pH, salinities and ectoine concentrations supplied were sequentially examined. Efficient and high yield of hydroxyectoine by recombinant *E. coli* BL21 strain was attained through the incubation of the cells in M63 medium containing 1 M NaCl, 1 mM ectoine, 1 mM isopropyl-β-D-thiogalactopyranoside, 10 mM glucose, and 50 mM xylose (pH 7.2) at 30°C, where uptake of ectoine favorably proceeded and at the same time, conversions of ectoine to hydroxyectoine in the cell cytoplasm resulted in 80 and 96 % at 48 and 72 h of incubations, respectively.

Chapter VI - Halophilic bacterium strain EG6 was isolated from Burg El Arab solar saltern lake in Alexandria, Egypt. Strain EG6 can grow in a wide range of NaCl concentrations up to 4 M. Phylogenetic position was established by 16S rDNA gene sequencing as a member of the genus *Halomonas*. This strain showed significant yield of ectoine, about 3.70 g/L, when subjected to the osmotic downshock treatment from 2 M to 0.3 M NaCl during 7 days. 0.3 M NaCl was the optimum concentration used for obtaining maximum yield of ectoine in comparison with other NaCl concentrations (0-0.7 M) that were used for osmotic downshock. In addition, same treatment of the type strain *Halomonas elongata* IFO 15536 had a lower yield of ectoine, about 2.89 g/L, than that of strain EG6. Release of ectoine from the cells was carried out within 10 min and its rates were increased by the dilution through downshock treatment. Significant differences were found in the amounts of ectoine synthesized and released by strain EG6 and *H. elongata* IFO 15536. Furthermore, growth of strain EG6 was highly increased in comparison with *H. elongata* IFO 15536 during downshock treatment from 2 M to 0.3 M NaCl.

Chapter VII - Fresh normal ejaculate clots by action of about 0.05 IU/ml thrombin on a fibrinogen-like compound, and subsequently lyses by action of about 0.1 IU/ml plasmin and about 300 % (of blood plasma norm) carboxypeptidase. Carboxypeptidases sequentially liberate amino acids, that were quantified in the present work: the most abundant amino acids in fresh seminal plasma are serine (6.09 mM), threonine (4.61 mM), lysine (3.45 mM), tyrosine (3.18 mM), glycine (2.79 mM), and arginine (2.29 mM). The concentration of the amino acids approximately doubles within 45 min (37°C), with an exceptional 12fold increase of glutamic acid. When compared to blood plasma, fresh seminal plasma contains 226fold gamma-glutamyl-transferase (γ-GT), 40fold ammonia, 21fold creatinine, and 18fold glutamate dehydrogenase (GLDH). Ammonia (NH_3), generated by glutamate dehydrogenase (GLDH), together with arginine or creatinine might protect sperm cells from female defence cells. Ammonia and natural guanidine derivatives might be physiologic immunsuppressors.

Chapter VIII - This chapter sets out to determine the effects of a) long-term abandonment of mown mountain meadows on Oxyaquic Hapludalf soil, and b) differing methods of thinning (2,044 versus 1,652 trees/ha) Norway spruce ecosystems on Typic Haplohumod and Entic Haplorthod soils. This was done by assessing the availability of particular forms of bioavailable nitrogen (N), including 14 proteinaceous monoamino acids (aspartic acid, glutamic acid, serine, histidine, glycine, threonine, alanine, tyrosine, valine, methionine, phenylalanine, isoleucine, leucine and proline) and three diamino acids (arginine, cystine and lysine). The research took place throughout the 2004 vegetation season and during the first half of the 2005 season. All of the study sites represent N-limited ecosystems with ammonium N dominant in plant nutrition.

Abandonment of meadows decelerated N-cycling and increased immobilization of N during the vegetation season. A significant ($P<0.05$) influence of meadow abandonment on concentrations of nitrates and glutamic acid in the Ap horizon was only found when data from 2004 and 2005 were pooled together. Of the amino acids, glutamic acid, arginine, aspartic acid and cystine were the most abundant. The three diamino acids studied composed <23% of total bioavailable N in the Ap horizon of both types of meadow (moderately mown *versus* abandoned). The amount of bioavailable N released in the top 7 cm of the Ap horizon in 2004 and 2005 was too low to supply the needs of the plant community. During the first half of the 2005 vegetation season, arginine and lysine were important as regards nutrition in the plant community, particularly in the moderately mown meadow. During 2004, amino acids were important for plant nutrition in the abandoned meadow at the end of summer. It appears that most N needed by plants is provided from the organic layer. The authors results indicate that the plant communities of moderately mown and abandoned meadows can be dependent on mineral only, or mineral plus mono- and diamino acid-N nutrition, in different vegetation seasons. Thinning of mountainous stands of Norway spruce by 392 trees/ha did not significantly affect concentrations of certain forms of bioavailable N, including diamino acids. Glutamic acid, alanine, aspartic acid and cystine were most abundant in the Oe horizon of these ecosystems.

Short Communication - Clinical hemostasis activation can be very dangerous, e.g. in pathologic disseminated intravascular coagulation (PDIC). Long transport times of this unstable plasma into the hemostasis laboratory or freezing/thawing of the samples induces artefactual hemostasis changes. 308 unselected EDTA plasmas of patients were analyzed for plasmatic amidolytic thrombin activity. EDTA-plasmas of 94 patients were stabilized with 1.25 M plasma concentration arginine and analyzed for plasmatic amidolytic thrombin activity directly or after freezing at -24°C/thawing at room temperature (RT). EDTA-plasmas were supplemented with 0-1100 mM arginine, stored for 24h at RT, and analyzed for plasmatic thrombin activity.

The normal range of amidolytic thrombin activity in arginine-stabilized EDTA plasma is 100 ± 20 % of normal, the overall patient range was here 131 ± 40 % of normal (mean value ± 1 standard deviation); the thrombin activity in patient plasmas distributed normally according to Gauss. Analysis of arginine-stabilized EDTA-plasma prior to or after freezing/thawing correlated with $r=0.980$, however thrombin activity in unstabilized frozen EDTA-plasma correlated only with $r=0.835$ to that of arginine stabilized EDTA-plasma. Addition of > 50 mM arginine to EDTA plasma increased the stability of thrombin activity after 24h storage at RT in individual plasmas, with an increase of correlation from $r=0.824$ to $r > 0.9$.

Arginine stabilization of EDTA-plasma enables the analysis of the true hemostasis activation state of an individual patient even in frozen plasma. This might be of great importance in both diagnosis and therapy monitoring of hemostasis disorders.

In: Diamino Amino Acids
Editors: H. A. Schäfer, L. M. Wohlbier, pp. 1-29
ISBN: 978-1-60456-892-9

Chapter I

Asymmetric Dimethyl L-Arginine (ADMA) and Delayed Cerebral Vasospasm after Aneurysmal Subarachnoid Hemorrhage (aSAH)

***Carla S. Jung*[1] *and Ryszard M. Pluta*[2,*]**
[1]Department of Neurosurgery, Ruprecht Karls-University Heidelberg, Germany;
[2]Surgical Neurology Branch, National Institute of Neurological Disorders and Stroke, National Institutes of Health, Bethesda, MD, USA.

Abstract

L-arginine is a semi-essential alpha diamino acid in which the second amino group is part of the guanidinium group at a distal carbon at the delta position. Because of the presence of this second amino group, L-arginine is a source of nitric oxide (NO) that is cleaved from the terminal guanidino nitrogen atom by nitric oxide synthase (NOS). NO, described by Robert F. Furchgott as an endothelium derived relaxing factor (EDRF) evokes, because of its free radical properties and affinity to heme, ferrous iron and cysteine, a wide spectrum of physiological and pathophysiological effects. For many years, different exogenous NOS inhibitors were used to elucidate the role of NOS and NO in health and disease.

In 1992, Patrick Vallance and colleagues described elevated levels of the endogenous NOS inhibitor, asymmetric dimethylarginine (ADMA) and associated its presence with kidney failure and essential hypertension. Recently, ADMA was also linked to cardiovascular diseases, erectile dysfunction, preeclampsia, hypercholesterolemia, and diabetes mellitus.

* Correspondence concerning this article should be addressed to: Ryszard M. Pluta, M.D., Ph.D., National Institutes of Health, 10 Center Drive, Room 5D39, Bethesda, MD, 20892-1414, tel: 001-301-594-8117; fax: 001-301-402-0380.

Di-methylarginines are produced by post-translational double methylation of L-arginine which is catalyzed by a family of protein N-methyltransferases (PRMT), using S-adenosylmethionine as a methyl group donor. Symmetric (SDMA) and asymmetric (ADMA) forms of di-methylarginine are derived during this process. SDMA has no known physiological effect, whereas ADMA acts as an endogenous inhibitor of NOS. ADMA is excreted in urine or eliminated by dimethylarginine dimethylaminohydrolase (DDAH I or II). One of the products of ADMA hydrolysis is L-citrulline, also a by-product and a marker of NO production by NOS, which in turn inhibits DDAH. ADMA hydrolysis increases NOS activity and NO production. ADMA-related endogenous inhibition of NO production by NOS can be modified by: 1) changing the availability of L-arginine; 2) changing the availability of L-citrulline; 3) stimulating or inhibiting DDAH activity; 4) modifying methylation via regulating availability of adenosylmethionine; or 5) modifying PRMT activity.

Since the pathophysiological and pharmacological effects of NO include cardiovascular, neurological, immunological, infectious, and tumorigenic processes, there is rapidly growing interest in the endogenous and exogenous modification of NO-controlled processes. This review summarizes research elucidating the role of NO, NOS and ADMA in regulating cerebral blood flow in health and in brain pathologies, focusing on development and prevention of delayed cerebral vasospasm after subarachnoid hemorrhage (SAH) evoked by a ruptured intracranial aneurysm.

Intracranial Aneurysms and Subarachnoid Hemorrhage (SAH)

Intracranial aneurysms are blood filled, often thin-walled bulges of cerebral vessels usually located at the bifurcations of major cerebral arteries of the circle of Willis. When the aneurysm ruptures, arterial blood fills out the cisterns of the subarachnoid space leading to a so-called SAH.

Almost a century ago, the first cases of aneurysmal subarachnoid hemorrhage (aSAH) were diagnosed (Cushing 1923; Symonds 1923). Since then, significant progress has been made in understanding the pathophysiology and treatment of aSAH. Substantially improved outcome and decreased fatality rates have been described within the past 50 years (Hop, Rinkel et al. 1997). However, despite the significant improvements, mortality following aneurysmal rupture remains unacceptably high at nearly 50%, with an additional one-third of patients requiring long–term care (Hop, Rinkel et al. 1997; Stapf and Mohr 2004) and less than one-third of all aSAH patients returning to functional life (Kassell and Drake 1982).

The incidence of intracranial aneurysms in the general population is difficult to estimate; its been reported to be as low as 2/100,000 in China and as high as 22. 5/100,000 in Finland but it is widely accepted as 10/100,000 population (Sudlow and Warlow 1997; Loch Macdonald and Weir 2001). The risk of developing aneurysms may differ among racial groups, genders, and the frequency of aneurysm detection increases with age. However, depending on the results of different studies, about 2 - 5% of the population in North America (10 to 15 million people) is believed to harbor cerebral aneurysms (McCormick and Acosta-Rua 1970; Rinkel, Djibuti et al. 1998; Burnett and Danish 2004). Approximately 30,000 Americans suffer aSAH yearly, with a peak incidence in the middle-aged population (Kassell and Drake 1982; Sudlow and Warlow 1997). Although, aSAH is relatively rare, compared to ischemic strokes (Sudlow and Warlow 1997), the relative youth of the patients (Kassell and Drake 1982) means that this event is responsible for one quarter of all years of life lost as a result of aneurysm rupture, complications of aSAH, and aneurysm treatment (Johnston, Selvin et al. 1998). Furthermore, even those patients who survive the initial subarachnoid bleed and have been successfully treated are not beyond danger since cerebral vasospasm leading to delayed cerebral neurological deficits (DIND) and cerebral ischemia remains the major cause for morbidity and mortality. However, despite many years of disappointing research, delayed cerebral vasospasm remains the most promising treatable factor affecting the outcome after aSAH (Bederson, Levy et al. 1998; Seiler and Binggeli 2001; van Gijn and Rinkel 2001).

Delayed Cerebral Vasospasm: A Historical Perspective

Delayed cerebral vasospasm is a reversible narrowing of the cerebral intradural arteries, following aSAH. In 1859, the symptoms and effects of cerebral vasospasm were first described by the British physician William M. Gull, who reported a case of DIND in the form of delayed dominant hemisphere stroke that led to the patient's death after SAH from a ruptured left middle cerebral artery aneurysm (Gull 1859). After the discovery of cerebral angiography by Egaz Moniz in 1927, almost thirty years passed before Reid, Johnson and Ollenshow demonstrated the presence of vasospasm during the 6th International Congress of Radiology in London in 1950, and Ecker and Riemenschneider first published angiographies confirming the presence of cerebral vasospasm in patients after SAH (Ecker and Riemenschneider 1951). Nevertheless, the existence of cerebral vasospasm was not widely accepted until the 1970´s when a relationship between vasospasm and DIND as well as the time course of development and resolution of cerebral vasospasm (Weir, Grace et al. 1978) were clearly established by Bryce Weir in his seminal paper. Usually, cerebral vasospasm develops after the third day post SAH, with a maximum incidence 7 to 9 days after the initial bleeding and resolves over the next 10 to 14 days (Weir, Grace et al. 1978; Black 1986; Adams, Kassell et al. 1987). Two forms of cerebral vasospasm have to be distinguished. Cerebral vasospasm detected radiologically using a "gold standard" technique, digital substraction arteriography, which is called angiographic vasospasm and neurological deficits attributed to angiographically proven cerebral vasospasm which are called "symptomatic vasospasm", "delayed cerebral ischemia" or "DIND". While arteriographic vasospasm is detectable in 30-70% of all patients during the first 7 days afterSAH, only 20-50% develop symptomatic vasospasm (Kassell, Sasaki et al. 1985; Sano 1994; Mayberg 1998; Weir, Macdonald et al. 1999). Thus, about 15% (Mayberg 1998) to 34% (Kassell, Sasaki et al. 1985; Dorsch 1995; Haley, Kassell et al. 1995; Mayberg 1998) of patients who survive an initial aSAH suffer DIND. DIND remains the single cause of 10% mortality and 11% severe morbidity after SAH (Dorsch 1995; Mayberg 1998). Symptoms of DIND may range from an increased headache to sudden paralysis, loss of consciousness or death (Newell, Eskridge et al. 1989; Mayberg 1998; Treggiari, Walder et al. 2003) Factors such as age, gender, cigarette smoking, genetic factors have been suggested as being predictive for developing vasospasm. All of them are controversial (Suzuki, Yasui et al. 1988; Inagawa 1992; Yoshimoto and Kwak 1995; Lanzino, Kassell et al. 1996), except the amount of blood or thickness of clot enveloping conductive arteries in the subarachnoid cisterns (Fisher, Kistler et al. 1980).

Since its first description more than 50 years ago, post-hemorrhagic delayed cerebral vasospasm, leading to DIND, remains a serious complication for patients who suffer aSAH (Kassell, Torner et al. 1990; Schievink 1997); its pathomechanism remains unclear and treatment is elusive.

PATHOPHYSIOLOGY OF VASOSPASM: IN A PURSUIT OF A HOLY GRAIL

Delayed cerebral vasospasm is a consequence of aSAH and is caused by a blood clot in the subarachnoid space in the vicinity of the large cerebral arteries (Loch Macdonald and Weir 2001; Loch Macdonald 2006). In various animal studies, vasospasm identical to that occurring in humans, could be reproduced by placing autologous blood next to the cerebral arteries or by injecting blood into the cerebrospinal fluid (CSF) (Liszczak, Varsos et al. 1983; Espinosa, Weir et al. 1984; Loch Macdonald and Weir 2001). The most powerful predictors of vasospasm in patients after SAH are the volume, density and prolonged presence of subarachnoid blood in the basal cisterns (Fisher, Kistler et al. 1980; Suzuki, Komatsu et al. 1980; Pasqualin, Rosta et al. 1984). Thus, it has been widely accepted that vasoactive / vasospastic substances are released during hemolysis from red blood cells (Allen, Henderson et al. 1974; Boullin, Mohan et al. 1976; Brandt, Ljunggren et al. 1981) (Barrows, Hunter et al. 1955; White, Macleod et al. 1987; Macdonald, Weir et al. 1991; MacAllister, Fickling et al. 1994), but not from platelets, cell membranes, or white blood cells which are also constituents of subarachnoid clots (Macdonald, Weir et al. 1991; Loch Macdonald and Weir 2001). Major players in this process are oxyhemoglobin (oxyHb) and deoxyhemoglobin (deoxyHb) (Ohta and Kajikawa 1980; White, Macleod et al. 1987; Copeland, Dyer et al. 1989; Pluta, Afshar et al. 1998).

There are several mechanisms by which oxy- and deoxy-hemoglobin can evoke vasoconstriction (Macdonald, Weir et al. 1991; Pluta, Afshar et al. 1998). The Nobel prize winner L. Ignarro (Ignarro 1990) proposed that a heme moiety of hemoglobin scavenges the potent vasodilator nitric oxide (NO) during SAH. This idea was based on the extremely high affinity of NO for the heme moiety (Keilin and Hartree 1937; Gow and Stamler 1998; Gladwin, Lancaster et al. 2003). Another widely accepted hypothesis was that oxyHb within the arterial wall (Foley, Kassell et al. 1993; Takenaka, Kassell et al. 1993) can stimulate endothelial cells to produce endothelin-1 (ET-1) (Machi and Stewart 1990; Cocks, Malta et al. 1991; Takenaka, Kassell et al. 1993), the most potent natural vascular vasoconstrictor (Yanagisawa, Kurihara et al. 1988). ET-1 levels increase after SAH in plasma and CSF and are thought to contribute significantly, despite some controversy (Ziv, Fleminger et al. 1992; Barone, Globus et al. 1994; Pluta, Boock et al. 1997; Pluta, Oldfield et al. 1997), to development of delayed cerebral vasospasm (Fujimori, Yanagisawa et al. 1990; Kobayashi, Hayashi et al. 1990; Sasaki and Kassell 1990; Yamaura, Tani et al. 1992; Gaetani, Rodriguez y Baena et al. 1994; Seifert, Loffler et al. 1995). Furthermore, oxidation of bilirubin, the product of hemoglobin degradation, produces bilirubin oxidation products (BOXes), which recently was shown to constrict cerebral arteries (Clark, Reilly et al. 2002). The mechanism of BOXes action is unclear but it is most likely via inhibition of hydrolysis of an asymmetric dimethylarginine (ADMA), an endogenous inhibitor of eNOS, leading to increased levels of asymmetric dimethylated L-arginine (Pluta, Jung et al. 2005), an endogeneous inhibitor of NOS, in CSF (Jung, Iuliano et al. 2004; Jung, Oldfield et al. 2007; Martens-Lobenhoffer, Sulyok et al. 2007). Therefore, decreased NO availability with increased ET-1 production may be responsible for delayed cerebral vasospasm (Hirose, Ide et al. 1995; Pluta, Boock et al. 1997; Pluta, Oldfield et al. 1997).

NO and NOS Pathways in Vasospasm

Endothelial derived relaxing factor (EDRF) (Furchgott and Zawadzki 1980), identified as NO (Ignarro, Byrns et al. 1987) is one of the few known gaseous messengers with a brief half-life of a few seconds (Stryer 1995). Because of its free radical properties and affinity for heme, ferrous iron and cysteine, NO evokes a wide spectrum of physiological and pathophysiological effects (Pluta 2005). Particularly, NO, via stimulation of soluable guanylyl cyclase, is a powerful vasodilator and plays a major role in regulating cerebrovascular tone and brain perfusion (Iadecola, Beitz et al. 1993; Thompson, Pluta et al. 1996; Toda, Tanaka et al. 2000; Pluta, Thompson et al. 2001).

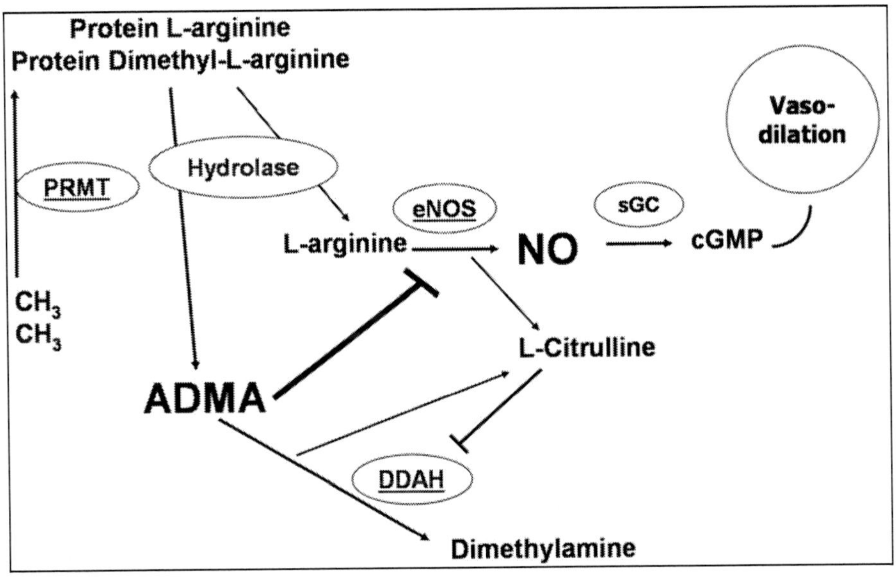

Figure 1. The graph depicts the in vivo synthesis and degradation mechanisms of asymmetric dimethylarginine (ADMA) and its influence on synthesis of nitric oxide (NO) and its effect on cerebrovascular tone via stimulation of soluable guanylyl cyclase leading to increased production of cyclic GMP (detailed description in the text). CH3: a methyl group; PRMT: protein-arginine-methyltransferase; DDAH: dimethylarginine dimethylamine hydrolase; eNOS: endothelial nitric oxide synthesis.

NO is synthesized by a family of NOS (Figure 1) and is derived from the terminal guanidino nitrogen atoms of L-arginine yielding L-citrulline as a by-product (Hibbs, Vavrin et al. 1987; Palmer, Ferrige et al. 1987; Hibbs, Taintor et al. 1988; Gold, Wood et al. 1989). After NO is synthesized and released from L-arginine by constitutive NOS in nanomolar concentrations (Davis, Martin et al. 2001), it can diffuse across the endothelial cell membrane and activate soluble guanylyl cyclase, which is a heterodimeric hemoprotein, converting 5-triphosphate to 3,5-cGMP. cGMP in turn activates a family of cGMP dependent protein kinases (Ignarro, Harbison et al. 1986; Ignarro 1990; Faraci and Heistad 1998) leading to smooth muscle relaxation and vasodilation (Figure 1) (Sobey and Faraci 1998; Vallance and Chan 2001). Recently, sodium nitrite in blood and saliva was reported as another possible source of NO under specific hypoxic and acidic conditions (Cosby, Partovi et al. 2003; Pluta 2005).

NO production is regulated by negative feedback between NO and NOS (Ignarro 2002; Mancardi, Ridnour et al. 2004) as well as by di-methylarginines (e. g. asymmetric dimethyl L-arginine, ADMA) which act as endogenous NOS-inhibitors (Fickling, Williams et al. 1993; MacAllister, Parry et al. 1996). There is growing evidence that cerebral vasospasm is associated with locally increased (in the CSF) levels of di-methylarginines (Jung, Iuliano et al. 2004; Jung, Oldfield et al. 2007; Martens-Lobenhoffer, Sulyok et al. 2007), evoking endothelial dysfunction due to decreased availability of NO resulting in constriction of cerebral vessels (Kim, Schini et al. 1992; Suzuki, Kajita et al. 1994; Pluta, Thompson et al. 2001; Sobey 2001; Iuliano, Pluta et al. 2004). Another source of NO is neuronal NOS (nNOS) in the cerebral vessel wall. nNOS immunoreactivity has also been reported to be affected and decreased after aSAH (Pluta, Thompson et al. 1996). Such a "double-hit" effect of SAH on production of NO leads to decreased cGMP levels in the arteries in spasm (Kim, Schini et al. 1992; Sobey and Faraci 1998). However, the observations on eNOS expression after SAH were controversial. Park et al. observed reduction of eNOS protein in rats 20 minutes after SAH (Park, Metais et al. 2001). A reduction of eNOS mRNA without a decrease in eNOS protein was reported at the time of highest vasospasm incidence 7 days after SAH by Hino et al. (Hino, Tokuyama et al. 1996) On the other hand, unaffected and unchanged protein expression was reported in spastic canine basilar arteries 7 days after SAH (Kasuya, Weir et al. 1995) and in mice 3 days after SAH (McGirt, Lynch et al. 2002). Furthermore, eNOS immunoreactivity and mRNA remain unchanged or increased in a primate model of delayed cerebral vasospasm (Sobey and Faraci 1998; Pluta, Thompson et al. 2001). In vivo, in a primate model after SAH, cerebral vessels in spasm lost their ability to dilate in response to intracarotid acetylcholine application (Iuliano, Pluta et al. 2004), similar to the endothelium-denuded arteries *in vitro* (Furchgott and Zawadzki 1980), suggesting dysfunction of eNOS in the intima of the arteries in spasm after SAH. Moreover, recently eNOS gene polymorphism has been hypothesized to be responsible for the post-SAH eNOS dysfunction (Khurana, Sohni et al. 2004). Additionally, there were decreased NO-metabolite levels as measured in the CSF of primates after SAH; these levels were negatively correlated with ADMA and the development of cerebral vasospasm after SAH (Jung, Iuliano et al. 2004; Jung, Oldfield et al. 2007; Martens-Lobenhoffer, Sulyok et al. 2007). A further, albeit indirect evidence that endothelial and neuronal NOS dysfunction is responsible for development of vasospasm, was that primate vessels in spasm after SAH maintained the ability to dilate when exposed to intracarotid infusion of exogenous NO or NO donors (Afshar, Pluta et al. 1995; Pluta, Boock et al. 1997). Together these findings support the hypothesis of NOS dysfunction developing after SAH and suggest that ADMA and suppressed NOS activity by ADMA may contribute to the development of delayed cerebral vasospasm after SAH.

ADMA Synthesis and Degradation

Methylation of the amino acid L-arginine was discovered in the 1970s, as a mechanism of post-translational modification of proteins (Kakimoto and Akazawa 1970; McDermott 1976). It has been shown that only L-arginine residues within proteins but no free amino acid can be methylated by a family of protein-arginine-methyltransferases (PRMT) (Kakimoto and Akazawa 1970; McDermott 1976; Ueno, Sano et al. 1992; Tang, Gary et al. 1998; Leiper and Vallance 1999). Until recently, very little was known about the physiological role of methylated arginine residues within proteins, but studies indicated that synthesis of asymmetric methylarginine residues is a highly regulated process which is thought to be comparable with protein phosphorylation in the regulation of protein function (Clarke 1993). Proteolysis of proteins containing methylarginine residues leads to release of free methylarginines into the cytoplasm (Kakimoto and Akazawa 1970).

In 1992, Patrick Vallance and colleagues described elevated levels of the endogenous NOS inhibitor, ADMA, and associated its presence with kidney failure and essential hypertension (Vallance, Leone et al. 1992; Vallance, Leone et al. 1992). Recently, ADMA has also been linked to endothelial dysfunction (Cooke 2000; Xiong, Yuan et al. 2001; Boger 2003), cardiovascular diseases (Boger 2003; Boger and Zoccali 2003; Mallamaci, Tripepi et al. 2005), hypertension (Azuma, Sato et al. 1995; Jin and D'Alecy 1996; Kielstein, Bode-Boger et al. 2003), erectile dysfunction (Kang, Yu et al. 2005), pre-eclampsia (Savvidou, Hingorani et al. 2003; Maas, Boger et al. 2004; Rytlewski, Olszanecki et al. 2005), hypercholesterolemia (Miyazaki, Matsuoka et al. 1999; Lu, Ding et al. 2004), atherosclerosis (Miyazaki, Matsuoka et al. 1999), diabetes mellitus (Paiva, Lehtimaki et al. 2003), and stroke (Valkonen, Laakso et al. 2003). Furthermore, delayed cerebral vasospasm was associated with elevated levels of ADMA in the CSF of patients after SAH (Jung, Oldfield et al. 2007; Martens-Lobenhoffer, Sulyok et al. 2007).

The initial step in the endogenous formation of di-methylarginines, especially ADMA, is the dimethylation of L-arginine residues by specific PRMTs (Tang, Frankel et al. 2000; Boisvert, Cote et al. 2003). Initially, a single PRMTenzyme was identified in calf thymus as being responsible for arginine methylation of all substrate proteins (Paik, Abou-Gharbia et al. 1983). Subsequently, seven genes encoding for seven different PRMT enzymes (PRMT1–7) were discovered (Lin, Gary et al. 1996; Katsanis, Yaspo et al. 1997; Tang, Gary et al. 1998; Chen, Ma et al. 1999; Branscombe, Frankel et al. 2001; Frankel, Yadav et al. 2002; Miranda, Miranda et al. 2004). Within individual cells the intracellular localization of different PRMT isoforms varies, with some (PRMT1, 2, 4 and 6) being predominantly nuclear while the remainder (3 and 5) being predominantly cytosolic (Lin, Gary et al. 1996). All PRMTs contain a conserved catalytic core region of about 310 amino acids and a variable sequence at the N-terminal. Further investigations revealed that there are two different types of PRMT enzyme activity (type I and type II) (Tang, Frankel et al. 2000; Boisvert, Cote et al. 2003). After enzymatic characterization of the seven different gene products, each PRMT enzyme could be assigned to either type I or type II activity. PRMT1–4 and 6 are classified as type I and PRMT5 and 7 as type II enzymes (Lin, Gary et al. 1996; Tang, Gary et al. 1998; Chen, Ma et al. 1999; Branscombe, Frankel et al. 2001; Frankel, Yadav et al. 2002; Miranda, Miranda et al. 2004). Differences between the two PRMT types are found in their substrate specificity and catalytic activities: Type I PRMT methylates a wide range of substrates including histones and non-histone nuclear proteins but excluding myelin basic protein, whereas Type II seems to methylate exclusively myelin basic protein (Ghosh, Paik et al. 1988). Furthermore, type II PRMTs catalyse formation of the asymmetric isomer monomethyl-L-arginine (L-NMMA) and the symmetric dimethylarginine (SDMA), while type I PRMTs are considered the predominant enzymes responsible for asymmetrical di-methylation (Tang, Frankel et al. 2000), producing L-NMMA and ADMA (Ghosh, Paik et al. 1988; Rajpurohit, Paik et al. 1992).

Both types of PRMT use S-adenosylmethionine, derived from L-methionine, as principal methyl group donor (Boger, Sydow et al. 2000), leading to methylated L-arginine and S-adenosylhomocysteine as a by-product. S-adenosylhomocysteine is further hydrolysed to L-homocysteine, which, in turn, can be re-methylated to L-methionine by methionine synthase utilizing vitamin B12 and 5-methyltetrahydrofolic acid as methyl donors. Elevation of S-adenosylmethionine stimulates PRMT activity and therefore, production of ADMA, whereas, S-adenosylhomocysteine inhibits PRMT activity (Lentz, Rodionov et al. 2003).

Methylarginines are endogenously present in the human body. The role of SDMA is unknown, but ADMA like L-NMMA, is a competitive endogenous NOS inhibitor (MacAllister, Parry et al. 1996). The concentration of ADMA in plasma is ten times greater than that of L-NMMA (Vallance, Leone et al. 1992). Due to its very low concentration, L-NMMA probably plays no or only a very limited role in regulating NO production in vivo (MacAllister, Parry et al. 1996). Methylarginines are eliminated from the body by a combination of renal excretion (Carnegie, Fellows et al. 1977; Yudkoff, Nissim et al. 1984) and metabolism. Therefore, competitive inhibition of NOS by ADMA can be reversed by L-arginine or limited by hydrolysis of ADMA by dimethylarginine dimethylamine hydrolases (Ogawa, Kimoto et al. 1989; Kimoto, Whitley et al. 1995; MacAllister, Parry et al. 1996; Leiper, Santa Maria et al. 1999; Vallance and Chan 2001). Two isoforms of ADMA hydrolases (DDAH) have been identified that can cleave ADMA into dimethylamine and L-citrulline (Kimoto, Whitley et al. 1995; MacAllister, Parry et al. 1996; Leiper and Vallance 1999): DDAH1 predominates in tissues which express nNOS (nNOS), (Ogawa, Kimoto et al. 1989; Tran, Fox et al. 2000) whereas, DDAH2, a 29644 DA protein, (Swiss Protein Database, Swiss-Prot accession no. O95865) is mainly found in tissues containing eNOS (Leiper and Vallance 1999; Leiper, Santa Maria et al. 1999). ADMA hydrolysis is regulated on one hand via negative feed-back from NO because S-nitrosylation of DDAH inactivates this enzyme (Leiper, Murray-Rust et al. 2002) and on the other hand via L-citrulline, an end-product of ADMA hydrolysis by DDAH as well as by NOS, which has also been shown to inhibit DDAH (Ito, Tsao et al. 1999; Vallance and Chan 2001; Jung, Iuliano et al. 2004).

ADMA, Cerebral Blood Flow and Cerebral Vasospasm

The central nervous system has the highest oxygen consumption in the body and the second highest blood flow of all organs. Therefore, under physiological conditions, a tight control of blood flow is in play to match the metabolic demand of the brain, which is known as blood flow-metabolism coupling of the neurovascular unit (Heistad and Kontos 1983; Khurana and Friedman 2004). Many preclinical studies showed that vascular NO synthesis plays a critical role in cerebral vascular tone (Faraci, Brian et al. 1995; Jin and D'Alecy 1996; Segarra, Medina et al. 1999) and cerebral blood flow, (Iadecola, Pelligrino et al. 1994) underlining the important role of NO as an endogenous vasodilator in regulating brain perfusion by influencing the tone of conductive and resistance arteries as well as of venous vessels (Iadecola, Zhang et al. 1993).

ADMA is an effective competitive inhibitor of NOS and is synthesized and released by endothelial cells in amounts that are sufficient to inhibit NO production by competing with the enzyme substrate L-arginine (McDermott 1976; Fickling, Williams et al. 1993; Chen, Xia et al. 1997; Cardounel, Cui et al. 2007). Serum levels of ADMA in healthy subjects range from 0. 5 to 1.0 μM (Vallance 2001). ADMA seems to be equipotent with the mono-methylated arginine L-NMMA; however, its concentration in plasma is significantly higher than that of L-NMMA (Vallance 2001). Furthermore, in healthy volunteers high doses of intravenously infused L-NMMA reduced cerebral blood flow associated with increased mean arterial pressure (Kamper, Spilt et al. 2004), while, low doses had no effect on cerebral blood flow (White, Hindley et al. 1999). In contrast, Kielstein et al. showed that already sub-pressor doses of ADMA increase vascular stiffness assessed non-invasively by pulse wave analysis and decreased cerebral perfusion measured by dynamic contrast-enhanced perfusion magnetic resonance imaging of the brain in healthy subjects (Kielstein, Donnerstag et al. 2006). Macrae and collaborators (Macrae, Dawson et al. 1993) confirmed that the effect of NOS inhibition in rats is primarily due to decreased cerebral blood flow rather than decreased metabolic activity.

In different neurological clinical settings, such as atherosclerosis (Mugge, Hanefeld et al. 2003), cerebral small-vessel disease (Khan, Hassan et al. 2007), transient ischemic attacks and stroke (Yoo and Lee 2001; Wanby, Brattstrom et al. 2003) as well as Alzheimer´s disease,(Selley 2003) increased ADMA levels have been reported and were a predictor of the progression of atherosclerosis, future vascular events and/or overall mortality (Vallance, Leone et al. 1992; Rawal, Lee et al. 1995; Das, Khan et al. 1996; Chen, Xia et al. 1997; Goonasekera, Rees et al. 1997; Herlitz, Petersson et al. 1997; Boger, Bode-Boger et al. 1998; Holden, Fickling et al. 1998; Usui, Matsuoka et al. 1998; Miyazaki, Matsuoka et al. 1999; Abbasi, Asagmi et al. 2001; Gorenflo, Zheng et al. 2001; Yoo and Lee 2001; Hermenegildo, Medina et al. 2002).

Experimentally, ADMA in concentrations of 1-300 μM causes a dose-dependent inhibition of NO synthesis. A concentration of 5 μM, as observed in plasma of patients with renal failure, was sufficient to inhibit isolated NOS, increase the tone of rat aortic rings, and raise the blood pressure of anesthetized guinea pigs (Vallance, Leone et al. 1992). Treatment of cerebral arteries, from several species, including humans, *in vitro* as well as *in vivo* in rats with exogenous ADMA causes concentration- and endothelium-dependent constriction of cerebral arteries comparable to vasospasm after SAH (Faraci, Brian et al. 1995; Segarra, Medina et al. 1999). Furthermore, intravenous infusion of ADMA into healthy humans decreased cerebral blood flow (Kielstein, Donnerstag et al. 2006).

In a primate model of delayed cerebral vasospasm after SAH, ADMA was detected in CSF and serum of control animals and in animals after SAH in a primate model of delayed cerebral vasospasm. Whereas ADMA levels in the serum remained unchanged, ADMA levels in CSF increased after SAH and were correlated with the degree of cerebral vasospasm (Jung, Iuliano et al. 2004). CSF levels of ADMA in monkeys with vasospasm after SAH peaked on day 7 and then decreased on days 14 to 21 (Jung, Iuliano et al. 2004) following closely the time-course of development and resolution of cerebral vasospasm in primates (Espinosa, Weir et al. 1984) and in humans (Weir, Grace et al. 1978). Furthermore, the increased CSF levels of ADMA were negatively correlated with nitrite and L-citrulline levels in CSF and were associated with decreased DDAH2 (the enzyme responsible for hydrolysis of ADMA) expression in vasospastic arteries (Jung, Iuliano et al. 2004), (Leiper, Santa Maria et al. 1999). On the other hand, PRMT1, the enzyme synthesizing ADMA, (Tang, Frankel et al. 2000) remained unchanged (Jung, Iuliano et al. 2004). These results strongly suggest that endogenous inhibition of eNOS by ADMA contributes to cerebral vasospasm development by decreasing NO availability (Jung, Iuliano et al. 2004). Experimental findings are further supported by similar observations made in patients after SAH. CSF levels of ADMA in patients who developed cerebral vasospasm significantly increased around day 4, peaked on days 7 to 9 after SAH, and gradually decreased through day 21 (Jung, Oldfield et al. 2007). Thus, they closely followed the time course for the development and resolution of cerebral vasospasm (Weir, Grace et al. 1978). Also, in patients after SAH, the degree of arteriographic vasospasm strongly correlated with CSF levels of ADMA in patients after SAH (Jung, Oldfield et al. 2007) and CSF ADMA levels negatively correlated with nitrite levels, which has been used as a measure of NO production, and which reflects endothelial NOS activity (Lauer, Preik et al. 2001). Although, a few controversial reports exist (Woszczyk, Deinsberger et al. 2003), many studies confirmed the observation of decreased nitrite levels in the CSF (Hirose, Ide et al. 1995; Schievink 1997; Sakowitz, Wolfrum et al. 2001). Similar results in patients after SAH were also reported by Martens-Lobenhoffer and coworkers, who showed, that on the day of admission, ADMA levels in CSF in patients with traumatic and aneurysmal SAH did not differ from those measured in healthy controls, but increased markedly during the course of the hospital stay (Martens-Lobenhoffer, Sulyok et al. 2007). Furthermore, a decreased L-arginine to ADMA ratio was observed and interpreted as a sign of impaired NO production rate in the brain after SAH being responsible for a simultaneous decrease of NO metabolites (nitrate and nitrite) in CSF (Martens-Lobenhoffer, Sulyok et al. 2007).

The cause of increased ADMA in CSF is unclear but was attributed to a loss of DDAH activity, the enzyme eliminating ADMA (Jung, Iuliano et al. 2004). Similar to the findings in monkeys (Jung, Iuliano et al. 2004), Martens-Lobenhoffer et al. reported a decreased SDMA/ADMA ratio (Martens-Lobenhoffer, Sulyok et al. 2007), as an indirect indicator of decreased DDAH activity (Achan, Tran et al. 2002) during vasospasm in patients after SAH.

Two major mechanisms may lead to elevated ADMA: 1) increased methylation of protein-bound L-arginine by up-regulation of PRMT; and 2) decreased hydrolysis of ADMA by DDAH. Boger et al. reported an up-regulation of PRMT mRNA expression in human endothelial cells by native and oxidized low-density lipoproteins (LDL) and suggested that elevated PRMT expression is responsible for ADMA elevation in hypercholesterolemia (Boger, Sydow et al. 2000). However, this mechanism is unlikely to be responsible for ADMA changes in vasospasm because no increased PRMT expression in arteries in spasm were observed (Jung, Iuliano et al. 2004). In an experimental shear stress model, Osanai et al. observed that the release of ADMA was increased up to 1.7-fold by shear stress at 15 dyne/cm^2. Parallel gene expression of PRMT I was enhanced. Shear stress at magnitudes higher than 25 dyne/cm^2 changed the gene expression pattern from PRMTI to DDAH facilitating the degradation of ADMA, thus returning ADMA to normal baseline levels (Osanai, Saitoh et al. 2003). The two isoforms of DDAH, DDAH1, which predominates in tissues which express nNOS (Ogawa, Kimoto et al. 1989; Tran, Fox et al. 2000), and DDAH2, which is mainly found in tissues containing eNOS (Leiper, Santa Maria et al. 1999) cleave ADMA into dimethylamine and L-citrulline (Kimoto, Whitley et al. 1995; MacAllister, Parry et al. 1996; Leiper, Santa Maria et al. 1999). Both are responsible for about 90% of ADMA degradation via hydrolysis (McDermott 1976; Ogawa, Kimoto et al. 1989; Vallance and Leiper 2004).

Pharmacological inhibition of DDAH increases ADMA levels in endothelial cells leading to blockage of NO-mediated relaxation of blood vessels (MacAllister, Parry et al. 1996). Also, a decline of DDAH activity occurs in pathological conditions associated with endothelial dysfunction (Ito, Tsao et al. 1999; Cooke 2000). Thus, it is likely that changes in DDAH activity or DDAH dysfunction play a role in ADMA increase in clinical settings of aSAH and vasospasm. Such a decreased endothelial DDAH2 activity in the arteries after SAH was reported and led to increased ADMA levels in CSF of monkeys with vasospasm, producing endothelial dysfunction via inhibition of eNOS (Jung, Iuliano et al. 2004).

The actual mechanism through which DDAH is inhibited after SAH is unknown, but may in part be caused by oxidative stress. The subarachnoid blood clot degrades and undergoes subsequent haemolysis releasing oxyHb, which is metabolized to substances such as bilirubin, carbon monoxide, Fe^{2+} and others, further generating free radicals and reactive oxygen species (for detailed review see (Pluta 2005)). DDAH is very susceptible to oxidative stress and a reduction in DDAH activity is frequently observed under this condition (Sydow and Munzel 2003). In the presence of oxidative stress, increased levels of bilirubin during haemolysis may be responsible for releasing bilirubin oxidation products (BOXes), potent vasoconstrictors for vascular smooth muscle cells (Kranc, Pyne et al. 2000; Clark, Reilly et al. 2002; Lyons, Shukla et al. 2004) and may link ADMA and SAH. Incubation of human umbilical vein endothelial cells (HUVEC) with BOXes lead to increased ADMA levels in HUVEC media confirming that BOXes can induce vascular constriction, also via eNOS inhibition (Pluta, Jung et al. 2005). Although, oxyHb was reported to evoke vasoconstriction in cerebral vessels (Macdonald, Weir et al. 1991; Mayberg 1998; Pluta 2005), *in vitro*, oxyHb did not increase ADMA levels in the media of HUVEC lysates (Pluta, Jung et al. 2005). Furthermore, increased intracellular protein turnover in endothelial cells and protein degradation of dimethylated arginine residues may contribute to elevated ADMA levels, especially as protein turnover is accelerated under oxidative stress. Additionally, the degradation of the subarachnoid blood clot itself may lead to ADMA release since recent studies measuring intracellular levels of ADMA in red blood cells showed that they contain free and bound ADMA at a combined concentration of up to 48 μmol/L (Billecke, Kitzmiller et al. 2006). Methyl-arginine uptake studies revealed that ADMA is taken up and can accumulate in endothelial cells in a way that intracellular levels exceed extracellular concentrations (Cardounel, Cui et al. 2007). A 5-fold increase in methylarginine concentration was observed in endothelial cells when they were exposed to methylarginines added to the culture medium (Pullamsetti, Kiss et al. 2005). Moreover, females have a higher incidence of cerebral vasospasm (Rosenorn, Eskesen et al. 1994). Because estrogen influences ADMA levels via activation of DDAH (Holden, Cartwright et al. 2003) as observed after estrogen therapy in postmenopausal women (Post, Verhoeven et al. 2003; Teerlink, Neele et al. 2003) there might be a link to a gender-dependent predisposition in DDAH activity. However, the exact mechanism of ADMA accumulation is unknown (Figure 2). The increase and accumulation of ADMA after SAH may not be the only reasons for the depletion of NO in the brain and the development of delayed cerebral vasospasm, but may be a possible factor strongly correlated with the risk of developing this severe complication. Thus, ADMA is an important endogenous modulator of cerebral vascular tone and may be involved in the pathogenesis of delayed cerebral vasospasm. However, further studies are needed to define clearly predisposition factors and the prognostic value of ADMA CSF concentrations for development of cerebral vasospasm and to the outcome of patients after aSAH.

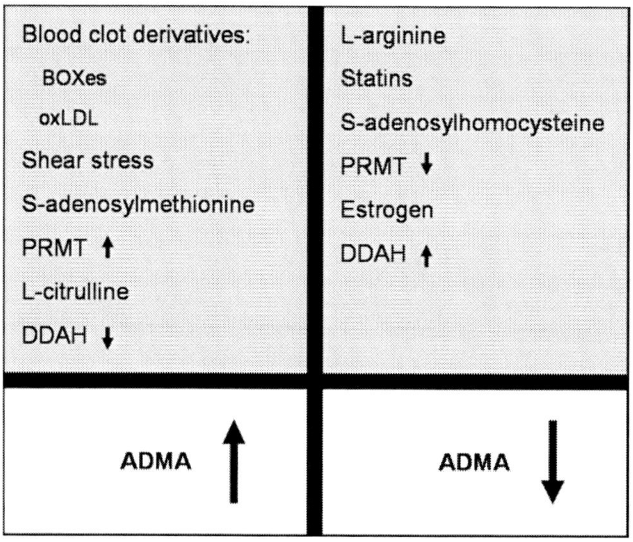

Figure 2. Biological factors regulating asymmetric dimethylarginine (ADMA) levels in vivo. BOXes: bilirubin oxidation products; PRMT: protein-arginine-methyltransferase; DDAH: dimethylarginine dimethylamine hydrolase.

THERAPEUTIC APPROACHES

In addition to mechanical measures such as a mechanical removal or intrathecal lyses of the subarachnoid blood clot, which present a difficult therapeutic endeavor with limited feasibility to decrease the amount of vasospastic agents, pharmaceutical approaches influencing ADMA levels are being developed.

Several agents have been proposed to decrease ADMA levels in vitro and in vivo (Figure 2) (Achan, Tran et al. 2002; Jiang, Li Ns et al. 2002; Holden, Cartwright et al. 2003). Decreased ADMA plasma levels were reported after treatment with simvastatin and pravastatin (Eid, Eritsland et al. 2003; Janatuinen, Laakso et al. 2003; Paiva, Laakso et al. 2003; Pereira, Bertolami et al. 2003; Jiang, Jiang et al. 2004; Lu, Ding et al. 2004). Prevention of vasospasm after SAH by statins, cholesterol-lowering drugs, has also been reported (McGirt, Lynch et al. 2002; Kosty 2005; Trimble and Kockler 2007; Tseng, Hutchinson et al. 2007) and suggests that NOS-related endothelial dysfunction may be corrected pharmacologically (McGirt, Lynch et al. 2002).

Probucol, a drug lowering cholesterol, (Barnhart, Sefranka et al. 1970; Sawayama, Shimizu et al. 2002) that should easily penetrate the blood-brain barrier revealed in earlier studies functional re-endothelialization after coronary angioplasty in rabbits (Lau, Leichtweis et al. 2003) and preservation of eNOS function by reducing ADMA serum levels in rats (Jiang, Li Ns et al. 2002). Thus, we hypothesized that eNOS dysfunction evoked by ADMA release into CSF in vasospasm could be corrected by probucol, and a blinded, placebo-controlled experiment was performed in primates to evaluate the effects of probucol on vasospasm after SAH. However, probucol failed to inhibit ADMA increases in CSF or prevent vasospasm in primates (Pluta, Jung et al. 2005), despite its effectiveness *in vitro* (Pluta, Jung et al. 2005) and *in vivo* (Jiang, Li Ns et al. 2002).

ADMA-related endogenous inhibition of NO production by NOS can be modified on different levels. In principle, each step of this methylation and hydrolysation cycle of ADMA as well as ADMA competitors for NOS constitute a potential target for therapeutic approaches: 1) increasing the availability of L-arginine; 2) changing the availability of L-citrulline; 3) modifying PRMT activity; 4) modifying methylation via regulating availability of adenosylmethionine; or 5) stimulating or inhibiting DDAH activity.

L-Arginine

L-arginine is a substrate for NOS for NO production. Furthermore, the competitive inhibition of NOS activity by ADMA or L-NMMA is reversible by adding L-arginine. In biological systems, however, the stoichiometery varied from a 1:1 ratio, requiring an excess of L-arginine to reverse the NOS inhibition (MacAllister and Vallance 1994). Nevertheless, it is tempting to speculate that increasing plasma L-arginine concentrations might also attenuate the detrimental effects of ADMA in patients with cerebrovascular disease. For example, the beneficial effects of raised L-arginine plasma concentrations were shown in atherosclerotic disease (Boger and Ron 2005). Furthermore, intravenous L-arginine infusion decreased ADMA and endothelin-1 levels and improved endothelial function in patients with angina pectoris and normal coronary arteriograms (Piatti, Fragasso et al. 2003). Moreover, there are several experimental studies in which L-arginine and NO donors reduce total cerebral infarct volume in permanent and transient models of ischemia, recently reviewed by Willmot et al. (Willmot, Gray et al. 2005). Furthermore, in a rabbit as well as in a double hemorrhage canine model, exogenous intracisternal administration of L-arginine lead to dilation of pre-constricted arteries (Kajita, Suzuki et al. 1994; Goksel, Ozum et al. 2001). In the primate model of delayed cerebral vasospasm after SAH, increased cerebral blood flow was observed after infusion of L-arginine; however, vasospasm was not prevented (Pluta, Afshar et al. 2000). Nevertheless, these observations support the hypothesis that increased concentrations and/or availability of L-arginine may help to overcome competitive inhibition of eNOS by ADMA and prevent development of vasospasm after SAH.

L-Citrulline

L-citrulline is a by-product of NO synthesis by NOS and of ADMA hydrolysis by DDAH (Kimoto, Whitley et al. 1995; MacAllister, Parry et al. 1996; Leiper and Vallance 1999). Furthermore, L-citrulline has been shown to inhibit DDAH (Ito, Tsao et al. 1999; Vallance and Chan 2001; Jung, Iuliano et al. 2004). Thus, hypothetically a decreased L-citrulline level may improve DDAH activity. In addition to the NO/NOS-pathway, L-citrulline plays additional roles in a variety of other biochemical pathways, e. g., the urea cycle, which also regulate the concentration of this amino acid. Therefore, small variations in L-citrulline concentrations might be adjusted by compensatory endogenous mechanisms. L-citrulline is converted to L-argininosuccinate by argininosuccinate synthase and subsequently to L-arginine by argininosuccinate lyase, (Curis, Nicolis et al. 2005) and might be a L-arginine precursor (Waugh, Daeschner et al. 2001). Schwedhelm et al. reported that high oral dose of L-citrulline induced dose-dependent elevations of L-arginine in healthy humans (Schwedhelm, Maas et al. 2008). Furthermore, the dose of 3g oral L-citrulline increased urinary exretion of cGMP, implying increased systemic NO production and bioactivity (Schwedhelm, Maas et al. 2008). L-citrulline levels in CSF were negatively correlated with ADMA CSF levels during cerebral vasospasm after SAH in primates, (Jung, Iuliano et al. 2004) whereas, patients revealed unchanged CSF and plasma levels of L-citrulline during the course of vasospasm after SAH (Jung, Oldfield et al. 2007). Further investigations of L-citrulline supplementation and its effect on endothelial function, ADMA expression and the NO/NOS-pathway are warranted.

PRMT

Asymmetric dimethylation of L-arginine residues of various cellular proteins by specific protein L-arginine methyltransferases (PRMT type I) is the initial step in the endogenous formation of ADMA (Tang, Frankel et al. 2000; Boisvert, Cote et al. 2003). In vitro, up-regulation of PRMT mRNA expression was observed in association with raised ADMA levels in models of hypercholesterolemia (Boger, Sydow et al. 2000) and shear stress

(Osanai, Saitoh et al. 2003). However, no increased PRMT expression was observed in arteries in spasm after SAH (Jung, Iuliano et al. 2004). Moreover, practical and clinical considerations make PRMT an unlikely target as post-translational methylation of L-arginine residues since methylation is a universal biological process affecting hundreds of proteins and (Boisvert, Cote et al. 2003) and the effects of this process in most cases remain unknown. Thus, interference with methylation may have unforeseen consequences (Maas 2005).

S-Adenosylmethionine

Elevation of S-adenosylmethionine stimulates PRMT activity and production of ADMA, whereas, S-adenosylhomocysteine inhibits PRMT activity (Lentz, Rodionov et al. 2003). Therefore, another potential target for therapeutic approaches seems to be lowering S-adenosylmethionine levels, thereby reducing substrate availability for asymmetric dimethylation of L-arginine residues, since both types of PRMT use S-adenosylmethionine, derived from L-methionine, as principal methyl group donor (Boger, Sydow et al. 2000), leading to ADMA and S-adenosylhomocysteine as a by-product. Comparable to PRMT targeting, S-adenosylmethionine and reduction of the methylation process at this stage is difficult and there may be unforeseen complications (Maas 2005).

Because of their role in the homocysteine– methionine cycle, folate and vitamin B12 supplementation was expected to increase synthesis of S-adenosylmethionine and subsequent ADMA formation. It was shown in patients with increased plasma homocysteine levels that folate lowered elevated plasma ADMA concentrations (Holven, Haugstad et al. 2003). This beneficial effect of folate was attributed to the reduction of homocysteine. But this effect remains controversial since another study showed that supplementation with folate and B vitamins had no impact on ADMA levels (Sydow, Schwedhelm et al. 2003).

DDAH

The major pathway of ADMA metabolism is hydrolysis by DDAH 1 and 2 (McDermott 1976; Ogawa, Kimoto et al. 1989; Vallance and Leiper 2004). Decreased activity of DDAH was observed in patients after SAH (Martens-Lobenhoffer, Sulyok et al. 2007) and attenuated expression of DDAH2 was described in spastic arteries (Jung, Iuliano et al. 2004). Therefore, it seems likely that changes in DDAH activity or DDAH dysfunction play a role in ADMA increase in the setting of cerebral vasospasm after SAH. Recently, the functionality of DDAH was further elucidated by DDAH-1 transgenic (Dayoub, Achan et al. 2003) and DDAH-1 deficient mice (Achan, Ho et al. 2005). In DDAH-1 deficient mice, plasma and tissue levels of ADMA were increased compared to control animals (Leiper, Nandi et al. 2007). Over-expression of human DDAH-1 in transgenic mice markedly decreased plasma levels of ADMA, thus selectively affecting endothelium-mediated responses without altering endothelium-independent responses (Dayoub, Achan et al. 2003). In addition, DDAH-1 over-expression protected against ADMA-induced endothelial dysfunction in the cerebral circulation of mice (Dayoub, Rodionov et al. 2008). Thus, NOS activity is reduced when DDAH activity is impaired (Cooke 2004) or when DDAH expression is reduced (Leiper,

Nandi et al. 2007). These potentially seminal findings may lead to future development of new, gene- and/or protein—based treatments against vasospasm.

Although, various pharmacological agents have been shown to reduce plasma ADMA levels (Kielstein and Zoccali 2005; Pluta 2005), no specific drug is available that increases degradation of ADMA by DDAH, thereby lowering ADMA levels.

References

Abbasi, F., T. Asagmi, et al. (2001). Plasma concentrations of asymmetric dimethylarginine are increased in patients with type 2 diabetes mellitus. *Am J Cardiol 88*(10): 1201-3.

Achan, V., H. K. Ho, et al. (2005). ADMA regulates angiogenesis: genetic and metabolic evidence. *Vasc Med 10*(1): 7-14.

Achan, V., C. T. Tran, et al. (2002). all-trans-Retinoic acid increases nitric oxide synthesis by endothelial cells: a role for the induction of dimethylarginine dimethylaminohydrolase. *Circ Res 90*(7): 764-9.

Adams, H. P., Jr., N. F. Kassell, et al. (1987). Predicting cerebral ischemia after aneurysmal subarachnoid hemorrhage: influences of clinical condition, CT results, and antifibrinolytic therapy. A report of the Cooperative Aneurysm Study. *Neurology 37*(10): 1586-91.

Afshar, J. K., R. M. Pluta, et al. (1995). Effect of intracarotid nitric oxide on primate cerebral vasospasm after subarachnoid hemorrhage. *J Neurosurg 83*(1): 118-22.

Allen, G. S., L. M. Henderson, et al. (1974). Cerebral arterial spasm. 2. In vitro contractile activity of serotonin in human serum and CSF on the canine basilar artery, and its blockage by methylsergide and phenoxybenzamine. *J Neurosurg 40*(4): 442-50.

Azuma, H., J. Sato, et al. (1995). Accumulation of endogenous inhibitors for nitric oxide synthesis and decreased content of L-arginine in regenerated endothelial cells. *Br J Pharmacol 115*(6): 1001-4.

Barnhart, J. W., J. A. Sefranka, et al. (1970). Hypocholesterolemic effect of 4,4'-(isopropylidenedithio)-bis(2,6-di-t-butylphenol) (probucol). *Am J Clin Nutr 23*(9): 1229-33.

Barone, F. C., M. Y. Globus, et al. (1994). Endothelin levels increase in rat focal and global ischemia. *J Cereb Blood Flow Metab 14*(2): 337-42.

Barrows, L. J., F. T. Hunter, et al. (1955). The nature and clinical significance of pigments in the cerebrospinal fluid. *Brain 78*(1): 59-80.

Bederson, J. B., A. L. Levy, et al. (1998). Acute vasoconstriction after subarachnoid hemorrhage. *Neurosurgery 42*(2): 352-60; discussion 360-2.

Billecke, S. S., L. A. Kitzmiller, et al. (2006). Contribution of whole blood to the control of plasma asymmetrical dimethylarginine. *Am J Physiol Heart Circ Physiol 291*(4): H1788-96.

Black, P. M. (1986). Hydrocephalus and vasospasm after subarachnoid hemorrhage from ruptured intracranial aneurysms. *Neurosurgery 18*(1): 12-6.

Boger, R. H. (2003). Association of asymmetric dimethylarginine and endothelial dysfunction. *Clin Chem Lab Med 41*(11): 1467-72.

Boger, R. H. (2003). The emerging role of asymmetric dimethylarginine as a novel cardiovascular risk factor. *Cardiovasc Res 59*(4): 824-33.

Boger, R. H., S. M. Bode-Boger, et al. (1998). Asymmetric dimethylarginine (ADMA): a novel risk factor for endothelial dysfunction: its role in hypercholesterolemia. *Circulation 98*(18): 1842-7.

Boger, R. H. and E. S. Ron (2005). L-Arginine improves vascular function by overcoming deleterious effects of ADMA, a novel cardiovascular risk factor. *Altern Med Rev 10*(1): 14-23.

Boger, R. H., K. Sydow, et al. (2000). LDL cholesterol upregulates synthesis of asymmetrical dimethylarginine in human endothelial cells: involvement of S-adenosylmethionine-dependent methyltransferases. *Circ Res 87*(2): 99-105.

Boger, R. H. and C. Zoccali (2003). ADMA: a novel risk factor that explains excess cardiovascular event rate in patients with end-stage renal disease. *Atheroscler Suppl 4*(4): 23-8.

Boisvert, F. M., J. Cote, et al. (2003). A Proteomic Analysis of Arginine-methylated Protein Complexes. *Mol Cell Proteomics 2*(12): 1319-30.

Boullin, D. J., J. Mohan, et al. (1976). Evidence for the presence of a vasoactive substance (possibly involved in the aetiology of cerebral arterial spasm) in cerebrospinal fluid from patients with subarachnoid haemorrhage. *J Neurol Neurosurg Psychiatry 39*(8): 756-66.

Brandt, L., B. Ljunggren, et al. (1981). Vasoconstrictive effects of human post-hemorrhagic cerebrospinal fluid on cat pial arterioles in situ. *J Neurosurg 54*(3): 351-6.

Branscombe, T. L., A. Frankel, et al. (2001). PRMT5 (Janus kinase-binding protein 1) catalyzes the formation of symmetric dimethylarginine residues in proteins. *J Biol Chem 276*(35): 32971-6.

Burnett, M. and S. Danish (2004). *Pathology and pathophysiology of aneurysmal subarachnoid hemorrhage. Management of cerebral aneurysms.* P. D. LeRoux, Winn, H. R., Newell, D. W., eds. Philadelphia, PA, Elsevier: 127-137.

Cardounel, A. J., H. Cui, et al. (2007). Evidence for the pathophysiological role of endogenous methylarginines in regulation of endothelial NO production and vascular function. *J Biol Chem 282*(2): 879-87.

Carnegie, P. R., F. C. Fellows, et al. (1977). Urinary excretion of methylarginine in human disease. *Metabolism 26*(5): 531-7.

Chen, B. M., L. W. Xia, et al. (1997). Determination of N(G),N(G)-dimethylarginine in human plasma by high-performance liquid chromatography. *J Chromatogr B Biomed Sci Appl 692*(2): 467-71.

Chen, D., H. Ma, et al. (1999). Regulation of transcription by a protein methyltransferase. *Science 284*(5423): 2174-7.

Clark, J. F., M. Reilly, et al. (2002). Oxidation of bilirubin produces compounds that cause prolonged vasospasm of rat cerebral vessels: a contributor to subarachnoid hemorrhage-induced vasospasm. *J Cereb Blood Flow Metab 22*(4): 472-8.

Clarke, S. (1993). Protein methylation. *Curr Opin Cell Biol 5*(6): 977-83.

Cocks, T. M., E. Malta, et al. (1991). Oxyhaemoglobin increases the production of endothelin-1 by endothelial cells in culture. *Eur J Pharmacol 196*(2): 177-82.

Cooke, J. P. (2000). Does ADMA cause endothelial dysfunction? *Arterioscler Thromb Vasc Biol 20*(9): 2032-7.

Cooke, J. P. (2004). Asymmetrical dimethylarginine: the Uber marker? *Circulation 109*(15): 1813-8.

Copeland, B. E., P. J. Dyer, et al. (1989). Hemoglobin by first derivative spectrophotometry: extent of hemolysis in plasma and serum collected in vacuum container devices. *Ann Clin Lab Sci 19*(5): 383-8.

Cosby, K., K. S. Partovi, et al. (2003). Nitrite reduction to nitric oxide by deoxyhemoglobin vasodilates the human circulation. *Nat Med 9*(12): 1498-505.

Curis, E., I. Nicolis, et al. (2005). Almost all about citrulline in mammals. *Amino Acids 29*(3): 177-205.

Cushing, H. (1923). Contributions to the clinical study of cerebral aneurysms. *Guys Hosp Rep 73*: 159-163.

Das, I., N. S. Khan, et al. (1996). Elevated endogenous nitric oxide synthase inhibitor in schizophrenic plasma may reflect abnormalities in brain nitric oxide production. *Neurosci Lett 215*(3): 209-11.

Davis, K. L., E. Martin, et al. (2001). Novel effects of nitric oxide. *Annu Rev Pharmacol Toxicol 41*: 203-36.

Dayoub, H., V. Achan, et al. (2003). Dimethylarginine dimethylaminohydrolase regulates nitric oxide synthesis: genetic and physiological evidence. *Circulation 108*(24): 3042-7.

Dayoub, H., R. Rodionov, et al. (2008). Overexpression of dimethylarginine dimethylaminohydrolase inhibits asymmetric dimethylarginine-induced endothelial dysfunction in the cerebral circulation. *Stroke 39*(1): 180-4.

Dorsch, N. W. (1995). Cerebral arterial spasm--a clinical review. *Br J Neurosurg 9*(3): 403-12.

Ecker, A. and P. A. Riemenschneider (1951). Arteriographic demonstration of spasm of the intracranial arteries with special reference to saccular arterial aneurysms. *J Neurosurg 8*: 660-667.

Eid, H. M., J. Eritsland, et al. (2003). Increased levels of asymmetric dimethylarginine in populations at risk for atherosclerotic disease. Effects of pravastatin. *Atherosclerosis 166*(2): 279-84.

Espinosa, F., B. Weir, et al. (1984). A randomized placebo-controlled double-blind trial of nimodipine after SAH in monkeys. Part 1: Clinical and radiological findings. *J Neurosurg 60*(6): 1167-75.

Faraci, F. M., J. E. Brian, Jr., et al. (1995). Response of cerebral blood vessels to an endogenous inhibitor of nitric oxide synthase. *Am J Physiol 269*(5 Pt 2): H1522-7.

Faraci, F. M. and D. D. Heistad (1998). Regulation of the cerebral circulation: role of endothelium and potassium channels. *Physiol Rev 78*(1): 53-97.

Fickling, S. A., D. Williams, et al. (1993). Plasma concentrations of endogenous inhibitor of nitric oxide synthesis in normal pregnancy and pre-eclampsia. *Lancet 342*(8865): 242-3.

Fisher, C. M., J. P. Kistler, et al. (1980). Relation of cerebral vasospasm to subarachnoid hemorrhage visualized by computerized tomographic scanning. *Neurosurgery 6*(1): 1-9.

Foley, P. L., N. F. Kassell, et al. (1993). Hemoglobin penetration in the wall of the rabbit basilar artery after subarachnoid hemorrhage and intracisternal hemoglobin injection. *Acta Neurochir (Wien) 123*(1-2): 82-6.

Frankel, A., N. Yadav, et al. (2002). The novel human protein arginine N-methyltransferase PRMT6 is a nuclear enzyme displaying unique substrate specificity. *J Biol Chem 277*(5): 3537-43.

Fujimori, A., M. Yanagisawa, et al. (1990). Endothelin in plasma and cerebrospinal fluid of patients with subarachnoid haemorrhage. *Lancet 336*(8715): 633.

Furchgott, R. F. and J. V. Zawadzki (1980). The obligatory role of endothelial cells in the relaxation of arterial smooth muscle by acetylcholine. *Nature 288*(5789): 373-6.

Gaetani, P., R. Rodriguez y Baena, et al. (1994). Endothelin and aneurysmal subarachnoid haemorrhage: a study of subarachnoid cisternal cerebrospinal fluid. *J Neurol Neurosurg Psychiatry 57*(1): 66-72.

Ghosh, S. K., W. K. Paik, et al. (1988). Purification and molecular identification of two protein methylases I from calf brain. Myelin basic protein- and histone-specific enzyme. *J Biol Chem 263*(35): 19024-33.

Gladwin, M. T., J. R. Lancaster, Jr., et al. (2003). Nitric oxide's reactions with hemoglobin: a view through the SNO-storm. *Nat Med 9*(5): 496-500.

Goksel, H. M., U. Ozum, et al. (2001). The therapeutic effect of continuous intracisternal L-Arginine infusion on experimental cerebral vasospasm. *Acta Neurochir 143*(3): 277-85.

Gold, M. E., K. S. Wood, et al. (1989). L-arginine causes whereas L-argininosuccinic acid inhibits endothelium-dependent vascular smooth muscle relaxation. *Biochem Biophys Res Commun 161*(2): 536-43.

Goonasekera, C. D., D. D. Rees, et al. (1997). Nitric oxide synthase inhibitors and hypertension in children and adolescents. *J Hypertens 15*(8): 901-9.

Gorenflo, M., C. Zheng, et al. (2001). Plasma levels of asymmetrical dimethyl-L-arginine in patients with congenital heart disease and pulmonary hypertension. *J Cardiovasc Pharmacol 37*(4): 489-92.

Gow, A. J. and J. S. Stamler (1998). Reactions between nitric oxide and haemoglobin under physiological conditions. *Nature 391*(6663): 169-73.

Gull, W. M. (1859). Cases of aneurysm of the cerebral vessels. *Guys Hosp Rep 5*: 281-304.

Haley, E. C., Jr., N. F. Kassell, et al. (1995). Phase II trial of tirilazad in aneurysmal subarachnoid hemorrhage. A report of the Cooperative Aneurysm Study. *J Neurosurg 82*(5): 786-90.

Heistad, D. and H. Kontos (1983). Cerebral circulation. Handbook of Physiology: The Cardiovascular System. F. Abboud and J. Shephard. Bethesda, MD, *American Physiological Society*: 137– 182.

Herlitz, H., A. Petersson, et al. (1997). The arginine-nitric oxide pathway in thrombotic microangiopathy. *Scand J Urol Nephrol 31*(5): 477-9.

Hermenegildo, C., P. Medina, et al. (2002). Plasma concentration of asymmetric dimethylarginine, an endogenous inhibitor of nitric oxide synthase, is elevated in hyperthyroid patients. *J Clin Endocrinol Metab 87*(12): 5636-40.

Hibbs, J. B., Jr., R. R. Taintor, et al. (1988). Nitric oxide: a cytotoxic activated macrophage effector molecule. *Biochem Biophys Res Commun 157*(1): 87-94.

Hibbs, J. B., Jr., Z. Vavrin, et al. (1987). L-arginine is required for expression of the activated macrophage effector mechanism causing selective metabolic inhibition in target cells. *J Immunol 138*(2): 550-65.

Hino, A., Y. Tokuyama, et al. (1996). Changes in endothelial nitric oxide synthase mRNA during vasospasm after subarachnoid hemorrhage in monkeys. *Neurosurgery 39*(3): 562-7; discussion 567-8.

Hirose, H., K. Ide, et al. (1995). The role of endothelin and nitric oxide in modulation of normal and spastic cerebral vascular tone in the dog. *Eur J Pharmacol 277*(1): 77-87.

Holden, D. P., J. E. Cartwright, et al. (2003). Estrogen stimulates dimethylarginine dimethylaminohydrolase activity and the metabolism of asymmetric dimethylarginine. *Circulation 108*(13): 1575-80.

Holden, D. P., S. A. Fickling, et al. (1998). Plasma concentrations of asymmetric dimethylarginine, a natural inhibitor of nitric oxide synthase, in normal pregnancy and preeclampsia. *Am J Obstet Gynecol 178*(3): 551-6.

Holven, K. B., T. S. Haugstad, et al. (2003). Folic acid treatment reduces elevated plasma levels of asymmetric dimethylarginine in hyperhomocysteinaemic subjects. *Br J Nutr 89*(3): 359-63.

Hop, J. W., G. J. Rinkel, et al. (1997). Case-fatality rates and functional outcome after subarachnoid hemorrhage: a systematic review. *Stroke 28*(3): 660-4.

Iadecola, C., A. J. Beitz, et al. (1993). Nitric oxide synthase-containing neural processes on large cerebral arteries and cerebral microvessels. *Brain Res 606*(1): 148-55.

Iadecola, C., D. A. Pelligrino, et al. (1994). Nitric oxide synthase inhibition and cerebrovascular regulation. *J Cereb Blood Flow Metab 14*(2): 175-92.

Iadecola, C., F. Zhang, et al. (1993). Role of nitric oxide synthase-containing vascular nerves in cerebrovasodilation elicited from cerebellum. *Am J Physiol 264*(4 Pt 2): R738-46.

Ignarro, L. J. (1990). Biosynthesis and metabolism of endothelium-derived nitric oxide. Annu *Rev Pharmacol Toxicol 30*: 535-60.

Ignarro, L. J. (2002). Nitric oxide as a unique signaling molecule in the vascular system: a historical overview. *J Physiol Pharmacol 53*(4 Pt 1): 503-14.

Ignarro, L. J., R. E. Byrns, et al. (1987). Endothelium-derived relaxing factor from pulmonary artery and vein possesses pharmacologic and chemical properties identical to those of nitric oxide radical. *Circ Res 61*(6): 866-79.

Ignarro, L. J., R. G. Harbison, et al. (1986). Activation of purified soluble guanylate cyclase by endothelium-derived relaxing factor from intrapulmonary artery and vein: stimulation by acetylcholine, bradykinin and arachidonic acid. *J Pharmacol Exp Ther 237*(3): 893-900.

Inagawa, T. (1992). Cerebral vasospasm in elderly patients treated by early operation for ruptured intracranial aneurysms. *Acta Neurochir (Wien) 115*(3-4): 79-85.

Ito, A., P. S. Tsao, et al. (1999). Novel mechanism for endothelial dysfunction: dysregulation of dimethylarginine dimethylaminohydrolase. *Circulation 99*(24): 3092-5.

Iuliano, B. A., R. M. Pluta, et al. (2004). Endothelial dysfunction in a primate model of cerebral vasospasm. *J Neurosurg 100*(2): 287-94.

Janatuinen, T., J. Laakso, et al. (2003). Plasma asymmetric dimethylarginine modifies the effect of pravastatin on myocardial blood flow in young adults. *Vasc Med 8*(3): 185-9.

Jiang, J. L., D. J. Jiang, et al. (2004). Effect of simvastatin on endothelium-dependent vaso-relaxation and endogenous nitric oxide synthase inhibitor. *Acta Pharmacol Sin 25*(7): 893-901.

Jiang, J. L., N. S. Li Ns, et al. (2002). Probucol preserves endothelial function by reduction of the endogenous nitric oxide synthase inhibitor level. *Br J Pharmacol 135*(5): 1175-82.

Jin, J. S. and L. G. D'Alecy (1996). Central and peripheral effects of asymmetric dimethylarginine, an endogenous nitric oxide synthetase inhibitor. *J Cardiovasc Pharmacol 28*(3): 439-46.

Johnston, S. C., S. Selvin, et al. (1998). The burden, trends, and demographics of mortality from subarachnoid hemorrhage. *Neurology 50*(5): 1413-8.

Jung, C. S., B. A. Iuliano, et al. (2004). Association between cerebrospinal fluid levels of asymmetric dimethyl-L-arginine, an endogenous inhibitor of endothelial nitric oxide synthase, and cerebral vasospasm in a primate model of subarachnoid hemorrhage. *J Neurosurg 101*(5): 836-42.

Jung, C. S., E. H. Oldfield, et al. (2007). Association of an endogenous inhibitor of nitric oxide synthase with cerebral vasospasm in patients with aneurysmal subarachnoid hemorrhage. *J Neurosurg 107*(5): 945-50.

Kajita, Y., Y. Suzuki, et al. (1994). Combined effect of L-arginine and superoxide dismutase on the spastic basilar artery after subarachnoid hemorrhage in dogs. *J Neurosurg 80*(3): 476-83.

Kakimoto, Y. and S. Akazawa (1970). Isolation and identification of N-G,N-G- and N-G,N'-G-dimethyl- arginine, N-epsilon-mono-, di-, and trimethyllysine, and glucosylgalactosyl- and galactosyl-delta-hydroxylysine from human urine. *J Biol Chem 245*(21): 5751-8.

Kamper, A. M., A. Spilt, et al. (2004). Basal cerebral blood flow is dependent on the nitric oxide pathway in elderly but not in young healthy men. *Exp Gerontol 39*(8): 1245-8.

Kang, K. K., J. Y. Yu, et al. (2005). The effect of DA-8159, a novel PDE5 inhibitor, on erectile function in the rat model of hypercholesterolemic erectile dysfunction. *Int J Impot Res*.

Kassell, N. F. and C. G. Drake (1982). Timing of aneurysm surgery. *Neurosurgery 10*(4): 514-9.

Kassell, N. F., T. Sasaki, et al. (1985). Cerebral vasospasm following aneurysmal subarachnoid hemorrhage. *Stroke 16*(4): 562-72.

Kassell, N. F., J. C. Torner, et al. (1990). The International Cooperative Study on the Timing of Aneurysm Surgery. Part 1: Overall management results. *J Neurosurg 73*(1): 18-36.

Kasuya, H., B. K. Weir, et al. (1995). Nitric oxide synthase and guanylate cyclase levels in canine basilar artery after subarachnoid hemorrhage. *J Neurosurg 82*(2): 250-5.

Katsanis, N., M. L. Yaspo, et al. (1997). Identification and mapping of a novel human gene, HRMT1L1, homologous to the rat protein arginine N-methyltransferase 1 (PRMT1) gene. *Mamm Genome 8*(7): 526-9.

Keilin, D. and E. F. Hartree (1937). Reaction of Nitric Oxide with Hæmoglobin and Methæmoglobin. *Nature 139*: 548-548.

Khan, U., A. Hassan, et al. (2007). Asymmetric dimethylarginine in cerebral small vessel disease. *Stroke 38*(2): 411-3.

Khurana, V. and J. Friedman (2004). *Biology of cerebral blood vessels and blood flow. Management of Cerebral Aneurysms*. D. Newell. Philadelphia, PA, Elsevier: 139– 168.

Khurana, V. G., Y. R. Sohni, et al. (2004). Endothelial nitric oxide synthase gene polymorphisms predict susceptibility to aneurysmal subarachnoid hemorrhage and cerebral vasospasm. *J Cereb Blood Flow Metab 24*(3): 291-7.

Kielstein, J. T., S. M. Bode-Boger, et al. (2003). Asymmetric dimethylarginine, blood pressure, and renal perfusion in elderly subjects. *Circulation 107*(14): 1891-5.

Kielstein, J. T., F. Donnerstag, et al. (2006). ADMA increases arterial stiffness and decreases cerebral blood flow in humans. *Stroke 37*(8): 2024-9.

Kielstein, J. T. and C. Zoccali (2005). Asymmetric dimethylarginine: a cardiovascular risk factor and a uremic toxin coming of age? *Am J Kidney Dis 46*(2): 186-202.

Kim, P., V. B. Schini, et al. (1992). Reduced production of cGMP underlies the loss of endothelium-dependent relaxations in the canine basilar artery after subarachnoid hemorrhage. *Circ Res 70*(2): 248-56.

Kimoto, M., G. S. Whitley, et al. (1995). Detection of NG,NG-dimethylarginine dimethylaminohydrolase in human tissues using a monoclonal antibody. *J Biochem (Tokyo) 117*(2): 237-8.

Kobayashi, H., M. Hayashi, et al. (1990). Effect of endothelin on the canine basilar artery. *Neurosurgery 27*(3): 357-61.

Kosty, T. (2005). Cerebral vasospasm after subarachnoid hemorrhage: an update. *Crit Care Nurs Q 28*(2): 122-34.

Kranc, K. R., G. J. Pyne, et al. (2000). Oxidative degradation of bilirubin produces vasoactive compounds. *Eur J Biochem 267*(24): 7094-101.

Lanzino, G., N. F. Kassell, et al. (1996). Age and outcome after aneurysmal subarachnoid hemorrhage: why do older patients fare worse? *J Neurosurg 85*(3): 410-8.

Lau, A. K., S. B. Leichtweis, et al. (2003). Probucol promotes functional reendothelialization in balloon-injured rabbit aortas. *Circulation 107*(15): 2031-6.

Lauer, T., M. Preik, et al. (2001). Plasma nitrite rather than nitrate reflects regional endothelial nitric oxide synthase activity but lacks intrinsic vasodilator action. *Proc Natl Acad Sci U S A. 98*(22)(Oct 23): 12814-9.

Leiper, J., J. Murray-Rust, et al. (2002). S-nitrosylation of dimethylarginine dimethylaminohydrolase regulates enzyme activity: further interactions between nitric oxide synthase and dimethylarginine dimethylaminohydrolase. *Proc Natl Acad Sci U S A 99*(21): 13527-32.

Leiper, J., M. Nandi, et al. (2007). Disruption of methylarginine metabolism impairs vascular homeostasis. *Nat Med 13*(2): 198-203.

Leiper, J. and P. Vallance (1999). Biological significance of endogenous methylarginines that inhibit nitric oxide synthases. *Cardiovasc Res 43*(3): 542-8.

Leiper, J. M., J. Santa Maria, et al. (1999). Identification of two human dimethylarginine dimethylaminohydrolases with distinct tissue distributions and homology with microbial arginine deiminases. *Biochem J 343* Pt 1: 209-14.

Lentz, S. R., R. N. Rodionov, et al. (2003). Hyperhomocysteinemia, endothelial dysfunction, and cardiovascular risk: the potential role of ADMA. *Atheroscler Suppl 4*(4): 61-5.

Lin, W. J., J. D. Gary, et al. (1996). The mammalian immediate-early TIS21 protein and the leukemia-associated BTG1 protein interact with a protein-arginine N-methyltransferase. *J Biol Chem 271*(25): 15034-44.

Liszczak, T. M., V. G. Varsos, et al. (1983). Cerebral arterial constriction after experimental subarachnoid hemorrhage is associated with blood components within the arterial wall. *J Neurosurg 58*(1): 18-26.

Loch Macdonald, R. (2006). Management of cerebral vasospasm. *Neurosurg Rev 29*(3): 179-93.

Loch Macdonald, R. and B. Weir (2001). *Cerebral Vasospasm*. San Diego, CA, Academic Press.

Lu, T. M., Y. A. Ding, et al. (2004). Effect of rosuvastatin on plasma levels of asymmetric dimethylarginine in patients with hypercholesterolemia. *Am J Cardiol 94*(2): 157-61.

Lyons, M. A., R. Shukla, et al. (2004). Increase of metabolic activity and disruption of normal contractile protein distribution by bilirubin oxidation products in vascular smooth-muscle cells. *J Neurosurg 100*(3): 505-11.

Maas, R. (2005). Pharmacotherapies and their influence on asymmetric dimethylargine (ADMA). *Vasc Med 10* Suppl 1: S49-57.

Maas, R., R. H. Boger, et al. (2004). Plasma concentrations of asymmetric dimethylarginine (ADMA) in Colombian women with pre-eclampsia. *Jama 291*(7): 823-4.

MacAllister, R. and P. Vallance (1994). Nitric oxide in essential and renal hypertension. *J Am Soc Nephrol 5*(4): 1057-65.

MacAllister, R. J., S. A. Fickling, et al. (1994). Metabolism of methylarginines by human vasculature; implications for the regulation of nitric oxide synthesis. *Br J Pharmacol 112*(1): 43-8.

MacAllister, R. J., H. Parry, et al. (1996). Regulation of nitric oxide synthesis by dimethylarginine dimethylaminohydrolase. *Br J Pharmacol 119*(8): 1533-40.

Macdonald, R. L., B. K. Weir, et al. (1991). Etiology of cerebral vasospasm in primates. *J Neurosurg 75*(3): 415-24.

Machi, T. and D. Stewart (1990). *Hemoglobin enhances release of endothelin from cultured bovine endothelial cells. Cerebral Vasospasm*. K. T. K. Sano, N. F. Kassell and T. Sasaki. Tokyo, University of Tokyo: 262–265.

Macrae, I. M., D. A. Dawson, et al. (1993). Inhibition of nitric oxide synthesis: effects on cerebral blood flow and glucose utilisation in the rat. *J Cereb Blood Flow Metab 13*(6): 985-92.

Mallamaci, F., G. Tripepi, et al. (2005). Prognostic value of combined use of biomarkers of inflammation, endothelial dysfunction, and myocardiopathy in patients with ESRD. *Kidney Int 67*(6): 2330-7.

Mancardi, D., L. A. Ridnour, et al. (2004). The chemical dynamics of NO and reactive nitrogen oxides: a practical guide. *Curr Mol Med 4*(7): 723-40.

Martens-Lobenhoffer, J., E. Sulyok, et al. (2007). Determination of cerebrospinal fluid concentrations of arginine and dimethylarginines in patients with subarachnoid haemorrhage. *J Neurosci Methods 164*(1): 155-60.

Mayberg, M. R. (1998). Cerebral vasospasm. *Neurosurg Clin N Am 9*(3): 615-27.

McCormick, W. F. and G. J. Acosta-Rua (1970). The size of intracranial saccular aneurysms. An autopsy study. *J Neurosurg 33*(4): 422-7.

McDermott, J. R. (1976). Studies on the catabolism of Ng-methylarginine, Ng, Ng-dimethylarginine and Ng, Ng-dimethylarginine in the rabbit. *Biochem J 154*(1): 179-84.

McGirt, M. J., J. R. Lynch, et al. (2002). Simvastatin increases endothelial nitric oxide synthase and ameliorates cerebral vasospasm resulting from subarachnoid hemorrhage. *Stroke 33*(12): 2950-6.

Miranda, T. B., M. Miranda, et al. (2004). PRMT7 is a member of the protein arginine methyltransferase family with a distinct substrate specificity. *J Biol Chem 279*(22): 22902-7.

Miyazaki, H., H. Matsuoka, et al. (1999). Endogenous nitric oxide synthase inhibitor: a novel marker of atherosclerosis. *Circulation 99*(9): 1141-6.

Mugge, A., C. Hanefeld, et al. (2003). Plasma concentration of asymmetric dimethylarginine and the risk of coronary heart disease: rationale and design of the multicenter CARDIAC study. *Atheroscler Suppl 4*(4): 29-32.

Newell, D. W., J. M. Eskridge, et al. (1989). Angioplasty for the treatment of symptomatic vasospasm following subarachnoid hemorrhage. *J Neurosurg 71*(5 Pt 1): 654-60.

Ogawa, T., M. Kimoto, et al. (1989). Purification and properties of a new enzyme, NG,NG-dimethylarginine dimethylaminohydrolase, from rat kidney. *J Biol Chem 264*(17): 10205-9.

Ohta, T. and H. Kajikawa (1980). *Cerebral vasospasm and hemoglobins: clinical and experimental studies. Cerebral Arterial Spasm*. R. H. Wilkins. Baltimore, Williams & Wilkins: 166– 172.

Osanai, T., M. Saitoh, et al. (2003). Effect of shear stress on asymmetric dimethylarginine release from vascular endothelial cells. *Hypertension 42*(5): 985-90.

Paik, W. K., M. Abou-Gharbia, et al. (1983). Metabolism of NG-monomethyl-L-arginine. *Can J Biochem Cell Biol 61*(8): 850-5.

Paiva, H., J. Laakso, et al. (2003). Effect of high-dose statin treatment on plasma concentrations of endogenous nitric oxide synthase inhibitors. *J Cardiovasc Pharmacol 41*(2): 219-22.

Paiva, H., T. Lehtimaki, et al. (2003). Plasma concentrations of asymmetric-dimethyl-arginine in type 2 diabetes associate with glycemic control and glomerular filtration rate but not with risk factors of vasculopathy. *Metabolism 52*(3): 303-7.

Palmer, R. M., A. G. Ferrige, et al. (1987). Nitric oxide release accounts for the biological activity of endothelium-derived relaxing factor. *Nature 327*(6122): 524-6.

Park, K. W., C. Metais, et al. (2001). Microvascular endothelial dysfunction and its mechanism in a rat model of subarachnoid hemorrhage. *Anesth Analg 92*(4): 990-6.

Pasqualin, A., L. Rosta, et al. (1984). Role of computed tomography in the management of vasospasm after subarachnoid hemorrhage. *Neurosurgery 15*(3): 344-53.

Pereira, E. C., M. C. Bertolami, et al. (2003). Effects of simvastatin and L-arginine on vasodilation, nitric oxide metabolites and endogenous NOS inhibitors in hypercholesterolemic subjects. *Free Radic Res 37*(5): 529-36.

Piatti, P., G. Fragasso, et al. (2003). Acute intravenous L-arginine infusion decreases endothelin-1 levels and improves endothelial function in patients with angina pectoris and normal coronary arteriograms: correlation with asymmetric dimethylarginine levels. *Circulation 107*(3): 429-36.

Pluta, R. M. (2005). Delayed cerebral vasospasm and nitric oxide: review, new hypothesis, and proposed treatment. *Pharmacol Ther 105*(1): 23-56.

Pluta, R. M., J. K. Afshar, et al. (1998). Temporal changes in perivascular concentrations of oxyhemoglobin, deoxyhemoglobin, and methemoglobin after subarachnoid hemorrhage. *J Neurosurg 88*(3): 557-61.

Pluta, R. M., J. K. Afshar, et al. (2000). Increased cerebral blood flow but no reversal or prevention of vasospasm in response to L-arginine infusion after subarachnoid hemorrhage. *J Neurosurg 92*(1): 121-6.

Pluta, R. M., R. J. Boock, et al. (1997). Source and cause of endothelin-1 release into cerebrospinal fluid after subarachnoid hemorrhage. *J Neurosurg 87*(2): 287-93.

Pluta, R. M., C. S. Jung, et al. (2005). In vitro and in vivo effects of probucol on hydrolysis of asymmetric dimethyl L-arginine and vasospasm in primates. *J Neurosurg 103*(4): 731-8.

Pluta, R. M., E. H. Oldfield, et al. (1997). Reversal and prevention of cerebral vasospasm by intracarotid infusions of nitric oxide donors in a primate model of subarachnoid hemorrhage. *J Neurosurg 87*(5): 746-51.

Pluta, R. M., B. G. Thompson, et al. (2001). Nitric oxide and vasospasm. *Acta Neurochir Suppl 77*: 67-72.

Pluta, R. M., B. G. Thompson, et al. (1996). Loss of nitric oxide synthase immunoreactivity in cerebral vasospasm. *J Neurosurg 84*(4): 648-54.

Post, M. S., M. O. Verhoeven, et al. (2003). Effect of hormone replacement therapy on plasma levels of the cardiovascular risk factor asymmetric dimethylarginine: a randomized, placebo-controlled 12-week study in healthy early postmenopausal women. *J Clin Endocrinol Metab 88*(9): 4221-6.

Pullamsetti, S., L. Kiss, et al. (2005). Increased levels and reduced catabolism of asymmetric and symmetric dimethylarginines in pulmonary hypertension. *Faseb J.*

Rajpurohit, R., W. K. Paik, et al. (1992). Enzymatic methylation of heterogeneous nuclear ribonucleoprotein in isolated liver nuclei. *Biochim Biophys Acta 1122*(2): 183-8.

Rawal, N., Y. J. Lee, et al. (1995). Urinary excretion of NG-dimethylarginines in multiple sclerosis patients: preliminary observations. *J Neurol Sci 129*(2): 186-91.

Ridnour, L. A., D. D. Thomas, et al. (2004). The chemistry of nitrosative stress induced by nitric oxide and reactive nitrogen oxide species. Putting perspective on stressful biological situations. *Biol Chem 385*(1): 1-10.

Rinkel, G. J., M. Djibuti, et al. (1998). Prevalence and risk of rupture of intracranial aneurysms: a systematic review. *Stroke 29*(1): 251-6.

Rosenorn, J., V. N. Eskesen, et al. (1994). [Prognosis after ruptured intracranial aneurysm in men and women]. *Ugeskr Laeger 156*(33): 4707-9.

Rytlewski, K., R. Olszanecki, et al. (2005). Effects of prolonged oral supplementation with l-arginine on blood pressure and nitric oxide synthesis in preeclampsia. *Eur J Clin Invest 35*(1): 32-7.

Sakowitz, O. W., S. Wolfrum, et al. (2001). Relation of cerebral energy metabolism and extracellular nitrite and nitrate concentrations in patients after aneurysmal subarachnoid hemorrhage. *J Cereb Blood Flow Metab 21*(9): 1067-76.

Sano, K. (1994). Acute ischaemic and delayed ischaemic neurological deficits as the causes of bad grading in aneurysmal subarachnoid haemorrhage. *Neurol Res 16*(1): 35-9.

Sasaki, T. and N. F. Kassell (1990). The role of endothelium in cerebral vasospasm. *Neurosurg Clin N Am 1*(2): 451-63.

Savvidou, M. D., A. D. Hingorani, et al. (2003). Endothelial dysfunction and raised plasma concentrations of asymmetric dimethylarginine in pregnant women who subsequently develop pre-eclampsia. *Lancet 361*(9368): 1511-7.

Sawayama, Y., C. Shimizu, et al. (2002). Effects of probucol and pravastatin on common carotid atherosclerosis in patients with asymptomatic hypercholesterolemia. Fukuoka Atherosclerosis Trial (FAST). *J Am Coll Cardiol 39*(4): 610-6.

Schievink, W. I. (1997). Intracranial aneurysms. *N Engl J Med 336*(1): 28-40.

Schwedhelm, E., R. Maas, et al. (2008). Pharmacokinetic and pharmacodynamic properties of oral L-citrulline and L-arginine: impact on nitric oxide metabolism. *Br J Clin Pharmacol 65*(1): 51-9.

Segarra, G., P. Medina, et al. (1999). Effects of some guanidino compounds on human cerebral arteries. *Stroke 30*(10): 2206-10; discussion 2210-11.

Seifert, V., B. M. Loffler, et al. (1995). Endothelin concentrations in patients with aneurysmal subarachnoid hemorrhage. Correlation with cerebral vasospasm, delayed ischemic neurological deficits, and volume of hematoma. *J Neurosurg 82*(1): 55-62.

Seiler, R. W. and R. Binggeli (2001). Is cerebral vasospasm still a clinical problem? *Acta Neurochir Suppl 77*: 1-4.

Selley, M. L. (2003). Increased concentrations of homocysteine and asymmetric dimethylarginine and decreased concentrations of nitric oxide in the plasma of patients with Alzheimer's disease. *Neurobiol Aging 24*(7): 903-7.

Sobey, C. G. (2001). Cerebrovascular dysfunction after subarachnoid haemorrhage: novel mechanisms and directions for therapy. *Clin Exp Pharmacol Physiol 28*(11): 926-9.

Sobey, C. G. and F. M. Faraci (1998). Subarachnoid haemorrhage: what happens to the cerebral arteries? *Clin Exp Pharmacol Physiol 25*(11): 867-76.

Stapf, C. and J. P. Mohr (2004). *Aneurysms and subarachnoid hemorrhage-epidemiology. Management of Cerebral Aneurysms*. P. D. LeRoux, Winn, H. R., Newell, D. W., eds. Philadelphia, PA, WB Saunders: 183-187.

Stryer, L. (1995). *Biochemistry*, W. H. Freeman and Company: 732.

Sudlow, C. L. and C. P. Warlow (1997). Comparable studies of the incidence of stroke and its pathological types: results from an international collaboration. International Stroke Incidence Collaboration. *Stroke 28*(3): 491-9.

Suzuki, A., N. Yasui, et al. (1988). [Cerebral vasospasm in aged patients with ruptured intracranial aneurysms]. *Neurol Med Chir (Tokyo) 28*(8): 786-90.

Suzuki, J., S. Komatsu, et al. (1980). Correlation between CT findings and subsequent development of cerebral infarction due to vasospasm in subarachnoid haemorrhage. *Acta Neurochir (Wien) 55*(1-2): 63-70.

Suzuki, Y., Y. Kajita, et al. (1994). Dysfunction of nitric oxide in the spastic basilar arteries after subarachnoid hemorrhage. *J Auton Nerv Syst 49* Suppl: S83-7.

Sydow, K. and T. Munzel (2003). ADMA and oxidative stress. Atheroscler Suppl 4(4): 41-51.

Sydow, K., E. Schwedhelm, et al. (2003). ADMA and oxidative stress are responsible for endothelial dysfunction in hyperhomocyst(e)inemia: effects of L-arginine and B vitamins. *Cardiovasc Res 57*(1): 244-52.

Symonds, C. P. (1923). Contributions to the clinical study of intracranial aneurysms. *Guys Hosp Rep 73*: 139-158.

Takenaka, K., N. F. Kassell, et al. (1993). Oxyhemoglobin-induced cytotoxicity and arachidonic acid release in cultured bovine endothelial cells. *Stroke 24*(6): 839-45; discussion 845-6.

Tang, J., A. Frankel, et al. (2000). PRMT1 is the predominant type I protein arginine methyltransferase in mammalian cells. *J Biol Chem 275*(11): 7723-30.

Tang, J., J. D. Gary, et al. (1998). PRMT 3, a type I protein arginine N-methyltransferase that differs from PRMT1 in its oligomerization, subcellular localization, substrate specificity, and regulation. *J Biol Chem 273*(27): 16935-45.

Teerlink, T., S. J. Neele, et al. (2003). Oestrogen replacement therapy lowers plasma levels of asymmetrical dimethylarginine in healthy postmenopausal women. *Clin Sci (Lond) 105*(1): 67-71.

Thompson, B. G., R. M. Pluta, et al. (1996). Nitric oxide mediation of chemoregulation but not autoregulation of cerebral blood flow in primates. *J Neurosurg 84*(1): 71-8.

Toda, N., T. Tanaka, et al. (2000). Cerebral vasodilatation induced by stimulation of the pterygopalatine ganglion and greater petrosal nerve in anesthetized monkeys. *Neuroscience 96*(2): 393-8.

Tran, C. T., M. F. Fox, et al. (2000). Chromosomal localization, gene structure, and expression pattern of DDAH1: comparison with DDAH2 and implications for evolutionary origins. *Genomics 68*(1): 101-5.

Treggiari, M. M., B. Walder, et al. (2003). Systematic review of the prevention of delayed ischemic neurological deficits with hypertension, hypervolemia, and hemodilution therapy following subarachnoid hemorrhage. *J Neurosurg 98*(5): 978-84.

Trimble, J. L. and D. R. Kockler (2007). Statin treatment of cerebral vasospasm after aneurysmal subarachnoid hemorrhage. *Ann Pharmacother 41*(12): 2019-23.

Tseng, M. Y., P. J. Hutchinson, et al. (2007). Biological effects of acute pravastatin treatment in patients after aneurysmal subarachnoid hemorrhage: a double-blind, placebo-controlled trial. *J Neurosurg 107*(6): 1092-100.

Ueno, S., A. Sano, et al. (1992). Distribution of free methylarginines in rat tissues and in the bovine brain. *J Neurochem* 59(6): 2012-6.

Usui, M., H. Matsuoka, et al. (1998). Increased endogenous nitric oxide synthase inhibitor in patients with congestive heart failure. *Life Sci 62*(26): 2425-30.

Valkonen, V. P., J. Laakso, et al. (2003). Asymmetrical dimethylarginine (ADMA) and risk of acute coronary events. Does statin treatment influence plasma ADMA levels? *Atheroscler Suppl 4*(4): 19-22.

Vallance, P. (2001). Importance of asymmetrical dimethylarginine in cardiovascular risk. *Lancet 358*(9299): 2096-7.

Vallance, P. and N. Chan (2001). Endothelial function and nitric oxide: clinical relevance. *Heart 85*(3): 342-50.

Vallance, P. and J. Leiper (2004). Cardiovascular biology of the asymmetric dimethylarginine:dimethylarginine dimethylaminohydrolase pathway. *Arterioscler Thromb Vasc Biol 24*(6): 1023-30.

Vallance, P., A. Leone, et al. (1992). Accumulation of an endogenous inhibitor of nitric oxide synthesis in chronic renal failure. *Lancet 339*(8793): 572-5.

Vallance, P., A. Leone, et al. (1992). Endogenous dimethylarginine as an inhibitor of nitric oxide synthesis. *J Cardiovasc Pharmacol 20*(Suppl 12): S60-2.

van Gijn, J. and G. J. Rinkel (2001). Subarachnoid haemorrhage: diagnosis, causes and management. *Brain 124*(Pt 2): 249-78.

Wanby, P., L. Brattstrom, et al. (2003). Asymmetric dimethylarginine and total homocysteine in plasma after oral methionine loading. *Scand J Clin Lab Invest 63*(5): 347-53.

Waugh, W. H., C. W. Daeschner, 3rd, et al. (2001). Oral citrulline as arginine precursor may be beneficial in sickle cell disease: early phase two results. *J Natl Med Assoc 93*(10): 363-71.

Weir, B., M. Grace, et al. (1978). Time course of vasospasm in man. *J Neurosurg 48*(2): 173-8.

Weir, B., R. L. Macdonald, et al. (1999). Etiology of cerebral vasospasm. *Acta Neurochir Suppl 72*: 27-46.

White, R. P., C. Hindley, et al. (1999). The effect of the nitric oxide synthase inhibitor L-NMMA on basal CBF and vasoneuronal coupling in man: a PET study. *J Cereb Blood Flow Metab 19*(6): 673-8.

White, R. P., R. M. Macleod, et al. (1987). Evaluation of the role hemoglobin in cerebrospinal fluid plays in producing contractions of cerebral arteries. *Surg Neurol 27*(3): 237-42.

Willmot, M., L. Gray, et al. (2005). A systematic review of nitric oxide donors and L-arginine in experimental stroke; effects on infarct size and cerebral blood flow. *Nitric Oxide 12*(3): 141-9.

Wink, D. A. and J. B. Mitchell (1998). Chemical biology of nitric oxide: Insights into regulatory, cytotoxic, and cytoprotective mechanisms of nitric oxide. *Free Radic Biol Med 25*(4-5): 434-56.

Woszczyk, A., W. Deinsberger, et al. (2003). Nitric oxide metabolites in cisternal CSF correlate with cerebral vasospasm in patients with a subarachnoid haemorrhage. *Acta Neurochir (Wien) 145*(4): 257-63; discussion 263-4.

Xiong, Y., L. W. Yuan, et al. (2001). Elevated serum endogenous inhibitor of nitric oxide synthase and endothelial dysfunction in aged rats. *Clin Exp Pharmacol Physiol 28*(10): 842-7.

Yamaura, I., E. Tani, et al. (1992). Endothelin-1 of canine basilar artery in vasospasm. *J Neurosurg 76*(1): 99-105.

Yanagisawa, M., H. Kurihara, et al. (1988). A novel peptide vasoconstrictor, endothelin, is produced by vascular endothelium and modulates smooth muscle Ca2+ channels. *J Hypertens Suppl 6*(4): S188-91.

Yoo, J. H. and S. C. Lee (2001). Elevated levels of plasma homocyst(e)ine and asymmetric dimethylarginine in elderly patients with stroke. *Atherosclerosis 158*(2): 425-30.

Yoshimoto, Y. and S. Kwak (1995). Age-related multifactorial causes of neurological deterioration after early surgery for aneurysmal subarachnoid hemorrhage. *J Neurosurg 83*(6): 984-8.

Yudkoff, M., I. Nissim, et al. (1984). Urinary excretion of dimethylarginines in premature infants. *Biochem Med 32*(2): 242-51.

Ziv, I., G. Fleminger, et al. (1992). Increased plasma endothelin-1 in acute ischemic stroke. *Stroke 23*(7): 1014-6.

In: Diamino Amino Acids
Editors: H. A. Schäfer, L. M. Wohlbier, pp. 31-102
ISBN: 978-1-60456-892-9

Chapter II

NEUROTROPIC AND IMMUNOTROPIC EFFECTS OF AMINO ACID L-LYSINE

Liudmila Severyanova and Maxim Dolgintsev
Pathophysiology Department, the Kursk State Medical University, Russia.

ABSTRACT

The aim of the study is to investigate L-lysine effects on the integrative brain activity and on the immunity as well as the interaction between the nervous and immune regulatory systems in the painful-emotional stress condition.

Methods: To assay the neurotropic amino acid effects on the following behavioral reactions of different biological significance were studied.

- Footshock aggression, tail-flick and hot plate tests (in Wistar rats) were used to compare peculiarities of L-lysine action on pain sensitivity and pain-induced behavior due to the stimulation of different nociceptor kinds.
- The learning in T-maze on the base of food reinforcement and exploratory activity of rats were also investigated.

To estimate the possible L-lysine influence on immunity the humoral immune response and delayed hypersensitivity phenomenon were induced by means of the corresponding ram erythrocytes injection to Wistar rats and CBA mice. Such congenital defensive mechanisms as bactericidal and phagocytic neutrophil activities were studied.

The intraperitoneal administration of L-lysine in the doses arranged from 0.15 up to 5000 mcg/kg was performed 15 min before experimental session. All control animals were injected with saline.

Results: L-lysine effects seemed to be algesic or analgesic in depending on the utilized pain model and administered dose. It should be stressed that the predominant role in the determination of pain-induced behavior features is played not by the painful stimulus intensity but by the biological significance of pain stressful situation and consequently by the supraspinal integrating mechanisms.

L-lysine administration promoted the formation and consolidation of the conditioned nutritional behavior in T-maze, but decreased the exploratory activity of rats.

It was shown that systemic L-lysine administration during the productive phase increases the humoral immune response value in rats but decreases the development of the delayed hypersensitivity reaction in mice.

In pain stress, provoked by electroshocks lasting 2 min per every day during productive phase immunosupression was revealed. Combination of pain stress and L-lysine administration during this period of immune reaction development is followed by the decrease in immune response.

The congenital defensive phenomena seemed to be less influenced by L-lysine. Thus, oxygen-dependent bactericidal neutrophil activity was enhanced only at the higher amino acid doses. However the phagocytic neutrophil activity was inhibited in the dose-dependent manner.

Conclusion: On the whole the obtained findings is the original knowledge about the amino acid incorporated into the active proteins the content of which is high in the upper brain structures particularly in the cerebral cortex.

According to these result the treatment strategies of L-lysine administration may be proved including the composition of L-lysine-containing diets and their using to modulate behavioral reactivity, to decrease emotional overstrain as well to correct immunodeficiency conditions.

Introduction

At present it is a generally proved that the brain activity depends on protein composition and function, in particular, the lysine-containing proteins. The direct dependence between these proteins content and both the species development level and morphological and functional peculiarities of brain nuclei level was found. It was determined that L-lysine-containing proteins content seemed to be the highest in the cerebral cortex neurons, especially in pyramidal cells, in a human (O. Terpilovskaya, V. Ivanov, 1990; R. Khudoerkov, 2001). It was also shown that L-lysine metabolism disorders caused the brain tissue degradation and mental deficiency (I. Ashmarin, P. Stukalov, N. Eschenko, 1999).

For the last years it was proved that L-lysine revealed the action an CNS-depressant with anticonvulsant effect that was realized by means of the enhance of the GABA-benzodiazepine-receptor complex affinity (Y.F. Chang, X.M. Gao, 1995). With use of radioligand binding it was also shown that L-lysine may be a particular serotonin receptor antagonist (M. Smriga, K. Torii, 2003).

Some attempts have been made to reveal L-lysine action on immune system activity but the concrete effects were not clarified (G. Belokrylov, O. Popova, I. Molchanova, 1991).

In the light of all the reviewed data the future study of L-lysine influence on the nervous and immune systems activity and their interaction may be considered as an actual modern problem.

This monograph contains the generalization of the modern information about L-lysine properties, metabolism and effects on an animal and human and analysis of the own study of the amino acid influence on the integrative brain activity and immune system responses as well as on the interaction between these regulatory systems in an emotional-stressful condition.

It was determined that in the used dose range parenteral administration L-lysine influenced different behavior kinds. The effects on the pain-induced behavior formation seemed to be dependent on the dose, pain sensitivity kind and biological significance of the stimulation situation. In the same doses and administration form L-lysine enhanced animal's reactions to unknown environment and decreased movements and exploratory activity. The amino acid administration improved the elaboration and maintenance of food conditioned reflex and, consequently, the learning process.

In parenteral administration L-lysine stimulated a humoral immune response to antigenic stimulation and decreased delayed hypersensitivity reaction type. The amino acid administration also weakened the immunosupression induced by emotional-painful stress. These effects might be estimated as immune system activating and stress-limiting action.

The congenital defensive mechanisms seemed to be less sensitive to L-lysine action. Thus, the oxygen-dependent bactericidal neutrophil capacity was enhanced only after the amino acid administration in the high doses and phagocytic activity was inhibited in dependence on the dose.

The presented reviewed and experimental data supplement the modern theoretical conception of the L-lysine role in a human and animal organism and may serve a ground for a practical recommendation, in particular, concerning L-lysine-containing diets.

1. Molecular Structure of L-Lysine, its Metabolism and Biologically Active Products in the Human and Animal Organisms

For the last two decades the study of amino acids effects has become an exclusively actual subdivision of neurochemistry. Until recently it was well-known that amino acids were constantly used as the substrate for the synthesis and resynthesis of proteins and biological regulators: hormones, enzymes, amines, pigments etc. In a human and mammal amino acids excess degradates up to end products and releases energy necessary for maintenance of the vital activity. However, it became obvious later that a number of amino acids played a role of the specific regulators of, in particular, brain functions. At present, glutamate- and aspartate-ergic neurotransmitter brain systems, their distribution, reception mechanisms and effects were studied in full details (K. Rajevski, 1989). Many of the last years studies were dedicated to discover the physiological role of L-arginine and NO as an active product of L-arginine catabolism. It was proved that NO accomplished gaseous neurotransmitter effect. The neurotropic and immunotropic L-arginine effects were studied (N. Kiryanova, 2006). There are grounds to consider that L-lysine can be included into the specially active amino acids group.

Metabolism of L-Lysine in the Organism, Its Biologically Active Products

Lysine is a diaminomonocarbonic essential acid separated from casein hydrolysate in 1889 and synthetized 1902 (H.-D. Yacubke, H. Yeshkait, 1985). It is well dissolved in water, acids and bases. Figure 1 illustrates the molecular composition of L-lysine.

$$\begin{array}{ccccccccc} CH_2 & - & CH_2 - CH_2 - CH_2 & - & CH & - & COOH \\ | & & & & | & & \\ NH_2 & & & & NH_2 & & \end{array}$$

Figure 1. Chemical formula of L-lysine.

The deficiency of L-lysine causes growth retardation in children and negative nitrogen balance and disorders of vital activity in adults. The day amino acid intake must be about 23 mg/kg of body mass. In industry it is manufactured by microbiological synthesis and included into diets for animals and humans.

The natural L-lysine is incorporated into almost all the proteins in animals and plants (H.-D. Yacubke, H. Yeshkait, 1985). The maximum amino acid concentration was found in histones and the minimum one – in wheat and other cereals proteins. In the neurochemical analysis of neuronal nuclear proteins with high lysine content the specific histone lysine methiltransferases were described which were closely connected with chromatin (O. Terpilovskaya, V. Ivanov, 1990).

In experiments when rats were fed amino acid-unbalanced diet with the consequent dietary treatment no differences were found in the amount of isoleucine, leucine, valine or methionine hydrolyzed from tRNA under any dietary treatment but a significantly higher level of lysine was determined ($p<0.001$) especially in the anterior piriform cortex (L.J. Magrum, P.S. Teh, M.R. Kreiter et al., 2002).

Until recently it was supposed that L-amino acid form was present in the mammals brain. However, with the development of stereochemical analysis certain D-lysine concentration was measured in body fluids and tissues (D.W. Armstrong, M.P. Gasper, S.H. Lee et al., 1993; I. Nagato, T. Akino, K. Ohno et al., 1987). The obtained data suggested that D-isomer could originate both from oral food intake and intestinal bacterial products (T. Fujita, M. Fujita, T. Kodama et al., 2003). In some proteins the L-lysine derivates were found, as namely: oxilysine (incorporated into connective tissue collagen) and N-methyllysine (in myosin composition).

In order to investigate the mechanisms of L-lysine physiological effects it was necessary to study catabolism pathways and reveal the functionally active metabolites. It was found that in mammals L-lysine degradation was predominantly realized via saccharopine pathway both in liver and kidney as well as through the pipecolic acid pathway in brain tissues, that was associated with L-alpha-aminoadipate synthesis (T. Fujita, M. Fujita, T. Kodama et al., 2003; C. Vianey-Liaud, P. Divry, C. Poinas, M. Mathieu, 1991; B.M. Wickwire, C.M. Harris, T.M. Harris et al., 1990; B.M. Wickwire, C. Wagner, H.P. Broquist, 1990). The increased content of the last two metabolites in the monkey brain and spinal cord was shown after the intracerebroventricular and intravenous administration of radioactive labeled L-lysine simultaneously their levels in plasma, liver and kidney were found to be less.

According to major L-lysine degradation pathways the enzymes lysine-ketoglutarate-reductase, saccharopine-dehydrogenase and saccharopine-oxido-reductase were defined as the major catabolism enzymes (J. Dancis, J. Hutzler, 1981; J. Dancis, J. Hutzler, 1982; J. Dancis, J. Hutzler, R.P. Cox, 1979; J. Dancis, J. Hutzler, N.C. Woody, R.P. Cox, 1976). In children with hereditary insufficiency of the above mentioned enzymes, the familial hyperlysinemia syndrome was described and manifested by speech development retardation, hyperactive behavior and certain neurological disorders (J. Dancis, J. Hutzler, R.P. Cox, 1979; S.D. Cederbaum, K.N. Shaw, J. Dancis et al., 1979).

Pipecolic acid is a cyclic imino acid (Figure 2). As a metabolite of L-lysine it was found in human physiological fluids and some plants (H.P. Broquist, 1991; T. Fujita, T. Hada, K. Higashino, 1999). Zellweger hyperpipecolatemia syndrome was described as the genetic disorder that manifested by the decreased oxidase activity and L-lysine catabolism in tissues (V.V. Rao, Y.F. Chang, 1992).

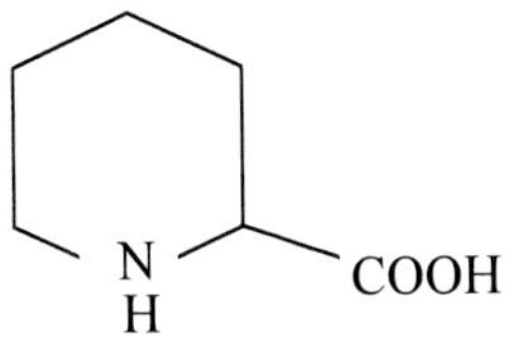

Figure 2. Chemical structure of L-pipecolic acid.

Special interest presented the study of the blood-brain transport of L[1-14C]pipecolic acid in the ray by single intracarotid injection and selective distribution in brain nuclei (A.K. Charles, Y.F. Chang, N.R. Myslinski, 1983). Brain uptake index was reported to be of maximum values for cerebral cortex, brain stem, and cerebellum. The kinetic study of L-pipecolic acid influx revealed a low- and high-capacity uptake mechanisms. A possible role of these mechanisms in the regulation of the neuronal function of L-pipecolic acid was suggested.

In some reports pipecolic acid was estimated as a neurotransmitter or neuromodulator controlling the central inhibitory GABA-system (T. Fujita, M. Fujita, T. Kodama et al., 2003). It was thought to be the precursor of a member of secondary metabolites (M. He, 2006). By using of a sensitive radioactive assay method and thin layer chromatography, it was proved that L-pipecolic acid oxidase was localized in both mitochondria and peroxisomes of rat liver (V.V. Rao, M.J. Tsai, X. Pan, Y.F. Chang, 1993). The 2-fold higher oxidation activity was found in cerebral cortex. In addition, the enzyme activity was shown to be inducable by glucagon (V.V. Rao, Y.F. Chang, 1990; V.V. Rao, Y.F. Chang, 1992).

L-alpha-aminoadipate was another L-lysine metabolite that was suggested to play a role of neurotransmitter. By investigation of superfused rat brain slices it was shown that the metabolite accumulation in slices was stereospecific and Na^+-dependent process that was stimulated by high K^+-concentration in Ca^{2+}-ions presence (A.K. Charles, Y.F. Chang, 1981).

It was also reported about L-alpha-aminoadipate activity as the weak competitive inhibitor of L-glutamate an L-aspartate intake by the cells of brain slices.

In addition to the above mentioned metabolites L-lysine seemed to be precursor of carnitin that was a biologically active component of the cardiac and skeletal muscles (V. Tanphaichitr, H.P. Broquist, 1973). It was found that carnitin took part in the control of fatty acids transport through mitochondria membrane. Carnitin liver synthesis dropped exclusively in rats fed with L-lysine deficient diet.

On the whole, the reviewed detailed data ground the aim to reveal the possible L-lysine and its active metabolites effects on the nervous and immune regulatory systems.

2. The Neurotropic Effects of L-Lysine

2.1. The Modern Concept of L-Lysine Influence on the Central Nervous System Activity

At present the idea does not demand for evidence that the functional brain activity depends on the biochemical peculiarities of neurons and neuroglia in particular of the qualitative composition and properties of proteins. The extensive histochemical investigation showed the high arginine- and lysine-rich proteins content of brain tissue (R. Khudoerkov, 2001). These proteins are built into the nucleoproteins and form 35% ribosomes mass. As it is known the ribosomes provide the over and over intensive RNA synthesis in neurons in comparison with other tissues. In that it seemed that lysine-rich proteins predominated in nervous cells and arginine-rich ones – in neuroglia.

The fact seems to be interesting that the brain nuclear proteins specifity is expressed in the greater degree than in other organs (O. Terpilovskaya, V. Ivanov, 1990). That is provided by the brain specific gene expression and exclusive RNA-processing. In the mammals brain neurons the histone methyl-transferase were identified. These enzymes catalyze the methyl groups translocation to the lysine residues in the N-terminal histone fragments. Thus, in the rat brain two lysine methyl-transferase were found. These were located exclusively in cell nuclei and closely connected to chromatin. These seemed to be specific for the histones H_3 and H_4. The neurospecific proteins were assessed as the base of structural and functional specifity of nervous cell genome that might be reflected in the unique ability to analytic and associative brain activity.

In the light of these data the attention should be paid to the neuroanatomical localization of lysine-containing proteins. The heightened content of these proteins was found in the evolutionally novel brain nuclei namely in cerebral cortex ceels and especially in the pyramidal ones (R. Khudoerkov, 2001). Other brain structures were arranged according to the lysine-rich proteins as follows: paleocortex, hippocampus, subcortical and brain stem nuclei. It was shown that the qualitative and quantitative characteristics of distinct brain nuclei were reflected in the peculiar emotional and behavioral reactions and learning (R. Khudoerkov, 1992; R. Khudoerkov, 2001).

High lysine residues content in the brain nuclear proteins was the ground to suggest that the amino acid inself might exert the modulatory action on the cellular physiological processes in particular on the proliferation, differentiation and apoptosis (N. Chalisova, V. Penniyaynen, 2003). The idea was experimentally studied using the cultivation of organotypic rat tissue fragments as follows: cerebral cortex, spleen and liver. The addition of L-lysine, L-arginine, L-asparagine and L-glutamate (0.05 ng/ml) into the tissue culture induced the changes of the growth zones in dependence on rat ageing and consequently on tissue differentiation and maturation stages. Thus, the less intensive growth zones were found in 1-day rats and an intensive zone was revealed in 21-day rats. So the modulatory amino acid effects (including L-lysine) on the growth of nervous, lymphoid and liver tissues was proved.

Method of fluorescent staining showed a negative correlation between growth zone size and apoptotic cell number (N. Chalisova, V. Penniyaynen, G. Haze, 2003). Thus it was concluded that the minimum concentration of lysine and other amino acids could influence on the programmed cell death mechanisms in tissues including nerve tissue.

The beginning of the systematized investigation of L-lysine influence on functional brain activity and behavior development might be named 70-th years of XX century. Since it was an essential amino acid the L-lysine dietary deficiency was the major idea to study and less investigation number the parenteral amino acid administration was used in the experimental and clinical investigations.

In a few studies for this period of time the increased [3H]lysine absorption and incorporation into brain mice proteins under influence of the electrical foot stimulation (A.J. Dunn, H.D. Rees, P.M. Iuvone, 1978; H.D. Rees, A.J. Dunn, 1977). It was concluded that the effect was mediated by ACTH since it was reproduced in spite of adrenalectomy and after $ACTH_{4\text{-}10}$ administration but inhibited by the preview dexamethazone injection. In contrast to this L-lysine incorporation into liver proteins seemed to be independent on the pituitary-adrenal system. The fact was interesting that lysine-vasopressin increased secretion in stress condition did not change [3H]lysine incorporation into brain and liver proteins.

In further investigations of the L-lysine neurotropic action in stress condition the anxiolytic effect was found and connected with neurotransmitter and hormonal responses (M. Smriga, M. Kameishi, H. Uneyama, K. Torii, 2002). Thus, rats fed lysine-deficient diet for 4 days exhibited the higher levels of anxiety and faecal excretion in stressful conditions namely, foot-shocks and "handling" stress.

Anxiolytic amino acid effect was also found in studies on humans. The intense anxiety and stress responses in particular in women because of menstrual blood bleeding were found in economically poor members of Syrian communities for whom wheat with low L-lysine content was the predominant food (M. Smriga, S. Ghosh, Y. Mouneimne et al., 2004). The double blind study showed that L-lysine addition to diet significantly decreased the extent of anxiety and sympathetic activation measured by means of the galvanic skin conductance.

The similar effect was shown in healthy subjects with relatively high trait anxiety in a randomized, double blind placebo controlled trial (D. Jezova, A. Makatsori, M. Smriga et al., 2005). In 29 healthy volunteers at the upper limit of the normal range of a trait anxiety scale, a mixture of L-lysine and L-arginine (3 g each/day) was administered for 10 days and resulted in decreased anxiety intensity. In addition to this it was determined that the amino acid mixture intake modified hormone levels in response to a psychoemotional stress. The psychoemotional stress model was based on public speech in front of an unknown audience and its main parts were 15 min of preparation and 15 min speech on an emotionally charged topic. It was revealed that the treatment with a mixture of amino acids resulted in higher neuroendocrine activation in response to psychoemotional stress in comparison with the control values. Concentrations of ACTH, cortisol and catecholamines in plasma were significantly increased during stress procedure. It should be noted that in the previous study the decreased hormone response to stress was revealed in depression patients (D. Jezova, A. Makatsori, R. Duncko et al., 2004). Thus, the treatment with amino acids promoted the normalization of the response to stress in particular hormone levels up to the pattern occurring in subjects with low trait anxiety.

In experiments on animals (pigs, broilers) the amino acids of L-lysine and L-arginine mixture supplementation resulted in the weakening of the anxiogenic action of transportation stress (S. Srinongkote, M. Smriga, K. Nakagawa, Y. Toride, 2003). Beside of this in rats the orally intaken L-lysine blocked the lowering urea synthesis and L-arginine blood level which were induced by "handling" stress (M. Smriga, K. Torii, 2003).

The attempt was undertaken to discover the mechanism of L-lysine anxiolytic effect. The prolong microdialisis (during 26 hours) by means of the tube implanted into the ventromedial hypothalamus was used in male rats fed L-lysine deficient diet. The significant decrease of circadian norepinephrine release from the nucleus was shown (M. Smriga, M. Mori, K. Torii, 2000). The neurotransmitter secretion was corrected up to the normal pattern after the supplemental lysine solution drinking (400 mmol/l).

The similar data were obtained in the study of L-lysine deficient food intake on the serotonin release in the central amygdalar nucleus that resulted on the consequent changes of psychobiological responses to stress (M. Smriga, M. Kameishi, H. Uneyama, K. Torii, 2002).

This idea development was continued in the experiments *in vitro* when L-lysine (0.07 and 0.7 mmol/dl) blocked the serotonin-induced contractions of guinea pig's ileum ($p<0.05$ and $p<0.01$ respectively). On using of radioligand method it was shown that L-lysine (0.8 mmol/dl) inhibited serotonin interaction with 5-HT_4-receptors (M. Smriga, K. Torii, 2003). It was concluded that amino acid might be the 5-HT_4-receptor antagonist and decreased stress-induced anxiety and faecal excretion that was mediated by these receptors. The conclusion was confirmed by the fact that L-lysine blocked arising of anxiety induced by the 5-HT_4-receptor agonist 5-hydroxy-tryptophane action.

In a series of studies with use of a pentylenetetrazol-induced clonic and tonic seizures model in mice L-lysine revealed CNS-depressant effect (Y.F. Chang, X.M. Gao, 1995; Y.F. Chang, N.R. Myslinski, 1985; Y.F. Chang, X.M. Gao, J.S. Chen, 1991; Y.F. Chang, Y. Wang, R.K. Cauley, X.M. Gao, 1993). Thus, intraperitoneal amino acid administration (2 mmol/kg and 10 mmol/kg) for 10 days resulted in the enhance of convulsant-induced (60 mg/kg) seizure latencies. This effect was also reproduced by the intracerebroventricular amino acid administration (0.1 mmol/kg) (Y.F. Chang, N.R. Myslinski, 1985; Y.F. Chang, Y. Wang, R.K. Cauley, X.M. Gao, 1993).

As for the mechanism of anticonvulsant L-lysine effect it might be based on the amino acid action on GABA-benzodiazepine receptor complex (Y.F. Chang, X.M. Gao, J.S. Chen, 1991; X.M. Gao, Y.F. Chang, 1989).

As the review of many studies which were performed in 80-th of XX century (P. Sergejev, N. Shimanovsky, 1987) showed GABA-receptors had rather complex molecular structure and functioning mechanisms (Figure 3). It was found that postsynaptic complex contained chloride channel permeability of which was regulated by two GABA-binding parts and at last by two subunits selectively binding benzodiazepines and picrotoxins. It is pointed that a released mediator activated GABA-receptor part that in its turn resulted in opening of chloride channel. Due to heightened chloride permeability the postsynaptic membrane hyperpolarization developed and its excitability decreased. It was determined that benzodiazepines prolonged the opened ion channel period in response to GABA-action. Thus, the molecular mechanisms of GABA-receptors functioning corresponded to receptor-ionophore complex.

In vitro experiments it was determined that L-lysine enhanced the specific [3H]flunitrazepam binding of bovine and mouse brain membranes in a dose-dependent manner (Y.F. Chang, X.M. Gao, J.S. Chen, 1991). Pretreatment with amino acid at 1.5, 10 & 20 mmol/kg not only reversed the inhibition of the specific [3H]flunitrazepam binding caused by pentylenetetrazol (0.46 mmol) at 50, 90 or 100 mmol/kg, but also enhanced the binding above control level and potentiated an anticonvulsant diazepam (0.2 mg/kg) effect.

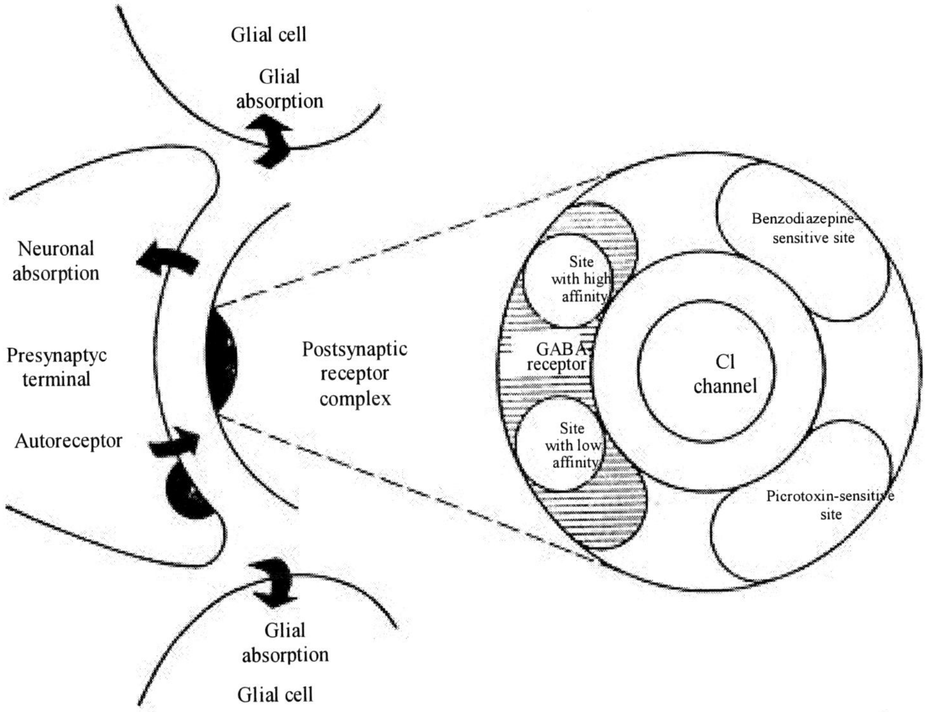

Figure 3. Schematic image of the central acsosomatic GABA-ergic synapsis (F.V. De Feudis, 1981).

Since the L-lysine-induced enhance of agonist binding to benzodiazepine receptors was inhibited with picrotoxin the conclusion was done that amino acid acted on the particular picrotoxin-sensitive sites which differed from the GABA-receptor binding site (X.M. Gao, Y.F. Chang, 1989). The Scatchard-analysis use showed that the L-lysine-enhanced binding of benzodiazepine-receptor agonist was due to the increased affinity of binding, but not to the change of receptor density. This conclusion corresponded to the dose-dependent stimulating influence of chloride ions on L-lysine effect. Thus, the possibility of amino acid action as the modulator of GABA-benzodiazepine receptors was proved. The effect was enhanced by the GABA administration and inhibited by pentobarbital as well as a benzodiazepine-receptor antagonist (Y.F. Chang, X.M. Gao, 1995).

On analysis of the mechanisms of L-lysine action as a CNS-depressant it was important to note that the intracerebroventricular injection of pipecolic acid (0.1 mmol/kg) caused the decrease of seizure latencies, did not enhance anticonvulsant diazepam activity, and even induced seizures at 0.6 mmol/kg. Thus, L-pipecolic acid effect might be not related to GABA-diazepine receptor (Y.F. Chang, V. Hargest, J.S. Chen, 1988).

To solve the problem completely the identification of pipecolic acid receptors was attempted in the solubilized membrane fraction from rat cerebral cortex (A.K. Charles, 1986). Specific bindings proteins for both pipecolic acid and a protein GABA-antagonist muscimol, were detected in the preparation. The labeled pipecolic acid and GABA binding proteins were separated by means of special centrifugation. It was shown that both GABA and pipecolic acid might bind to the same receptor complex, if not to the same binding site. However, the possible presynaptic interrelationship between pipecolic acid and GABA-ergic transmission was not excluded (Y. Kase, K. Takahama, T. Hashimoto et al., 1980).

In addition it was shown that L-pipecolic acid was found to inhibit GABA-binding to bovine synaptic membranes especially in the presence of hexobarbital or pentobarbital (P. Feigenbaum, Y.F. Chang, 1986). Thus, the heterogeneity of L-pipecolic acid-displaceable GABA binding sites was confirmed. It was concluded that this brain-lysine metabolite might be an endogenous ligand acting as a neuromodulator on the GABA receptor ionophore complex, or it might act on its own membrane binding sites due to allosteric effect on the GABA receptor complex.

The data deserve to be noted that L-lysine might accelerate the development of receptor desensitization kinetics in the presence of GABA receptor agonist (S.J. Fountain, R.A. North, 2006).

The possibility is not excluded that L-lysine influenced neuronal excitability and activation mechanisms. The primary studies of the problem were undertaken to clarify the mechanisms of nutrient deficiency recognition and homeostasis maintenance. In rats feeding L-lysine deficient diet the electrical neuronal activity of the medial and lateral hypothalamus as well as cerebral blood flow and brain oxygenation determined by means of magnetoresonance method, were investigated (K. Torii, T. Yokawa, E. Tabuchi et al., 1996). L-lysine deficiency for 4 days in young Wistar male rats resulted in decreased amino acid content in plasma, anorexia development and loss of body weight. The electrical activity of the mentioned hypothalamus parts decreased and increased blood oxygenation appeared after intraperitoneal L-lysine administration (0.2 mmol per 10 ml/kg) as well as after ionophoresis or solution intake. The similar changes were not found in other brain regions such as hippocampus, thalamus etc. These findings were the ground to conclude that the hypothalamic neurons possessed the specific sensitivity to L-lysine level and played the role in the recognition of its deficiency.

As for the cellular mechanisms of amino acids effects the experiments on anesthetized cats equipped with cranial windows were performed. It was determined that L-lysine and L-arginine influenced the opening of ATP-sensitive potassium channels in response to vasodilator (H.A. Kontos, E.P. Wei, 1998).

In summary, it may be considered enough proved the existence of two main aspects of L-lysine action on the functional brain activity. In the first place L-lysine effects performed at the cellular level due to the close connection of lysine-containing proteins with chromatin, namely the effects on the proliferation, differentiation processes as well as excitability and neuronal apoptosis. In the second place, L-lysine influenced the intercellular interactions by means of the neurotransmitter systems activity including the formation of anxiolytic and anticonvulsant effects.

However, the third possible aspect of L-lysine neurotropic action was virtually unstudied. This is the integrative brain activity manifesting itself in the formation of the behavior of different biological significance. One of the main tasks of our study is the investigation of this aspect of L-lysine neurotropic effects.

2.2. The Investigation of L-Lysine Effects on the Behavior of Different Biological Significance

To study the neurotropic L-lysine action the following experimental models were used.

1. The animals (rats) behavior in the different kinds of painful simulation:
   a) the thermoalgesic reactions in "hot plate" test;
   b) behavioral responses to the electric skin stimulation of tail;
   c) the aggressive-defensive behavior (foot-shock aggression) induced by the inescapable electric painful stimulation of paws.
2. The exploratory activity and emotionality the "glass arena" and "open field" tests.
3. The food conditioned reflex elaboration in T-maze.

The experiments were performed on 217 adult mail Wistar rats which were obtained from the nursery "Stolbovaya" of the Russian Academy of Medical Sciences. The rats were in quarantine in the Kursk State Medical University vivarium and no diseases were revealed. Animals were housed in group cages (10 to a cage) in temperature-controlled environments (20-23 °C), maintained on 12:12 h light-dark cycle and provided with the standard granular feed and water *ad libitum*. Experiments were performed on the same time period from 9 a.m. till 2 p.m. Control and experimental rat groups included the animals which were simultaneously obtained from nursery.

L-lysine ("ICN") was dissolved in saline and intraperitoneally injected in corresponding dose in dependence on an experimental session aim. Control rats were injected with the equivalent saline volumes in account 1 ml per 1 kg of body mass.

All animals handling and testing were performed in accordance with the declaration of Helsinki and Animal Care Guidelines. The study protocol was approved by the Ethical Committee of the Kursk State Medical University (Russia).

Any statistically significant differences between obtained data for all tests were determined using a Student's paired t-test ($p<0.05$). In addition, non-paired Mann-Whitney and z-criteria were also used when it was needed according to the analysis task (E. Gubler, A. Genkin, 1973; N. Plokhinsky, 1970).

2.2.1. L-Lysine Action on Nociceptive Sensitivity and Pain-Induced Behavior in Rats

L-Lysine Action on Thermoalgesic Reactions

To study pain sensitivity to thermal stimulation, the rats were individually placed on the plate in a restraint plexiglas cylinder (diameter and height were equal 25×25 cm). The temperature of the plate was maintained at 53.0±0.3 °C. Four trials lasting 1 min were carried out with 1 min interval between them. The latency of the first sign of hindpaw licking or jumping to avoid heating pain was estimated as an index of the nociceptive threshold. The paw licking reactions were observed in the 1st and 2nd exposures and jumping to the margin of the limiting cylinder usually arised in the 3rd and 4th trials (L. Sernov, V. Gatsura, 2000). L-lysine was administered in the doses 0.15, 0.5, 1.5, 5.0, 15.0, 50.0, 500.0, 1500.0 and 5000.0 mcg/kg 12 min before the experiment onset.

Table 1. The latencies (s, M±m) of the pain reactions of rats in the thermal stimulation (the experimental session 1)

Reaction / Group	Licking		Jumping	
	1st exposure	2nd exposure	3rd exposure	4th exposure
Control (n=10)	11.1±1.1	18.0±3.3	5.5±0.5	6.7±1.5
L-lysine: 0.15 mcg/kg (n=10)	8.1±0.7*	19.3±5.3	17.5±4.8*	7.2±0.9
0.5 mcg/kg (n=12)	11.0±3.1	28.3±13.8	14.6±3.5*	13.6±2.2*
50.0 mcg/kg (n=10)	6.1±0.4*	10.7±1.0*	16.4±2.4*	12.6±2.3*

Note: * – p<0.05-0.001 in comparison with the control values.

Table 2. The latencies (s, M±m) of the pain reactions of rats in the thermal stimulation (the experimental session 2)

Reaction / Group (n=10)	Licking		Jumping	
	1st exposure	2nd exposure	3rd exposure	4th exposure
Control	7.9±1.0	22.7±3.5	16.6±2.4	13.6±2.9
L-lysine: 1.5 mcg/kg	12.3±1.3*	21.8±3.9	13.6±3.2	22.2±6.5
5.0 mcg/kg	11.4±1.2*	23.0±3.0	23.1±4.5	18.1±3.0
15.0 mcg/kg	10.9±1.7	18.9±2.3	23.1±6.4	8.6±2.3
500.0 mcg/kg	8.9±1.4	22.2±4.0	18.5±5.6	13.7±5.0
1500.0 mcg/kg	8.4±1.6	19.2±3.3	18.3±4.4	11.7±2.4
5000.0 mcg/kg	8.6±0.7	17.4±3.4	11.8±3.1	8.7±3.0

Note: * – p<0.05 in comparison with the control values.

Each series included saline and L-lysine-treated rat groups and the results of the thermoalgesic sensitivity investigation of the groups on the time courses are summarized the Tables 1 and 2 as absolute values of latencies.

In the first exposure of L-lysine-treated rats (the doses 0.15 and 50.0 mcg/kg) the latencies of paw licking decreased in comparison with the control group values by 27% ($p<0.05$) and 45% ($p<0.001$) respectively. This permits to conclude that the pain sensitivity is increased (Figure 4).

On the contrary, amino acid administration in the doses 1.5 and 5.0 mcg/kg resulted in the prolonged latencies of the reactions, and consequently in analgesic effect (by 56% and 45%, $p<0.05$). The same character of L-lysine effect was also found in the rat group treated with L-lysine in the dose 15 mcg/kg (38%). However, the statistically significant difference from the control group values was not obtained. In the rest of the doses the administration of an amino acid induced either less expressed change of the paw licking latencies or did not induce any changes.

In the 2nd trial L-lysine effects were expressed in less degree (Figure 5). Thus the statistically significant L-lysine effect – decrease of pain sensitivity threshold – were reproduced only in the rat group that received amino acid in the dose 0.5 mcg/kg. In the rest of the arrangements of the administered doses the changes of thermoalgesic sensitivity have no statistically significant differences from the control findings.

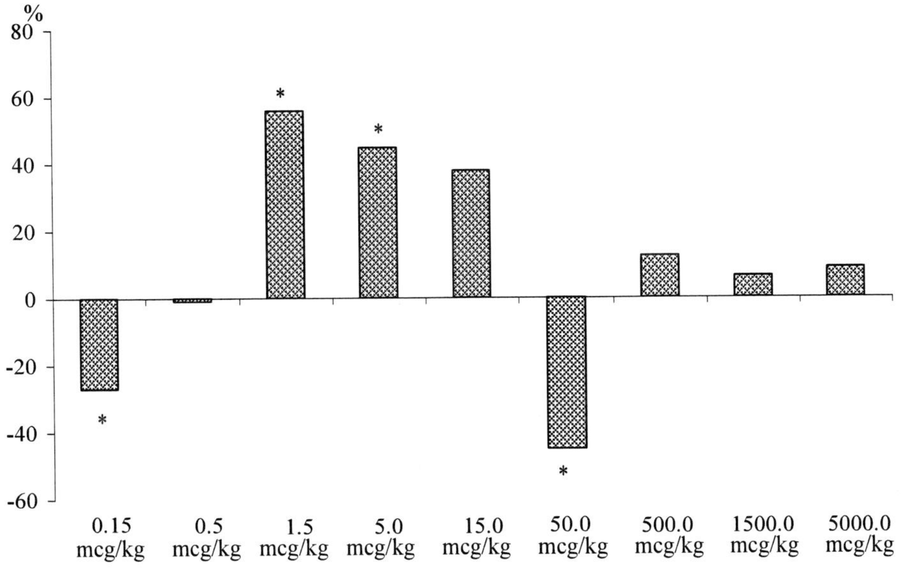

Figure 4. The changes in the latencies of the paw licking reactions of rats during the 1st exposure after L-lysine administration. Designations: columns – the changes of latencies in percent in relation to the control group indices. * – $p<0.05-0.001$ in comparison with the control values.

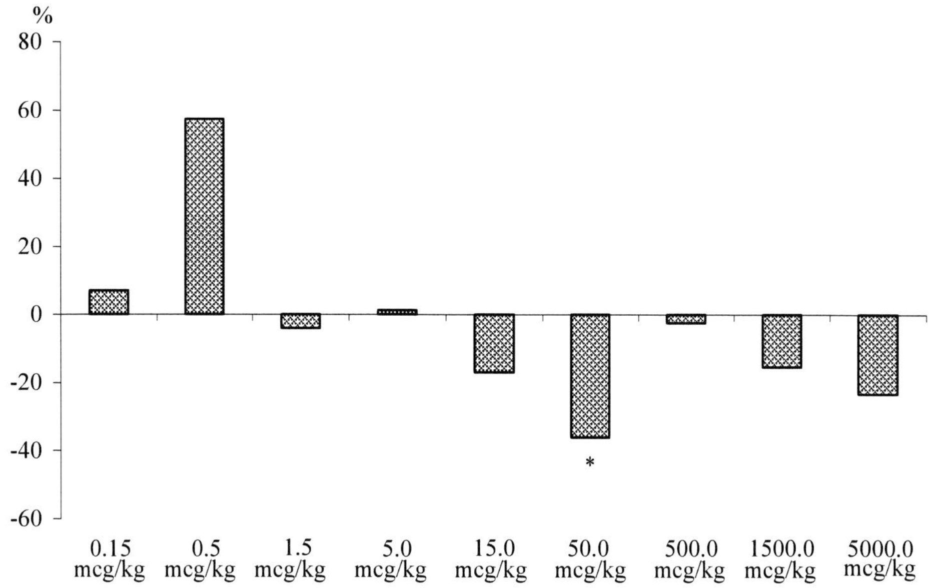

Figure 5. The changes in the latencies of the paw licking reactions of rats during the 2nd exposure after L-lysine administration. Designations as in Figure 4.

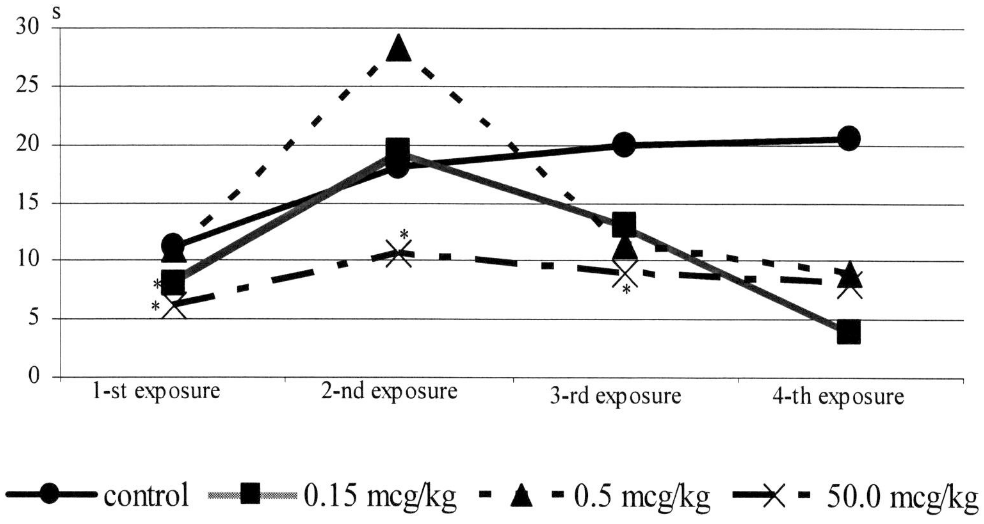

Figure 6. The dynamic changes in the latencies of the paw licking reactions of the control rats and after L-lysine administration. Designations: * – p<0.05-0.001 in comparison with the control values.

To detalize the peculiarities of L-lysine action on the pain reactions in the hot plate test it seemed to be important to study the dynamics of thermoalgesic sensitivity estimated due to paw licking sign in the same rats during the experimental session. Thus, amino acid administration in the dose 50 mcg/kg produced notable decrease of the paw licking latencies in all 4 trials (to 45-55%, p<0.05-0.001) in comparison with the control group findings. These changes were the most marked in the first two trials (Figure 6).

L-lysine administered in a dose 0.15 mcg/kg produced the similar tendency in effects. The analysis of the summarized number of paw licking reactions in the 1st and 2nd trials did not reveal statistically significant differences of the indices from the control values (Figure 9, Table 3 and 4). In the 3rd and 4th trials the changes of the avoidance behavior were investigated and estimated according to the latencies of jumping to the margin top of the limiting cylinder.

Table 3. The number of the paw licking reactions during the 1st and 2nd exposure and jumping reactions during the 3rd and 4th exposure (the experimental session 1)

Group \ Reaction	Licking 1st and 2nd exposure	Jumping 3rd and 4th exposure
Control (n=10)	75% (15)	85% (17)
L-lysine:		
0.15 mcg/kg (n=10)	70% (14)	100%$^{+}$ (20)
0.5 mcg/kg (n=12)	50% (11)	100%$^{+}$ (22)
50.0 mcg/kg (n=10)	80% (16)	100%$^{+}$ (20)

Note: percentage is determined in relation to the total number of trials; $^{+}$ – p=0.06-0.07 in comparison with the control values; in brackets – absolute meanings.

Table 4. The number of the paw licking reactions during the 1st and 2nd exposure and jumping reactions during the 3rd and 4th exposure (the experimental session 2)

Group (n=10) \ Reaction	Licking 1st and 2nd exposure	Jumping 3rd and 4th exposure
Control	85% (17)	75% (15)
L-lysine:		
1.5 mcg/kg	95% (19)	90% (18)
5.0 mcg/kg	80% (16)	100%* (20)
15.0 mcg/kg	85% (17)	90% (18)
500.0 mcg/kg	70% (14)	85% (17)
1500.0 mcg/kg	95% (19)	90% (18)
5000.0 mcg/kg	85% (17)	85% (17)

Note: percentage is determined in relation to the total number of trials; * – p<0.05 in comparison with the control values; in brackets – absolute meanings.

After L-lysine administration in the doses 0.15, 0.5 and 50.0 mcg/kg in the 3rd exposure the statistically significant increase of the thermoalgesic sensitivity thresholds was shown (to 166-218%, p<0.05-0.001) (Figure 7). Amino acid administration in the rest of rat groups produced slightly expressed statistically insignificant changes of the jumping latencies as compared with the control group findings. The above mentioned tendency to the decrease of thermoalgesic sensitivity (to 8-104%) was reproduced in the 4th trial (Figure 8) when the amino acid doses 0.15, 0.5, 1.5, 5.0 and 50.0 mcg/kg were administered.

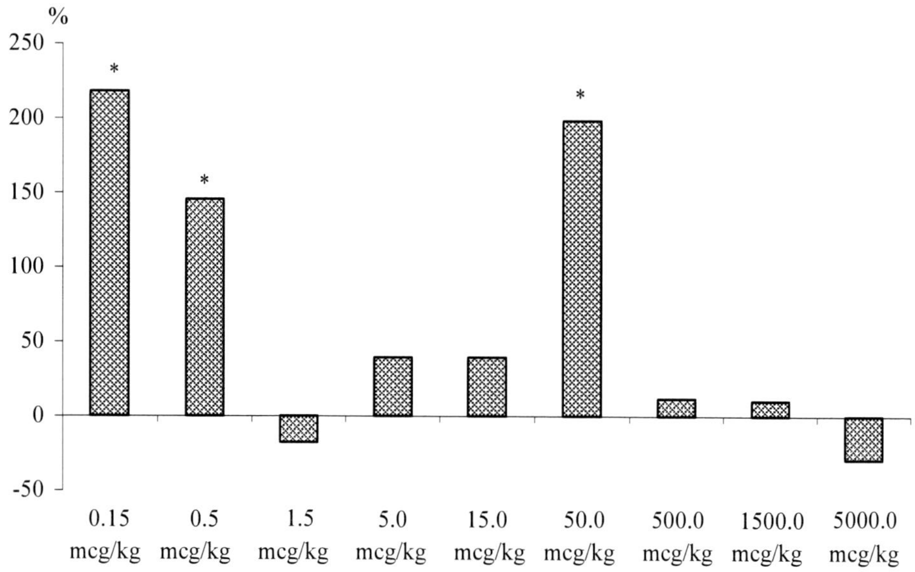

Figure 7. The changes of the jumping reaction latencies in rats during the 3rd exposure after L-lysine administration. Designations are the same as in Figure 4. Note: * – $p<0.05$-0.001 in comparison with the control values.

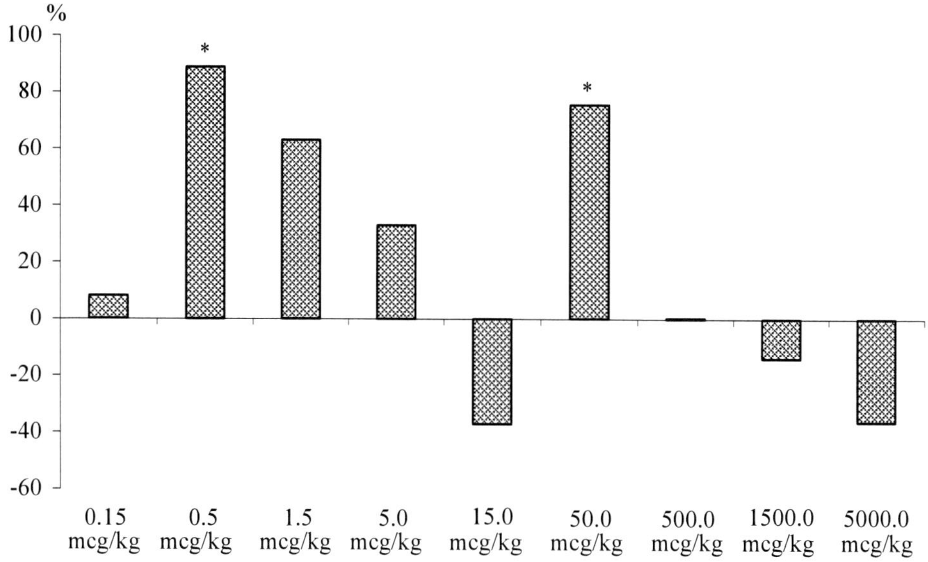

Figure 8. The changes of the jumping reaction latencies in rats during the 4th exposure after L-lysine administration. Designations as in Figure 4.

The statistically significant increase of jumping thresholds was found only in the two groups of rats that received L-lysine in the doses 0.5 mcg/kg (to 104%, $p<0.05$) and 50.0 mcg/kg (to 80%, $p<0.05$). Further increase of the amino acid doses resulted in statistically insignificant decrease of the pain-induced behavior thresholds (to 14-36%). These effects do not permit to conclude about the maintenance of the algesic amino acid action in the 4th trial.

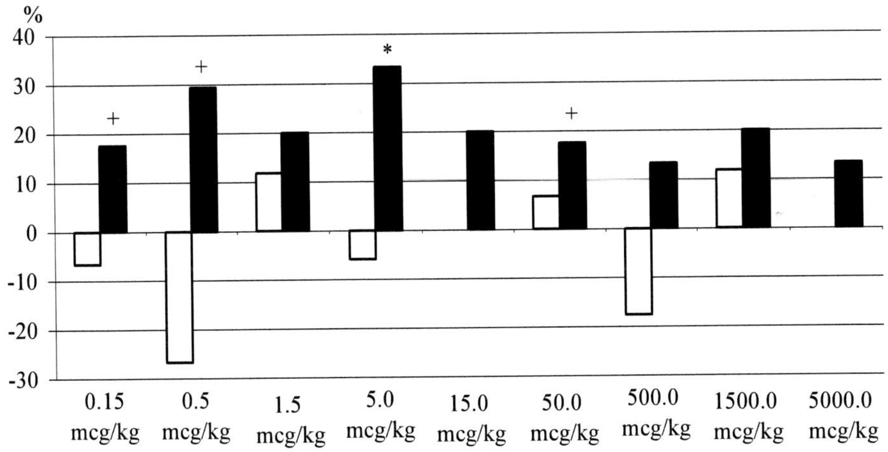

Figure 9. The changes of the paw licking reactions number during the 1st and 2nd exposure and jumping reactions number during the 3rd and 4th exposures after L-lysine administration. Designations: columns – the changes of paw licking reactions number during the 1st and 2nd exposures (white) and jumping reactions number during the 3rd and 4th exposures (dark) in percent in relation to the control values. * – $p<0.05$; + – $p=0.06-0.07$ in comparison with the control values.

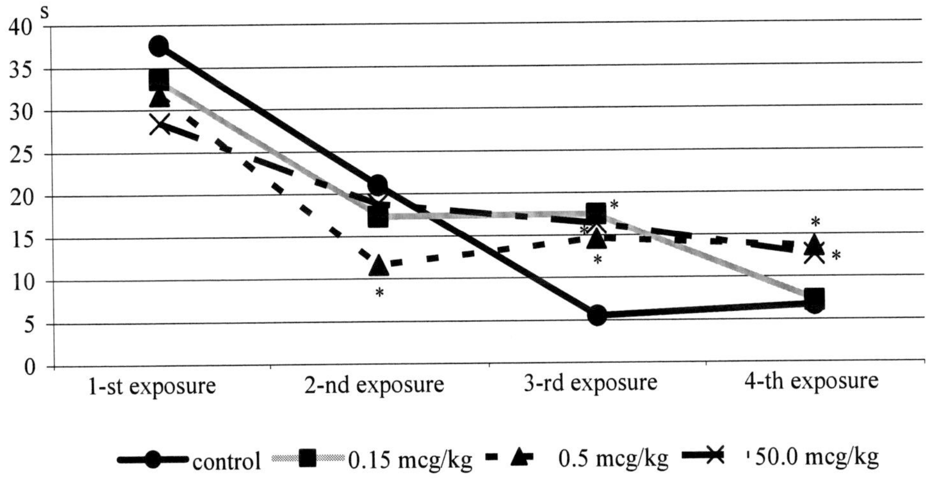

Figure 10. The dynamic changes of the jumping reaction latencies in the control rats and after L-lysine administration. Designations: * – $p<0.05-0.001$ in comparison with the control values.

Analysis of the jumping threshold changes in the same animals during experimental session showed the essential differences between the dynamics of paw licking latencies in the 1st and 4th trials. Thus, L-lysine administration in the doses 0.15, 0.5 and 50.0 mcg/kg resulted in the increased thresholds of the pain-induced avoidance (by 88-218%, $p<0.05-0.001$) in the final trials (Figure 10). However, the estimation of summarized avoidance number in the 3rd and 4th trials revealed that the index in all the groups essentially exceeded the control values (Figure 9, Tables 3 and 4). The statistically significant differences were noted when L-lysine was administered in a dose 5.0 mcg/kg (to 33%, $p<0.05$), and amino acid administration in the doses 0.15, 0.5 and 50.0 mcg/kg induced the changes of jumping number which were close to the statistically significant value (by 18-29%, $p=0.06-0.07$).

Thus, L-lysine in the used arrangement of doses induced the changes of the thermoalgesic sensitivity in rats in the hot plate test. On the whole these changes were characterized by the parabolic dependence on the administered doses and, consequently, the thermoalgesic and thermoanalgesic amino acid effects were found. For example the statistically decreased latencies of paw licking response in the 1st exposure on the hot plate confirmed the conclusion about the enhanced pain sensitivity. The most steady amino acid effect was associated with the dose 50.0 mcg/kg when the decrease of the pain response threshold was reproduced in the final trials (the full period of observation lasted about 7 min). In the control condition the final trials were characterized by the decrease of the primary response to the painful stimuli. On the contrary L-lysine administration in the average doses (1.5, 5.0 and 15.0 mcg/kg) was followed by the statistically significant increase of the threshold values of the primary painful response and consequently with the pain sensitivity weakening.

As for the changes of jumping response under L-lysine influence (this is predominant reaction in the two final exposures to the thermoalgesic stimulation) it was found that the arising of jumping was slowered but frequent in all the doses. On the whole these changes may be estimated as the enhanced intensity of avoidance behavior.

L-Lysine Influence on the Nociceptive Sensitivity and Pain-Induced Behavior in Tail Electric Stimulation

To determine the pain sensitivity thresholds in rats the lamella electrodes were attached to the tail beginning part and this site was irritated by electric current. The arising pain reaction was estimated due to the threshold measures (mA) of the following responses: the tail tension and head turn in direction to electrodes. The emotional-affective pain response thresholds were also measured as follows: the thresholds of body rotation, vocalization and electrode site biting. Two trials in each rat were carried out with 1 min inter-trial interval.

The experiments were performed in the free behavior condition of rats in the special chamber. So the observation could be realized by an researcher without any influence on rats (S. Tikhomirov, S. Nikolayev, 1988). L-lysine was administered in the doses 0.15, 0.5, 1.5, 5.0, 15.0, 50.0 mcg/kg 12 min prior to the experiment onset. The control animals were injected with the equivalent volume of saline.

As it is shown in Table 5 and Figure 11 L-lysine administration in the above mentioned doses provoked the statistically significant reduction of all the components of pain-induced behavior thresholds with the exception of the vocalization response. In other words this was the notable algesic action. On using the doses 0.15 and 0.5 mcg/kg the decrease of pain response thresholds (such as tail tension and head turn in direction to the electrode sites) reached 16-28% ($p<0.05-0.01$) in comparison with the control values. The degree of body rotation and electrode site biting reactions as the signs of emotional-affective behavior were changed and exceeded the control values in less degree (up to 15-20%, $p<0.05-0.001$).

Table 5. The thresholds of the painful and emotional-affective reactions of rats (mA, M±m) in the tail electric stimulation test after L-lysine administration

Index \ Groups	Control	0.15 mcg/kg	0.5 mcg/kg	1.5 mcg/kg	5.0 mcg/kg	15.0 mcg/kg	50.0 mcg/kg
Tail tension	0.78±0.04	0.66±0.03*	0.64±0.03 *	0.50±0.03*	0.61±0.02*	0.62±0.06*	0.49±0.03*
Head turn in direction to electrodes	1.10±0.09	0.91±0.06	0.80±0.07 *	0.62±0.03*	0.89±0.04*	0.82±0.08*	0.63±0.05*
Vocalization	0.95±0.11	0.85±0.07	0.81±0.11	0.76±0.05	0.84±0.07	0.93±0.11	0.66±0.8*
Body rotation	1.49±0.06	1.28±0.05*	1.20±0.07 *	1.12±0.06*	1.26±0.03*	1.17±0.09*	0.93±0.04*
Electrode site biting	1.51±0.07	1.27±0.05*	1.22±0.06 *	1.12±0.05*	1.23±0.03*	1.22±0.1*	0.94±0.06*

Note: * – p<0.05-0.001 in comparison with the control values.

The algesic effect was virtually the similar one in the L-lysine administration in the doses 5.0 and 15.0 mcg/kg. The tail tension and head turn reaction thresholds became less (by 19-26%, p<0.05-0.01) in comparison with the control means and the degree of emotional-affective reaction activation was characterized by 19-38% threshold reduction (p<0.05-0.001).

The most intensive amino acid action was found after its administration in the doses 1.5 and 50.0 mcg/kg. It should be stressed that these dose effects essentially differed. Thus, after L-lysine injection in the dose 1.5 mcg/kg the pain sensitivity enhanced in the greater degree (up to 36-44%, p<0.001) than the affective reactivity (up to 25-26, p<0.05-0.001).

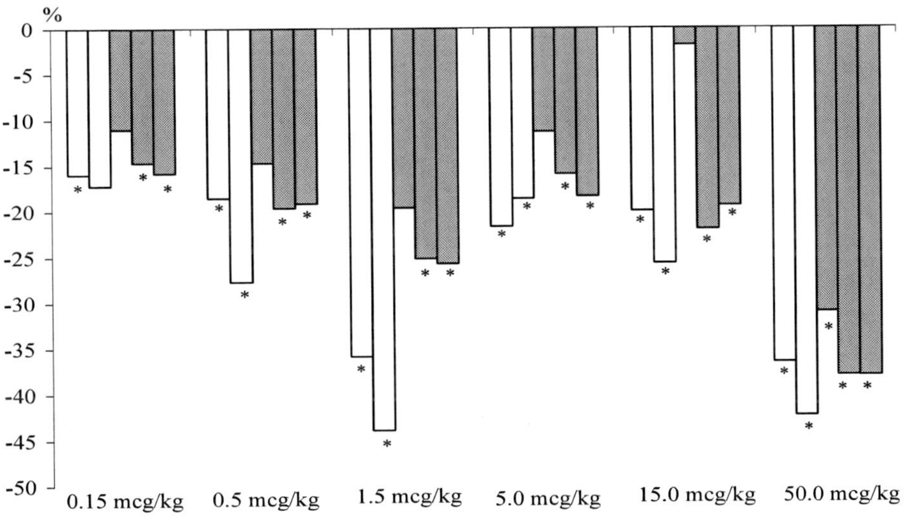

Figure 11. The changes of the pain reactions thresholds in rats in electrical tail stimulation after L-lysine administration. Columns: the changes in the thresholds of the tension tail reaction, head turn to electrodes (white), vocalizations, body rotations and electrode site biting (dark) in percent in relation to their values in the control condition (the left scale); * – p<0.05-0.001 in comparison with the control values.

In summary, the electric stimulation of rat tail provoked very expressed defensive behavior. The electrode site biting was the attack to a physical thing that served the cause of pain and might be considered as the sign of affective defence. In such situation the tail tension as the spinal reflex seemed to be more sensitive to L-lysine action. With the progressive dose the activating amino acid effect on the supraspinal structures controlling the defence behavior formation was also increased.

L-Lysine Influence on the Aggressive-Defensive Behavior of Rats in Unavoidable Painful Electric Stimulation

The pair grouped male rats of the identical age were stimulated by the action of gradually increasing alternative electric current (1 V per 1 s) that was conducted through the special chamber floor. The arising of aggressive-defensive reaction was estimated by the thresholds of the consequently developing behavior components as follows: flinching, vocalization, rising up, running and fighting. The frequency of attacks (percentage of the number of exposures) during the period of time when the voltage reached the critical value 70 V. The electric current action was stopped on reaching of the above mentioned intensity of stimulation and aggression arising. Two trials were performed with 1 min inter-trial interval (R.J. Rodgers, J.M. Semple, 1978). The amino acid was administered in the doses 0.15, 0.5, 1.5, 5.0, 15.0 and 50.0 mcg/kg 12 min before the experimental session.

The results of L-lysine effect on the foot-shock aggression development in rats are represented in the Table 6. The peculiarity of the amino acid action in this condition was the U-form curve of the dose-effect dependence way (Figure 12). Thus, the administration of the mean arranged amino acid doses 1.5, 5.0 and 15.0 mcg/kg resulted in the reduction of flinching and vocalization response thresholds up to 18% of the control values but the enhance of pain perception was statistically significant when the dose 1.5 mcg/kg ($p<0.05$) was injected. The changes of defensive running threshold in this rat group reached 13% ($p<0.01$), and aggression increase was revealed due to the greater number of fighting (by 20-50% in comparison with control values). The dose 5.0 mcg/kg effect was the most close to the statistically significant value ($p<0.05$).

Table 6. The thresholds values of rat behavior (V, M±m) and fighting rate (% in relation to the group trials number) in the foot-shock-aggression test after intraperitoneal L-lysine administration

Index \ Groups	Control	0.15 mcg/kg	0.5 mcg/kg	1.5 mcg/kg	5.0 mcg/kg	15.0 mcg/kg	50.0 mcg/kg
Flinching	28.4±2.0	26.5±0.9	30.0±1.6	28.1±1.7	27.4±1.7	26.4±2.7	37.7±3.8*
Vocalization	35.8±1.6	35.3±1.6	36.7±2.4	29.2±2.2*	34.0±1.8	31.4±2.7	37.7±5.5
Rising up	31.0±3.2	34.4±1.3	40.6±2.2*	35.6±2.9	33.9±2.7	30.0±7.0	39.3±8.3
Running	47.7±1.4	46.2±1.2	60.2±2.2*	41.7±1.4*	45.9±1.8	49.1±2.0	57.3±3.8*
Fighting	56.6±3.1	58.0±4.4	67.0±0	59.7±3.1	61.0±3.1	54.0±2.8	53.0±0.0
Frequency of attacks	30	40	10	70	80	50	10

Note: * – p<0.05-0.001 in comparison with the control values.

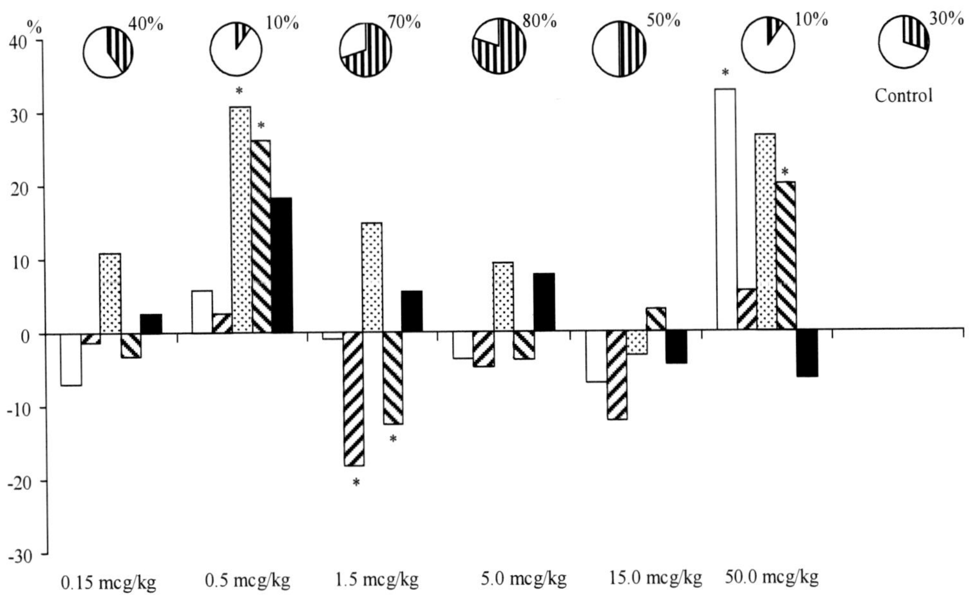

Figure 12. The changes in the indices of an aggressive-defensive reaction after L-lysine administration. Columns: the changes of the thresholds of flinching (white), vocalization, rising up, running (shaded), and fighting (black) in percent in relation to their control values (the left scale). The round diagrams: the fighting rate (shaded) in relation to the total number of trials. * – $p<0.05$-0.001 in comparison with the control values.

The administration of L-lysine in the doses 0.5 and 50.0 mcg/kg resulted in the opposite – analgesic and antiaggression effects. The increase of pain threshold as the pain criteria (by 3-33% in comparing with the control ones) in the rat group-treated with amino acid in the dose 50.0 mcg/kg as well as the en of the defensive behavior were found (decrease of thresholds by 20-31%, $p<0.05$-0.001). The changes of fighting reaction thresholds were unidentical in tendency and insignificant never the less the attack frequency essentially decreased (to 10%).

The significant between-group difference of the foot-shock-aggression was not found if the L-lysine administered dose was 0.15 mcg/kg. The affective aggression intensity estimated according to the attack frequency was also close to the control value.

On the whole, the painful and defensive responses arised simultaneously in paired rats because of paw painful irritation in the electric floor chamber. The reactions developed in the conditions when the animal communication and perception of visual and smelling stimuli were actual. The inescapble character of progressing intense stimulation resulted in the consummatory aggression attacks to another animal. In such situation the L-lysine effect depended on a dose and was either activating or inhibitory one.

2.2.2. L-Lysine Effects on the Exploratory Animal Behavior

L-Lysine Influence on the Rat Activity in the "Glass Arena" Test

The exploratory rat activity was investigated in the glass cylinder with diameter 30 cm and the height of wall – 30 cm that was named "glass arena" (L. Severyanova, I. Bobyntsev, 1995). The frequency of risings as the vertical component of exploratory rat activity as well as groomings and emotionality (due to the number of boluses and urination) were estimated. L-lysine was administered in the doses 0.15, 0.5, 1.5, 5.0, 15.0, 50.0 mcg/kg 10-12 min before the experiment onset.

The behavior parameters changes of the control animal group and L-lysine-treated groups in "glass arena" test are represented in Table 7 and Figure 13.

Table 7. The indices of rat behavior in the "glass arena" test after L-lysine administration (M±m)

Indices / Groups	Vertical activity	Grooming	Emotionality
Control	8.10±0.97	0.80±0.25	1.30±0.37
L-lysine:			
0.15 mcg/kg	4.30±0.98*	0.40±0.27	1.70±0.47
0.5 mcg/kg	3.10±1.20*	0.30±0.21	1.60±0.31
1.5 mcg/kg	5.20±0.83*	1.0±0.37	1.0±0.30
5.0 mcg/kg	8.20±0.80	1.80±0.39*	1.90±0.50
15.0 mcg/kg	3.80±0.96*	0.10±0.10*	2.40±0.27*
50.0 mcg/kg	3.90±1.05*	0.50±0.27	1.20±0.49

Note: * – $p<0.05-0.01$ in comparison with the control group values.

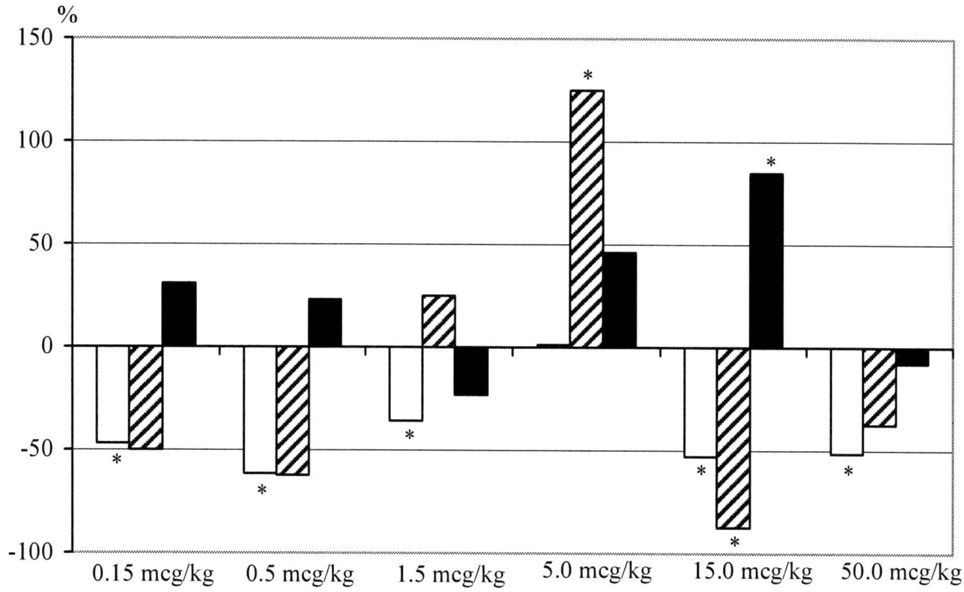

Figure 13. The changes of behavior indices in "glass arena" test after L-lysine administration. Note: columns – the changes of vertical activity (white), grooming (shaded), and emotionality (black) in relation to their values in the control group (the left scale), * – $p<0.05-0.01$ in comparison with the control values.

As it was shown the predominant L-lysine effect in "glass arena" was the weakening of the vertical exploratory behavior. An amino acid in the most used doses essentially decreased the frequency of rising (by 36-62%, $p<0.05$-0.01). Only in the rat group that received L-lysine in the dose 5.0 mcg/kg the intensity of vertical activity virtually corresponded to the control values.

The changes of grooming in the greater degree depended on the injected amino acid dose. Thus, in the rat groups which were L-lysine-treated ones in the doses 0.15 and 0.5 mcg/kg the tendency to decrease this behavior kind was revealed (by 50-62%, $p>0.05$). On the contrary, the subsequent raising of the used amino acid dose up to 1.5 mcg/kg provides increase of grooming activity (by 25%, $p>0.05$). This L-lysine effect became especially expressed in the case of the dose 5 mcg/kg. For this group the grooming activity enhance was significant in comparison with the control values (by 112%, $p<0.05$). Nevertheless, L-lysine effects on the exploratory behavior showed the opposite tendency in the further increase of the administered doses. Thus, the significant grooming weakening was found in the use of the dose 15.0 mcg/kg (by 87%, $p<0.05$). Such tendency of amino acid action was also reproduced in the case of the dose 50.0 mcg/kg (decrease by 37%, $p>0.05$).

The L-lysine influence on the animal emotionality changes exhibited in the prevalent enhance of this behavioral component. This effect reached the significant difference from the control value in the case of the used amino acid dose 15.0 mcg/kg (by 85%, $p<0.05$). The attention must be payed to the fact that the enhanced emotionality usually was not followed by the increase of vertical exploratory activity and grooming.

L-Lysine Action on the Exploratory Activity and Emotionality of Rats in the "Open Field" Test

The experiments were carried out in the strictly limited period of time, namely: from 9 a.m. till 3 p.m. in the dark room that was maximum isolated from external noise. The open field presented floor space 90×90 cm with surrounding height 30 cm walls. The floor was divided into 100 squares with the side 9 cm. The central 16 squares were lighted by the electric lamp light. The characteristics of the lamp were reproduced as follows: the height of location was equal to 60 cm and the power – 150 Wt.

Each exposure lasted 5 min. In the experimental session the following behavior criteria were measured: a) in the central squares – immobility duration, movements, the number of acrossed squares; b) in the peripheral squares – movements, the number of acrossed squares, the number and duration of rising on hind paws, the duration of local within-square movements, immobility duration, the number and duration of groomings, the number of faecal boluses (M. Dolgintsev, 2007).

All the above mentioned parameters were recorded due to the special computer program that permitted to measure the frequency duration and the beginning time of every considered reaction which subsequently arised for 5 min of an animal presence in the open field.

L-lysine was intraperitoneally injected in the doses 0.15, 0.5, 1.5, 5.0, 15.0 and 50.0 mcg per 1 kg of body mass 15 min prior the beginning of investigation session. The control rats were injected with saline in the equivalent volumes. All the rats were grouped 15 animals per one group.

The findings obtained in the "open field" test were represented in absolute values in Tables 8, 9 and 10.

Table 8. The changes of the rat behavior in the central squares (M±m)

Index / Groups	Immobility (s)	Movement (s)	The number of acrossed squares
Control	3.13±0.86	4.13±1.12	2.73±0.28
L-lysine: 0.15 mcg/kg	3.53±0.97	2.80±0.64	2.33±0.27
0.5 mcg/kg	2.87±1.05	2.40±0.39	2.47±0.22
1.5 mcg/kg	0.73±0.21*	1.60±0.33*	2.0±0.0*
5.0 mcg/kg	2.67±0.77	2.93±0.99	2.53±0.32
15.0 mcg/kg	3.40±0.97	2.40±0.41	2.07±0.07*
50.0 mcg/kg	2.93±0.98	2.33±0.41	2.40±0.27

Note: * – $p<0.05$ in comparison with the control values.

The changes of horizontal motor rat activity in the central squares acquired a quite certain tendency. Thus, the latencies of immobility in central squares were significantly decreased under the L-lysine treatment in the dose 1.5 mcg/kg (by 77%, $p<0.05$) (Figure 14). The changes of this index in other groups were insignificant ones.

The essential decrease (by 7-61%) of the movement duration and the number of acrossed central squares in comparison with the control values were found under the influence of L-lysine in all the arranged doses (Figures 15 and 16). The significant changes of the between control group – L-lysine-treated group difference of the movement duration were reached in the case of amino acid dose 1.5 mcg/kg (by 61%, $p<0.05$), and the changes of acrossed square number were significant ones in the both doses 1.5 mcg/kg (by 27%, $p<0.05$) and 15 mcg/kg (by 24%, $p<0.05$). The obtained results confirmed the fact that L-lysine administration provided the shortening of their presence in the lighted central open field squares.

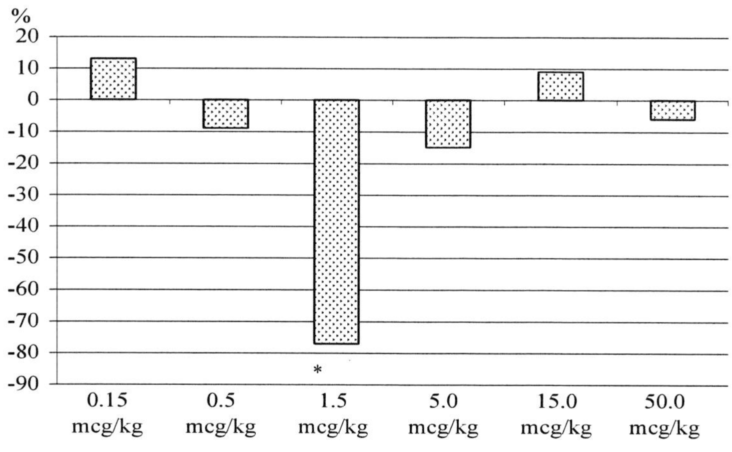

Figore 14. The changes of the immobility duration in the central squares in the "open field" test after L-lysine administration. Columns: the changes of the immobility duration in the central squares in percent in relation to the control values (the left scale). * – p<0.05 in comparison with the control values.

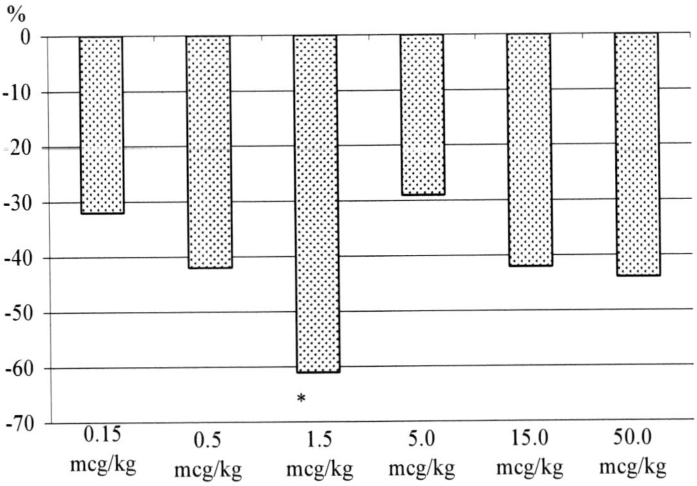

Figore 15. The changes of the movement duration in the central squares in the "open field" test after L-lysine administration. Columns: the changes of the movement duration in the central squares in percent in relation to the control values (the left scale). * – p<0.05 in comparison with the control values.

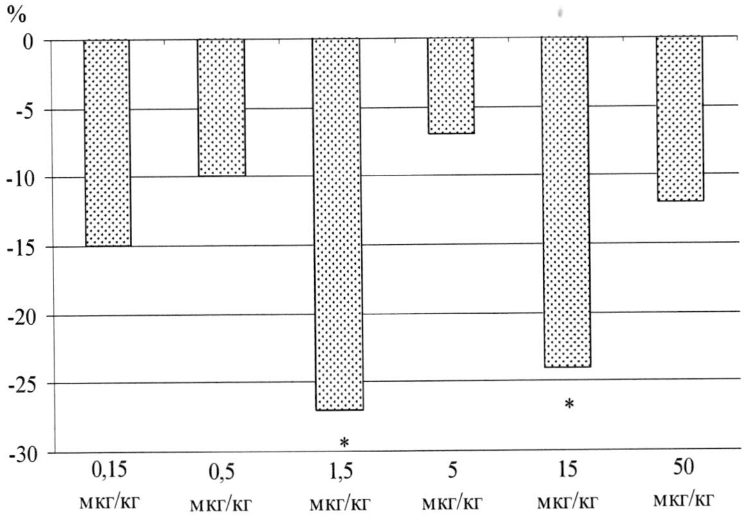

Figure 16. The changes of the acrossed central squares number in the "open field" test after L-lysine administration. Columns: the changes of the acrossed central squares number in percent in relation to the control values (the left scale). * – p<0.05 in comparison with the control values.

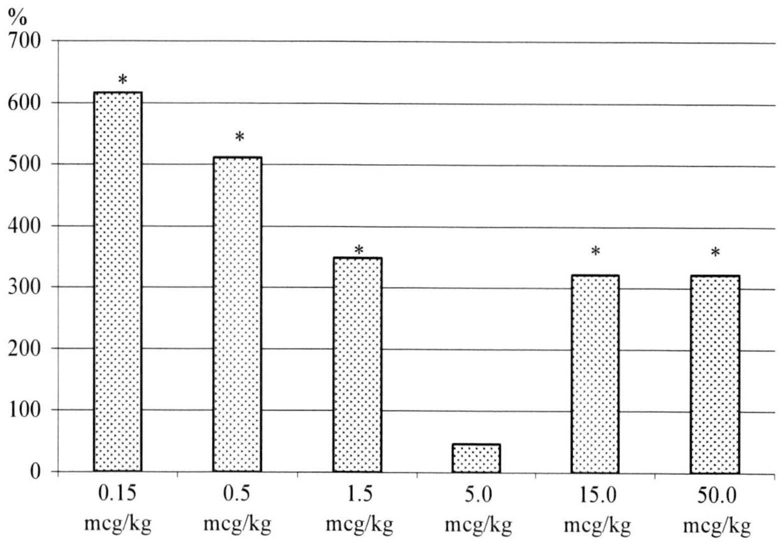

Figure 17. The changes of the immobility duration in the peripheral squares in the "open field" test in rats after L-lysine administration. Columns: the changes of the immobility duration in the peripheral squares in percent in relation to the control values (the left scale). * – p<0.05-0.001 in comparison with the control values.

Since the rats were present in the nonlighted peripheral squares for a longer period of the total experiment duration the findings which were obtained in these conditions played the most important role (Tables 9 and 10).

Table 9. The changes of the horizontal motor activity and emotionality of rats in the peripheral squares (M±m)

Group \ Index	Immobility (s)	Movement in the same place (s)	Movement (s)	The number of acrossed squares	The number of faecal boluses
Control	11.07±6.22	122.73±6.36	87.80±7.94	143.0±17.10	0.87±0.34
L-lysine:					
0.15 mcg/kg	79.20±21.03*	84.20±10.35*	49.27±9.38*	69.60±15.80*	0.93±0.40
0.5 mcg/kg	67.67±13.31*	81.93±10.87*	56.93±10.32*	77.73±16.92*	1.73±0.28*
1.5 mcg/kg	49.53±13.42*	129.60±8.20	54.93±10.46*	85.13±17.92*	2.0±0.39*
5.0 mcg/kg	16.13±5.09	122.47±11.35	84.47±10.90	136.80±22.03	2.73±0.48*
15.0 mcg/kg	46.73±9.19*	152.87±9.08*	32.73±5.29*	58.93±11.68*	3.07±0.54*
50.0 mcg/kg	46.73±14.91*	142.40±11.33	53.53±10.47*	95.93±19.22*	2.60±0.44*

Note: * – p<0.05-0.001 in comparison with the control values.

The increased movement duration (by 46-616%) in comparison with the control one was found in the application of all the used amino acid doses (Figure 17). The most expressed changes of this index were revealed in the rat groups treated with the L-lysine minimum doses: 0.15 and 0.5 mcg/kg (by 511-616%, p<0.01-0.001). Some weakening of the effect was found on increasing L-lysine dose. Thus, the index values of the treated with the dose 1.5 mcg/kg rat group exceeded the control ones by 348% (p<0.05). The administration of the dose 5.0 mcg/kg did not result insignificant effect, however the tendency to the increased immobility duration maintained (by 46%, p>0.05). The further increase of dosage up to 15.0 and 50.0 mcg/kg was followed by the behavioral component enhance by 322% (p<0.05) in both groups.

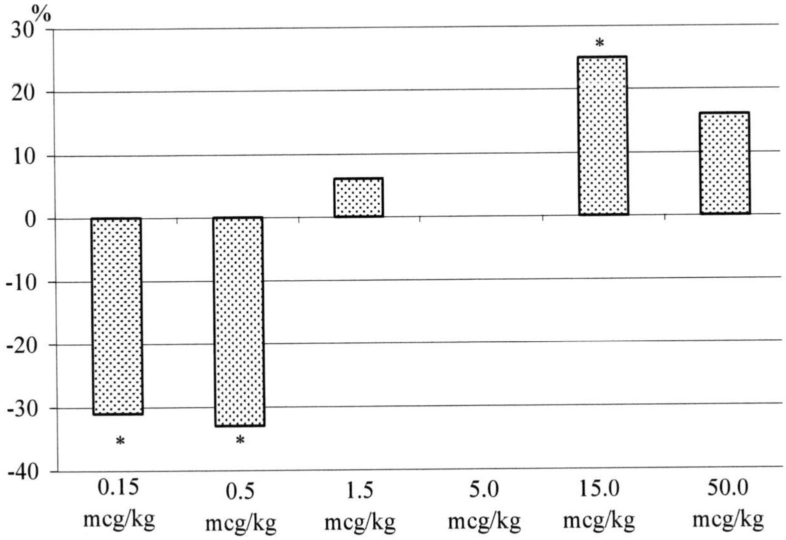

Figure 18. The changes of the movement duration in the same place in the peripheral squares in rats in the "open field" test after L-lysine administration. Columns: the changes of the movement duration in the same place in the peripheral squares in percent in relation to the control values (the left scale). * – p<0.05-0.01 in comparison with the control values.

Table 10. The changes of the rizing on hind paws and grooming in rats in the peripheral squares (M±m)

Index / Group	Rizing on hind paws (s)	The number of rizing on hind paws	Grooming (s)	The number of groomings
Control	34.53±4.06	19.33±2.17	25.13±3.72	3.73±0.69
L-lysine: 0.15 mcg/kg	35.0±7.78	11.20±2.20*	46.73±8.19*	5.33±0.74
0.5 mcg/kg	33.20±5.91	12.87±2.53*	55.67±7.09*	7.67±0.98*
1.5 mcg/kg	33.20±5.45	15.33±2.38	31.47±7.24	3.20±0.68
5.0 mcg/kg	40.20±5.11	21.60±2.95	31.33±6.01	3.93±0.55
15.0 mcg/kg	26.60±4.24	13.33±2.37*	30.60±5.36	4.73±0.72
50.0 mcg/kg	31.80±5.32	14.53±2.49	20.67±4.88	3.27±0.82

Note: * – p<0.05-0.001 in comparison with the control values.

The changes of the movements in the same place were unequal ones after L-lysine injections (Figure 18). The decrease of the component (by 31-33%, p<0.01) was found after amino acid administration in the minimum doses 0.15 and 0.5 mcg/kg. The component indices in the rat groups which were treated with L-lysine doses 1.5 and 5.0 mcg/kg were close to the control values i.e. amino acid effect was not found. However, the tendency to the prolonged movements at site (by 25%, $p<0.05$) appeared on the increasing of the amino acid dose up to 15.0 mcg/kg and also maintained when the dose value was 50 mcg/kg (by 16%, $p>0.05$).

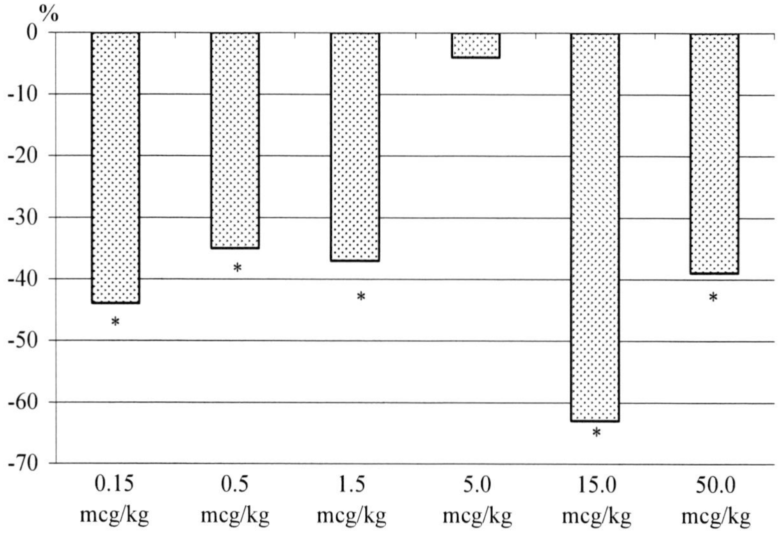

Figure 19. The changes of the movement duration in the peripheral squares in the "open field" test in rats after L-lysine administration. Columns: the changes of the movement duration in the peripheral squares in percent in relation to the control values (the left scale). * – $p<0.05$-0.001 in comparison with the control values

The horizontal motor activity according to the locomotion period duration and the number of acrossed squares manifested the inverse dependence on the immobility duration. The essential shortening of the locomotion period was shown in the majority of groups (by 35-63%, $p<0.05$-0.001) (Figure 19). The significant difference from the control values was not revealed only in the case of dose 5.0 mcg/kg.

The similar changes were also found when the number of accrossed squares was estimated (Figure 20). L-lysine administration in the most used doses with the exception of the dose 5.0 mcg/kg was followed by the weakened horizontal motor activity that manifested by the decreased number of the acrossed squares (by 33-59%, $p<0.05$-0.001).

The changes of feacal boluses number (Figure 21) that was one of the important behavior components and reflected the emotionality of animals exhibited the notable dose-dependent character. The L-lysine effect in the minimum dose 0.15 mcg/kg was close to the control one i.e. it was not sufficiently expressed. However, on increasing of amino acid dose in the rest groups the significant increase of bolus number was revealed (by 100-254%, $p<0.05$-0.01). In this connection it is possible to conclude about the enhanced emotional behavioral component in the open field test under L-lysine influence.

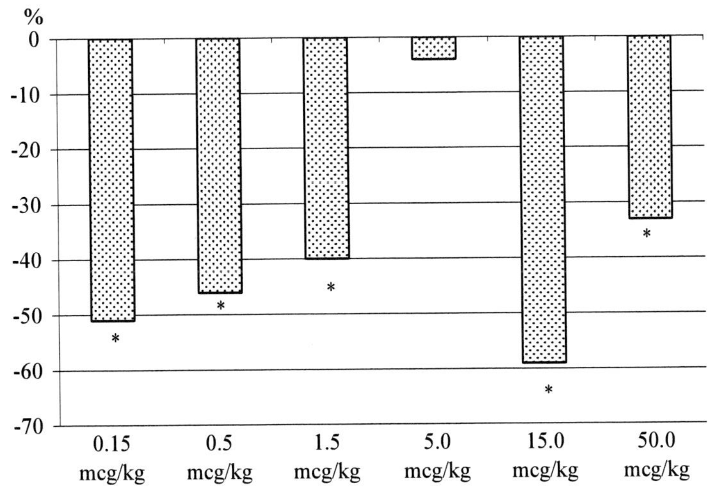

Figurer 20. The changes of the acrossed peripheral squares number in the "open field" test in rats after L-lysine administration. Columns: the changes of the acrossed peripheral squares number in percent in relation to the control values (the left scale). * – p<0.05-0.001 in comparison with the control values

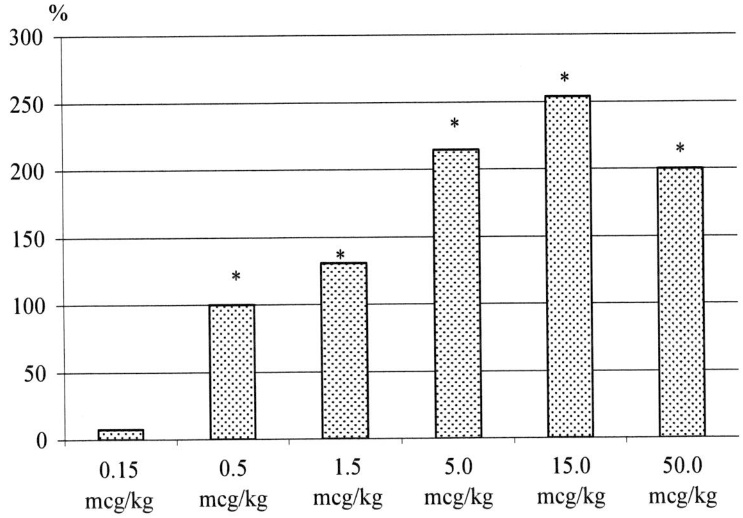

Figure 21. The changes of the feacal boluses number in the peripheral squares in the "open field" test in rats after L-lysine administration. Columns: the changes of the feacal boluses number in the peripheral squares in percent in relation to the control values (the left scale). * – p<0.05-0.01 in comparison with the control values.

The studying of the vertical motor rat activity was carried out by means of the estimation of the frequency and duration of rising on hind paws. The changes of this activity duration seemed to be insignificant (Figure 22). In the most rat groups the essential decrease of the rising number in comparison with control values was noted (Figure 23). The significant difference was reached in the cases of doses 0.15 mcg/kg (by 42%, p<0.05), 0.5 mcg/kg (by 33%, p<0.05) and 15.0 mcg/kg (by 31%, p<0.05).

The frequency and duration of groomings were also investigated. The significant changes of this behavior duration were determined on using of minimum L-lysine doses 0.15 and 0.5 mcg/kg. This was increase of grooming duration by 86% ($p<0.05$) and 121% ($p<0.001$) respectively (Figure 24). The amino acid administration in the greater doses produced insignificant results. The grooming frequency changed in the similar manner (Figure 25). The only difference was the significant change in the case of the only dose 0.5 mcg/kg (by 105%, $p<0.01$).

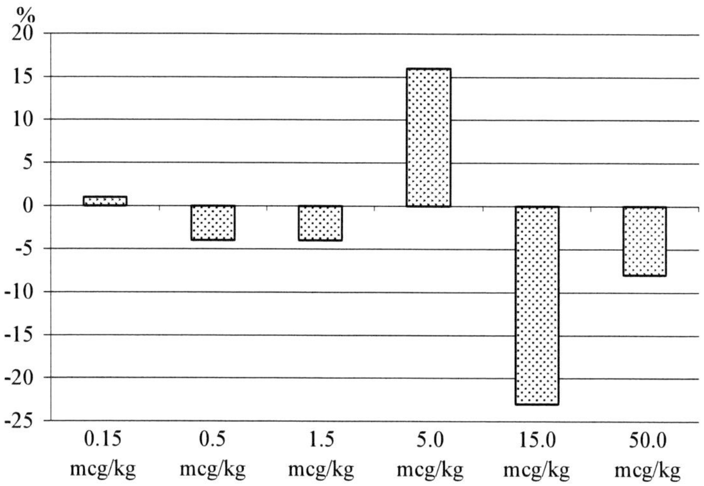

Figure 22. The changes in the duration of the rizing on hind paws in the peripheral squares in the "open field" test in rats after L-lysine administration. Columns: the changes in the duration of the rizing on hind paws in the peripheral squares in percent in relation to the control values (the left scale).

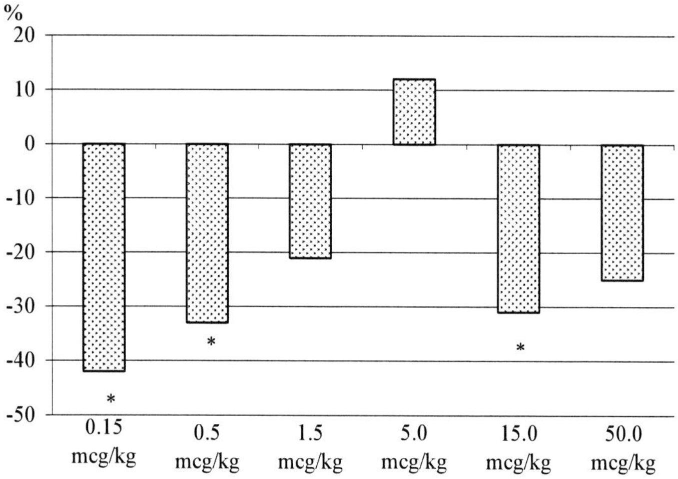

Figure 23. The changes in the number of the rizing on hind paws in the peripheral squares in the "open field" test in rats after L-lysine administration. Columns: the changes in the number of the rizing on hind paws in the peripheral squares in percent in relation to the control values (the left scale). * – $p<0.05$ in comparison with the control values.

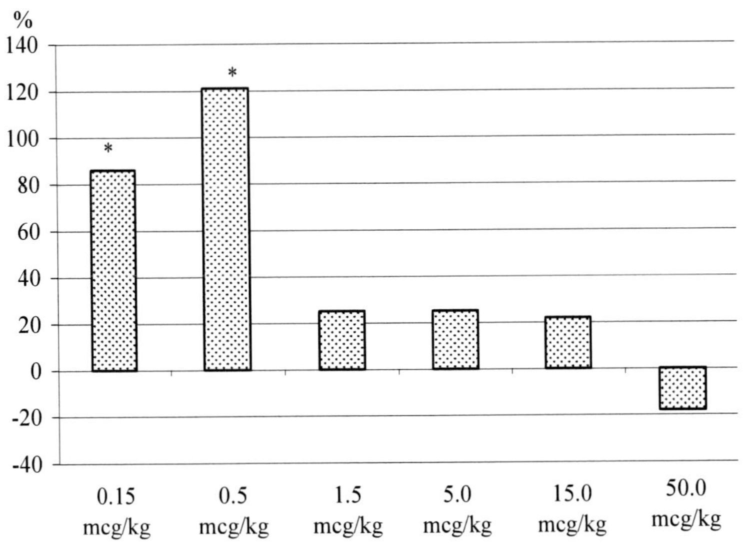

Figore 24. The changes of the grooming duration in the peripheral squares in the "open field" test in rats after L-lysine administration. Columns: the changes of the grooming duration in the peripheral squares in percent in relation to the control values (the left scale). * – p<0.05-0.001 in comparison with the control values.

In summary, the obtained findings showed that the L-lysine administration in the chosen dosage arragement inhibited the horizontal and vertical locomotion activity in the association with the emotionality and grooming reactions. This designated that L-lysine injections enhanced the animal response to the unknown environment.

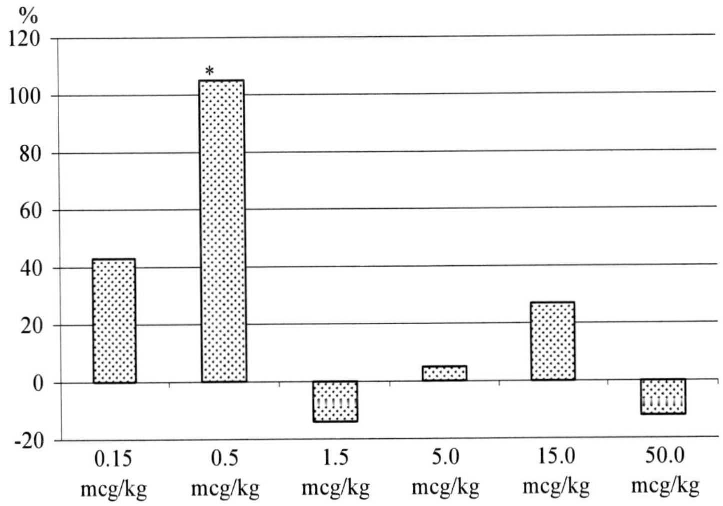

Figure 25. The changes of the groomings in the peripheral squares in the "open field" test in rats after L-lysine administration. Columns: the changes of the groomings in the peripheral squares in percent in relation to the control values (the left scale). * – p<0.01 in comparison with the control values.

2.2.3. The Amino Acid Influence on the Feeding Behavior in T-maze

The experiments were carried out on Wistar male rats with body mass 250-280 g. The food conditioned reflex was elaborated due to the certain site in T-maze and 9 days later after finishing of learning session the maintenance of the conditioned reflex was checked (I. Ashmarin, A. Kamensky, S. Melekhov, 1978). The experiments were carried out in T-maze. The recording of the necessary indices was realized by the special computer program.

Before the learning session the rats were placed in T-maze for 30 min during 2 days to become adapted to the condition. The animals took the limited food in the bread balls form, which were put at the corresponding T-maze site. Food deprivation lasted 48 hours.

The conditioning was being conducted from 9 a.m. till 2 p.m. per day for 4 days. The rats might fulfill 5 trials one-by-one with an inter-trial interval 1 min. The duration of each trial was 3 min. The bread balls were put on the T-maze branch shelves and played a role of refreshment. L-lysine was administered in the doses 0.5 and 5.0 mcg/kg 15 min before the experimental session.

During each experimental session the following findings were recorded: behavior latency (the period from the moment of a rat putting into T-maze till its removing from the start-chamber), reaction time (that was necessary for a rat to come into the corresponding T-maze branch and eat a bread ball on a shelf), the member of mistakes (coming into improper T-maze branch), the number of nonrealized reactions (a rat could not find bread balls for 3 min when it was present in T-maze). The performance of 5 responses in 5 trials and during the specified time was considered the learning criterion.

The maintenance of the elaborated conditioned reflex was investigated 9 days after learning session due to the similar method. The animals were injected intraperitoneally with saline in the volume 2 ml per 1 kg of body mass. During the experiment the following findings were recorded: the number of correct reactions (the estimation of the conditioning level), latency of each response, the duration of reaction, the number of mistakes. The degree of conditioned reflex maintenance was determined due to the correct responses percentage. Only the trained rats were included into statistical assessment.

L-lysine was administered in the doses 0.5 and 5.0 mc/kg 15 min before the experiment onset. The control rats were injected with the equivalent saline volume. All the animals were grouped, 26-27 rats per one group.

The preliminary analysis of L-lysine influence on food conditioned reflex behavior in T-maze showed that the percentage of trained amino acid-treated rats was greater in the comparison with the control value. Thus, on the 4th day of leaning 42-46% L-lysine-treated rats performed the conditioned reflex behavior while the correct performance was observed in 15% trained control animals (Table 11).

Table 11. The distribution of rats into the groups in the investigation of L-lysine influence on a feeding behavior in T-maze

Groups / Numbers of rats	Control	0.5 mcg/kg	5.0 mcg/kg
The initial number	53	26	26
The number of the trained rats	8 (15%)	12 (46%)	11 (42%)

Table 12. The latencies during the elaboration of a food conditioned reflex (s, M±m)

Series of experiments	The days of learning			
	1	2	3	4
Control	72.5±18.1	75.9±24.4	69.5±21.3	21.9±5.6
L-lysine: 0.5 mcg/kg	43.7±11.2	47.7±9.9	36.3±10.2	13.8±5.8
5.0 mcg/kg	72.0±17.8	67.3±17.4	39.3±12.7	13.3±4.4

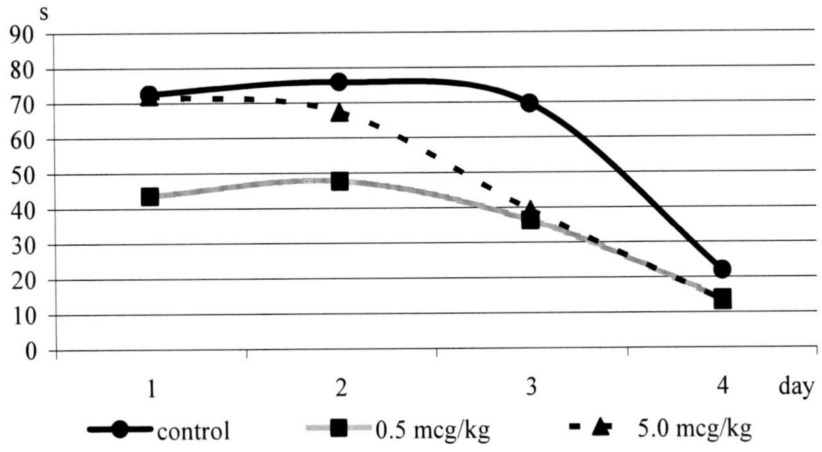

Figure 26. The changes of the latencies during the elaboration of a food conditioned reflex after L-lysine administration.

L-lysine administration resulted in the dose-dependent activating effect on the food conditioning process. However, the process of rat learning performed unindectically in the use of distinct doses. The amino acid administration in the dose 0.5 mcg/kg induced shortening of the response duration (Figure 27, Table 13) and reaching the significant difference from control values on the 4th learning day (by 46%, $p<0.05$). The dynamic changes of this index might be noted, namely: in the first 3 days of learning session the tendency to prolong reaction time (by 11-87%) was found in contrast to the 4th day value. The changes of mistake number in the learning process were identical to the reaction time ones (Figure 29 and Table 15) but did not reach the significant threshold. The latencies and number of unperformed reactions also did not reach the significant inter-group difference (Figure 26, 28, Tables 12, 14).

The reproduction of elaborated conditioned food reflex on the 13th day was revealed in 67% of the trained rats that exceeded by 50% ($p>0.05$) the index value of the control animals (Table 16 & Figure 30). The latencies of the rat group which received L-lysine in the dose 0.5 mcg/kg decreased by 37% ($p>0.05$). The changes of the reaction time and number of mistakes were insignificant.

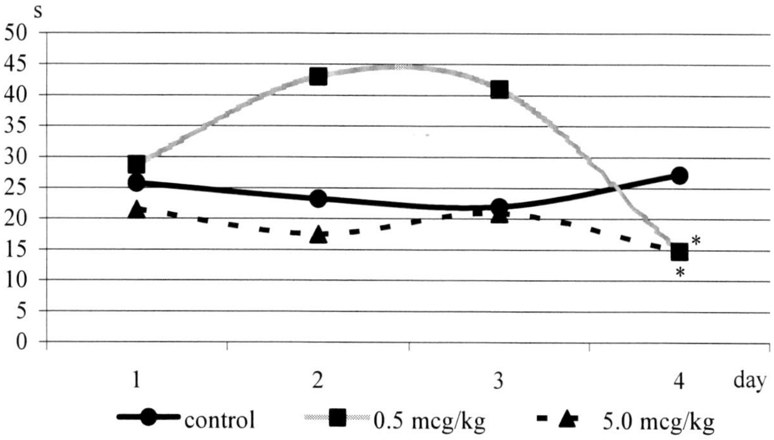

Figure 27. The changes in the duration of the conditioned reflex response during the elaboration of a food conditioned reflex after L-lysine administration. Designations: * – p<0.05 in comparison with the control values.

Table 13. The duration of the reflex response for the period of the food conditioned reflex elaboration (s, M±m)

Series of experiments	The days of learning			
	1	2	3	4
Control	25.7±6.4	23.2±6.6	21.9±3.5	27.1±5.6
L-lysine: 0.5 mcg/kg	28.7±5.9	43.0±10.4	41.0±13.5	14.8±1.7*
5.0 mcg/kg	21.5±3.0	17.5±3.6	20.9±4.9	14.8±3.2*

Note: * – p<0.05 in comparison with the control values.

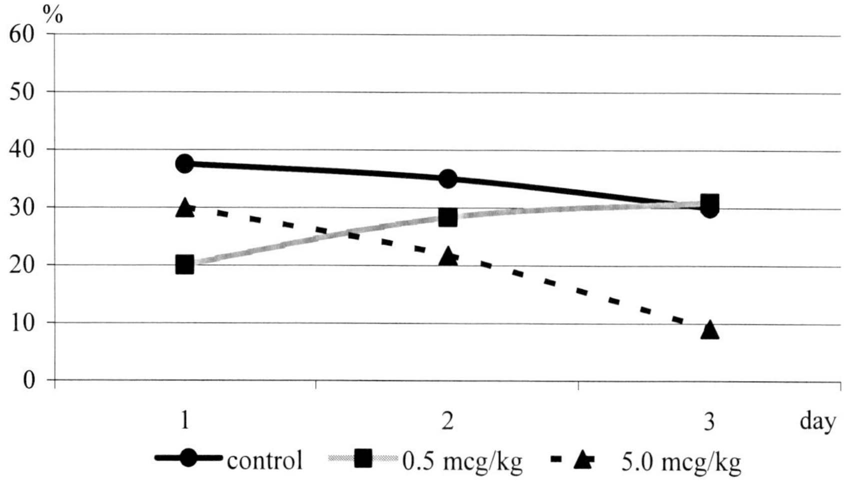

Figure 28. The changes in the unfulfilled reactions number during the elaboration of a food conditioned reflex after L-lysine administration.

Table 14. The number of the unfulfilled reactions during the elaboration of a food conditioned reflex (%, M±m)

Series of experiments	The days of learning		
	1	2	3
Control	37.5±12.8	35.0±13.5	30.0±13.1
L-lysine: 0.5 mcg/kg	20.0±8.2	28.3±10.3	21.7±10.0
5.0 mcg/kg	36.4±12.3	30.9±10.2	9.1±6.2

Table 15. The number of mistakes during the elaboration of a food conditioned reflex (M±m)

Series of experiments	The days of learning			
	1	2	3	4
Control	1.4±0.3	1.1±0.3	1.1±0.4	1.4±0.3
L-lysine: 0.5 mcg/kg	1.6±0.2	1.5±0.3	1.6±0.4	1.0±0.2
5.0 mcg/kg	1.0±0.1	0.8±0.2	1.2±0.2	0.5±0.2*

Note: * – $p<0.05$ in comparison with the control values.

L-lysine administration in the greater dose – 5.0 mcg/kg – changed the conditioning and maintenance of feed conditioned reflex in the similar manner. The leaning level of the mentioned rat group that was determined due to the latencies and reaction time, seemed to be greater than one of the control group. In the L-lysine-treated group the percentage of leaning was shown as follows: on the first day – 46% rats and for the first 3 days – 82% while in the control group the conditioned reflex was elaborated only in 50%.

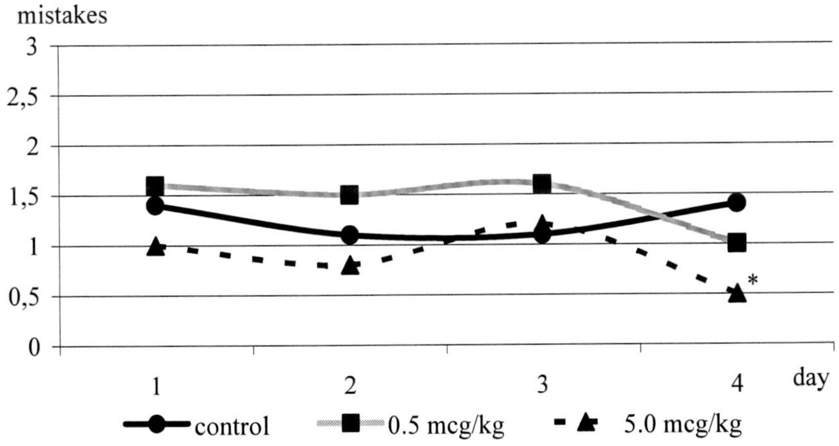

Figure 29. The changes of the mistakes number during the elaboration of a food conditioned reflex after L-lysine administration. Designations are the same as in Figure 27.

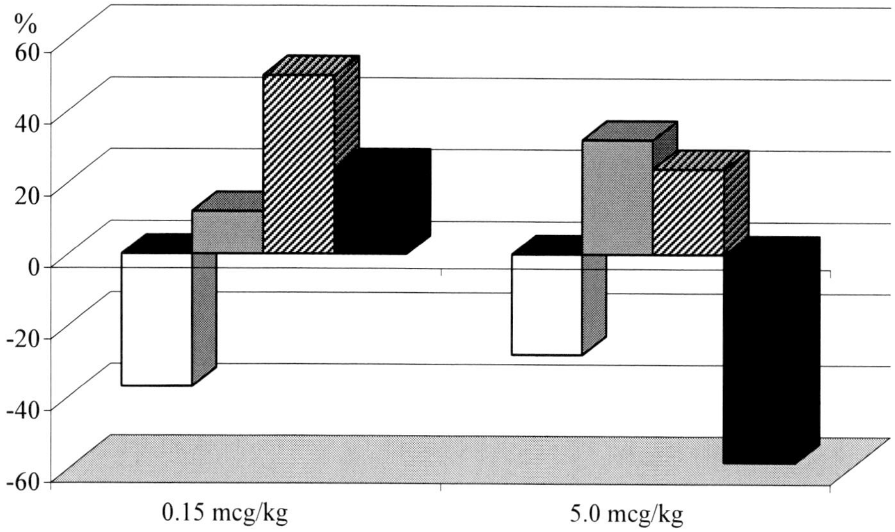

Figure 30. The changes of the indices which are used to estimate the maintenance of the feeding conditioned behavior. Designations: columns – the changes of the latencies (white), the reaction duration (grey), the number of unfulfilled reactions (shaded), the number of mistakes (black) in percent in relation to the control values (the left scale).

As shown in Figure 27 the significant decrease of reaction time (by 46%, $p<0.05$) was found on the 4th day in the rats which received an amino acid in the dose 5 mcg/kg and reproduced for the whole period of conditioned reflex elaboration. It was less than in the control group by 5-25% ($p>0.05$).

The number of runnings to the uncorresponding T-maze-branch was less in the majority of the amino acid-treated rats (by 27-60%) up to the reaching of significance on the 4th learning day (Figure 29, Table 15). The changes of latencies and the number of missed responses were insignificant (Figures 26, 28, Tables 12, 14). The elaborated conditioned reflex was maintained in 46% rats of the group. Exceeding the control values under L-lysine influence was equal to 24% ($p>0.05$) (Figure 30, Table 16). The number of mistakes on the 13th day decreased by 58% ($p>0.05$) in comparison with control values. The significant changes of latencies and reaction time were not revealed. The tendency to improvement of food conditioned reflex maintenance under L-lysine action could be noted.

In summary, L-lysine effects on learning and memory on the base of positive emotional conditioning showed clear dose-dependent character. The improvement of feeding conditioned reflex elaboration was determined on increasing amino acid dose.

Table 16. The indices of the food conditioned reflex maintenance (M±m)

Series of experiments	Latencies	Reaction duration	The number of the correct responses	The number of mistakes
Control	86.0±22.4	21.0±4.0	50.0±15.6	0.9±0.3
L-lysine: 0.5 mcg/kg	54.4±17.5	23.5±4.3	75.0±11.6	1.1±0.3
5.0 mcg/kg	61.6±17.1	27.7±8.0	61.8±11.6	0.4±0.1

2.3. The Analysis of L-Lysine Influence on the Integrative Brain Activity

To analyze the obtained data it should be noted first of all that in pain-induced behavior formation the amino acid effects depended on the dose, pain sensitivity kind and biological significance of the situation of stimulus action. Thus, L-lysine action on thermoalgesic sensitivity was characterized by parabolic dose-dependence curve: it was algesic after the minimum and maximum dose administration (0.15 and 50.0 mcg/kg) and analgesic one after intermediary doses (1.5 and 5.0 mcg/kg). However, it showed be noted that amino acid action in all the doses resulted in the increase of avoidance reaction frequency in relation to thermoalgesic stimulation.

Concerning L-lysine influence on "hot plate" test it is necessary first of all to take into consideration the biological significance of animal behavior components during the full period – 7 min, when they experienced thermoalgesic stimulation 4 times. The primary animal reaction – hind paw licking which arised on the first exposure would be estimated as the most important component that was the pain sensitivity measure. It weakened on the next exposures being replaced with the avoidance reaction, namely with jumping to the cylinder margin top. This kind of behavior had more complex organization and depended not only on the nociceptive perception but on the superspinal interactions which determined avoidance behavior. On the whole, the obtained findings may be estimated as the L-lysine-provided enhance of the thermal pain-evoked behavior.

The amino acid effects on the behavior induced by the electric painful skin irritation seemed to be unidentical depending on the biological significance of the concrete situation.

Thus, L-lysine administration in the used doses enhanced the pain sensitivity and affective defense manifestations in the electrical skin tail stimulation. Meanwhile the amino acid action after systemic administration was characterized by U-curved-dependence on dose concerning aggressive-defensive behavior that was induced by the gradually increasing inescapable stimulation of paired rats in electric floor chamber. In this situation L-lysine administration in the intermediary doses evoked algesic and aggressogenic effects but the action was opposite in the cases of the minimum (0.5 mcg/kg) and maximum (50.0 mcg/kg) doses administration.

To assess the obtained findings it is necessary to stress that the investigated behavior kinds were provoked by the electric current painful stimulation of equivalent intensity but in the situation of different biological significance. Namely that determined the behavior peculiarities. Thus, the defensive behavior seemed to be the greater expressed on the electrical tail stimulation. The electrode site biting might be considered as the affective defense manifestation namely: that was the attack on the physical thing which played a role of pain reason. In this situation the spinal pain reflex seemed to be more sensitive to L-lysine action. The activating amino acid effect on the supraspinal brain regions which were involved into the defense behavior formation, enhanced with dose increasing.

Quite another situation was reproduced when paired rats were put on the electric chamber floor and their communication, visual and smell perceptions were possible simultaneously with painful paw irritation. In such situation the inescapable increasing stimulation and pain resulted in the consummatory outbursts of aggression in the form of typical species attacks against a partner. In this condition L-lysine action was either activating or inhibitory depending on the dose.

It is known that the defense and aggression as different behavior kinds are also provided by different neurochemical mechanisms which depend on certain neurotransmitter systems activity, in particular, serotonin-, catecholamine-, choline- and GABA-ergic systems (V. Poshivalov, 1986). The limbic, striatum and cortex functions provide integrative mechanisms of aggression and defense and, in their turn, are realized due to the neurotransmitter systems activity. It is naturally to suggest that L-lysine effects may be realized due to the activity and interactions of neuromediator systems with a certain neuroanatomical localization. This suggestion is confirmed by the findings that serotonin release from the central amygdale nucleus (M. Smriga, M. Kameishi, H. Uneyama, K. Torii, 2002) and norepinephrine secretion in the ventro-medial hypothalamus (M. Smriga, K. Torii, 2000) decrease in the L-lysine deficiency in rats. Besides the amino acid could manifest its properties as the competitive serotonin antagonist which were determined *in vitro* in guinea pigs ileum (M. Smriga, M. Mori, K. Torii, 2000). Thus, L-lysine could exert influence on the brain structures which play an important role in the agonistic-behavior formation.

As for the mechanisms of two kinds of L-lysine action on the foot-shock aggression the "dose – effect" law may be its base since the neurotransmitter system reactions are characterized by this law. Thus, it is known that the intraspecies animal aggression is as a rule enhanced under action of the small doses of GABA, and benzodiazepines as well as the minimum brain serotonin decrease while it is inhibited on increasing of doses (V. Poshivalov, 1986). In our study this regularity may be manifested in the algesic and aggressogenic L-lysine action in the doses 1.5, 5.0, and 15.0 mcg/kg but by the opposite effect of the greater dose – 50.0 mcg/kg. The amino acid administration in the dose 0.5 mcg/kg activated defensive behavior and inhibited aggression that may be the result of its action as the serotonin receptor agonist. The block of these receptors could provide the defensive behavior in the electric painful stimulation, while the cholinergic and adrenergic mechanisms were important for the manifestation of threatening and attacks (V. Poshivalov, 1986). It is possible to suppose that the special interactions of neurotransmitter systems were formed in the electric painful tail stimulation due to which the defensive behavior became prevalent.

At last, the obtained in our study data confirmed the fact that the biological significance of the situation of painful stimulation played the prevalent role in the formation of pain-induced behavior which depended on the supraspinal integrative interactions. This role is greater than the pain stimulation intensity.

In that relation the information of the last years about the brain structures which are activated in nociceptive stimulation, is of particular interest (Y.F. Chang, Y. Wang, R.K. Cauley, X.M. Gao, 1993; A. Hess, M. Sergejeva, L. Budinsky et al., 2007; J.N. Wood, A.N. Akopian, P. Cesare et al., 2000; E. Torebjörk, 2000; L. Villanueva, P.W. Nathan, 2000). The magnetic nuclear resonance was used to investigate anesthetized rats brain areas involved into the response to thermal pain stimulation of hind paw or to the zimozan subcutaneous injection-induced inflammation (A. Hess, M. Sergejeva, L. Budinsky et al., 2007). It was found that cingular, retrosplenal, sensomotor and insular cerebral cortex zones, medial and lateral thalamic nuclei, prethecal area, hypothalamus and periaqueductal grey matter were the most constantly and frequently bilaterally activated. The increased blood oxygenation of these brain regions was observed as the response to the contralateral intact paw stimulation. Thus, the conception of peripheral and central sensitization was added with the discovery of the cerebral sensitization phenomenon.

Table 17. The comparative characteristics of the nociceptive systems under thermal and mechanical stimulation

The components of nociceptive system	Stimulation kinds	
	Thermal	Mechanical
Primary afferents: nociception	blocked by capsaicin	independence on capsaicin
Signal transmission (the neurons of spinal cord posterior horn)	neurokinin A, calcitonin, somatostatin receptors	neurokinin$_1$-receptors
Central sensitization	excitatory amino acids at NMDA-receptors	substance P, cooperative activation of the glutamate- and AMPA-receptors

As for the difference of L-lysine effects during the thermal and electric current painful stimulation these may be provided by the particular mechanisms of each pain kind and pain-induced behavior formation. It should be noted first of all that according to modern concept the low threshold mechanical nociceptors are involved into the response to skin electric current stimulation (C. Weidner, M. Schmelz, R. Schmidt et al., 1999; R. Schmidt, M. Schmelz, C. Forster et al., 1995). For the last years the existence of differentiated thermal and mechanical nociceptive systems was confirmed by the sufficient facts (Table 17). The differentiation begins already on the peripheral nociceptors level and continues in the central nociceptive process (L. Monconduit, L. Bourgeais, J.-F. Bernard et al., 2000; R.D. Treede, W. Margel, 1995). In particular, the difference of excitation conduction along thermoalgesic and mechanical nociceptive pathways in the posterior spinal cord horns is known as follows: the excitation conduction along thermoalgesic pathways involves neurokinin$_2$-receptors, neurokinin A and somatostatin and the conduction along mechanical pathways involves substance P and neurokinin$_1$-receptors.

On activation of both nociceptive systems certain peculiarities of the descending inhibitory influences were also found. The aim of study was to investigate the thermal and mechanical hypersensitivity in the inflammation induced by carrageenan injection. This drug is the inhibitor of serotonin and norepinephrin reuptake that induces analgesic effect due to the enhance of descending antinociceptive influence along the corresponding neurotransmitter inputs into the spinal cord. Simultaneous administration of the drug-inhibitor of the peripheral and central cycloxygenase and consequent involvement of the peripheral and central sensitization phenomena resulted in a different action on the mentioned nociceptive systems though the investigation sessions were performed on the same animals at the same time. The synergic blocking effect on the thermal hyperalgesia and the simple summation effect on the mechanical one were found. It was suggested that the between-effects difference was based on certain peculiarities of the descending inhibitory modulatory influences on the thermal and mechanical nociceptive pathways. It should be noted that the 5-

HT_{1A}-receptors activation is considered the most effective mechanism of the serotoninergic antinociceptive system (F.C. Colpaert, M. Hamon, Z. Wiesenfeld-Hallin, 2006; L. Villanueva, P.W. Nathan, 2000).

Certain L-lysine influence on the brain neurotransmitter systems might be the base of its effects on the other investigated behavior kinds. Thus, systemic amino acid administration depressed exploratory activity and enhanced emotionality manifestations in the "open field" and "glass arena" tests. In other words, the amino acid intensified the animal reaction to the unknown environment. The predominant tendency to enhanced grooming under L-lysine influence in these conditions might be assessed as the behavioral "shunting" mechanism that decreased the excessive strain induced by the anxiety and fear experience (R. Ilyuchenok, 1972).

There was the evidence to suppose that the mentioned amino acid effects were due to its action on the GABA-benzodiazepine complex which activity influenced the development of anxiety-depression disorders (A. Kalueff, 2006). Common for humans and animals map-scheme was composed to illustrate the brain GABA-ergic regions involvement in the anxiety and depression pathogenesis. The following brain structures were included in the map: amygdala, hippocampus, hypothalamus, prefrontal cortex and mesencephalon. In the studies the exploratory activity inhibition in "the forced novelty" condition (the "open field" test) and "vegetative" behavior (defecations, urinations, grooming) were presented as the experimental anxiety models.

As for the L-lysine activating influence on the conditioning and maintenance of the food conditioned reflex in rats the mechanisms of the effect might be explained as follows: the short-term and long-term memory improvement and some drive increase. The last effect was found after L-lysine administration in the dose 5.0 mcg/kg and manifested by the reaction-time shortening.

The described L-lysine effects might also exert due to its action on the neurotransmitter systems involved into the memory tracing reproduction. These were known as: choline-, epinephrine-, dopamine-, serotonin-, GABA- and histamine-ergic systems (Yu. Borodkin, P. Shabanov, 1986). The important role is played by the metabolic background on which the specific neurochemical and topographical changes developed that was closely connected with the conditioning and brain peptides activity (R. Kruglikov, 1981). In this connection it is necessary to stress the specific L-lysine role in the brain peptides modifications that modulated their functions (J.L. Spidel, R.C. Craven, C.B. Wilson et al., 2004; A.M. Martin, D.J. Pouchnik, J.L. Walker, J.J. Wyrick, 2004). Thus, it was found that the L-lysine residues acetylation controlled the activity of histones, highly mobile proteins groups, nuclear and transcription factors and the methylation of lysine specific residues in histones further regulated the chromatin structure and gene expression (J.L. Spidel, R.C. Craven, C.B. Wilson et al., 2004). Besides, these residues modification in the small protein ubiquitin or ubiquitin-like molecules promoted the multiple functions of different proteins.

In summary, in our study the neurotropic L-lysine effects, found by us, manifested in the influence on the integrative brain activity. These effects could be connected with both neurotransmitter and neurochemical mechanisms.

3. The Immunotropic Effects of L-Lysine

3.1. The Modern Concept of L-Lysine Influence on the Immune System Activity and Congenital Defensive Mechanisms

At present the considerable number of data was collected concerning amino acids capacity to influence the immunological reactivity.

The some studies L-lysine was proposed to serve a modulator of the immune system. It was shown that L-lysine intake was followed by its fast absorption by tissues so 5-6 hours later the tissue L-lysine content exceeded the level of other amino acids (A.M. Uhe, G.R. Collier, K. O'Dea, 1992).

In clinical studies it was indicated that L-lysine administration essentially influenced the cellular immune mechanisms in patients with traumatic disease (D. Vologzhanin, N. Kalinina, A. Sosyukin et al., 2005). If the disease development was followed by complications, considerable decrease of plasma L-lysine content was found and explained as the evidence of its increased utilization. In the whole patients group with different traumatic disease development L-lysine level correlated with the absolute account of CD3+ lymphocytes (especially of CD8+ subclass) and spontaneous γ-interferon production (D. Vologzhanin, N. Kalinina, A. Sosyukin et al., 2005). In the group of patients with complicated trauma disease development the analogous analysis revealed the higher correlation connections with the CD4+ cells number. It was concluded that the decrease of serum lysine content correlated with the enhanced cellular immune response (D. Vologzhanin, N. Kalinina, A. Sosyukin et al., 2005). The increased utilization of the amino acid was suggested as the mechanism of this phenomenon. In addition, insufficient lysine content can stimulate the immune system activity.

The considerable lysine influence on the inflammation intensity was revealed in different periods after trauma (D. Vologzhanin, N. Kalinina, A. Sosyukin et al., 2005). During the first week after traumatic injury the increased intensity of inflammation was associated with the elevated amino acid level. However, the interrelation changed. During the second week after trauma the enhancement of inflammation was followed by the decrease of lysine content because of its increased utilization. The special analysis of the rate of inflammation and complications development showed unfavorable prognosis if low lysine content arised during the first week of posttraumatic period.

The investigations were performed to study the influence of L-lysine-hydrochloride on the adaptive immune transplantation of murine splenocytes and macrophages in the presence of BCG-antigen and HBsAg to the intact mice (S. Dasgupta, V. Chandran, A. Bhinge et al., 2004; S. Dasgupta, A. Bhinge, V. Chandran et al., 2003). The elevation of antibody titre in rat groups which received the cell with both antigen kinds, was found in comparison with control values. The immune response changes were the most expressed to the HBsAg-containing cells. It was suggested that L-lysine could act as the biocombined adjuvant and enhanced immune response by means of nonspecific activation and the modulation of cell proliferation.

According to some findings L-lysine stimulated neutrophil phagocytic activity without notable changes of the specific immune response (G. Belokrylov, I. Molchanova,

E. Sorochinskaya, 1986; G. Belokrylov, O. Popova, I. Molchanova et al., 1991). *In vitro* experiments no detoxificating amino acid property in the relation to toxic action of benzene or aflatoxin B_1 on murine splenocytes was shown (G. Belokrylov, O. Popova, E. Sorochinskaya, 1999; G. Belokrylov, O. Derevnina, O. Popova et al., 1996).

It was revealed that in the absence of a specific mitogen L-lysine caused moderate proliferation of healthy donors lymphocytes in vitro (I. Rakityanskaya, A. Kucher, T. Abramova, 2000). Meanwhile, in the presence of mitogen the amino acid inhibited the mononuclear proliferation due to suppression of tritium-thymidine utilization for DNA-synthesis.

In a special series of experiments the membrane transport of cationic amino acids seemed to be essential for the cells which were actively metabolizing L-arginine or L-lysine (D.H. Crawford, S. Chen, C.A.R. Boyd, 1994). It was shown that phytohaemoagglutinin stimulation of peripheral blood T-lymphocytes specifically activated lysine transport though system y+. Both phytohaemoagglutinin-activated CD4+, CD8+ T-cells increased lysine transport through this system. It was also found that y+ activity was markedly increased in CD45RA+ (naive T-cells) proliferation, but not in the CD45RO+ (memory) cells. CD45RA+ lymphocytes response was manifested by the enhanced proliferation and cytotoxicity in relation to alloantigens and interleukin-2 production. CD8+ cell population demonstrated the maximum y+ transport system activity and suppressed autoimmune reactions.

There are findings concerning the changes of the histochemical characteristics of typical tissue basophiles in the majority of the murine lymph nodes (V. Pokrovsky, L. Shulzhenko, 2000). The chronobiological method was used, so it was determined that in winter these cells manifested the lysine residues presence in proteins and putting in good order contact with the marginal zone of lymphoid follicles. In summer the basophile chains disappeared.

The findings are of interest as the metal- and SH-dependent aminopeptidases were discovered in leukemic cells and involved into the realization and regulation of specific cellular functions. Among these enzymes the aminopeptidase was found that caused lysine catabolism (Y. Gureeva, N. Golubeva, O. Lubkova, 1999).

Against the background of lysine deficiency the hepatocyte microsome enzymes induction by xenobiotics (phenobarbital type) caused amino acid mobilization from different organs and tissues to provide detoxificating liver function (T. Nurmagambetov, B. Amirov, T. Kuanysheva, T. Sharmanov, 1991).

The virusological investigations permitted to find that L-lysine was capable to inhibit the Herpes simplex virus replication in cell and to shorten the disease duration (E. Ayala, D. Krikorian, 1989; J.G. Beauman, 2005; R.S. Griffith, D.C. De Long, J.D. Nelson, 1981; R.S. Griffith, A.L. Norins, C. Kagan, 1978; B.B. Singh, J. Udani, S.P. Vinjamury et al., 2005; R.S. Griffith, D.E. Walsh, K.H. Myrmel et al., 1987). The inhibitory effect of L-lysine-alpha-oxidase in relation to the reproduction of Herpes simplex virus type 1 (HSV-1) *in vitro* was proved due to oxidative desamidization (I. Smirnova, S. Diorditsa, S. Alexejev, I. Zaytsev, 1998).

In a special set of experiments L-lysine-alpha-oxidase that was released from Trichoderma sp. was added into infected by HSV-1 cell culture of the Green monkey kidneys in the doses 0.0007, 0.007, 0.07, 0.7 and 17.5 mcg/ml. The L-lysine-alpha-oxidase administration in all the doses with the exception of the maximum one caused antiviral effect.

The highest dose action seemed to be cytotoxic. It was suggested that antivirus enzyme activity was realized inside cells since the synthesis of nuclear and DNA-connecting proteins of HSV-1 virus was inhibited. In addition to these data it was proved that the effective L-lysine-alpha-oxidase concentration seemed to be 100-times less in comparison with the concentration of traditional antiherpes drugs.

The further studies of different enzyme forms determined its noticeable activity in relation to the synthesis of HSV-1 and HSV-2 virus antigens (I. Smirnova, S. Alexejev, S. Diorditsa et al., 1999; A. Sclischeva, S. Alexejev, I. Smirnova, V. Podboronov, 2003). By use of morphological histological and immunoblotting methods the effects of soluble and gel L-lysine-alpha-oxidase forms were estimated due to application to the HSV-1 affected sites of rabbit skin and eye cornea. The enzyme gel application seemed to be the preferable form of treatment. In addition, the combined administration both gel and water dissolved forms inhibited the most effectively the HSV-2-induced affection in guinea pigs.

For the first time, L-lysine-oxidase and its antitumour, antivirus and antibacterial effects were discovered by Japanese researchers (E. Ishikawa, M. Imagawa, S. Hashida et al., 1983). L-lysine-alpha-oxidase is a glycopeptide with molecular mass 120 000 D (N. Gogichaeva, E. Lukasheva, E. Gavrilova et al., 2000). It catalyzes the oxidative lysine desamidization which resultant in α-keto-α-aminocapronic acid and hydrogen peroxide formation. The enzyme was proved to act only on L-lysine and in quite low degree (6%) on L-ornitine and L-arginine which were considered as the structural analogous matters. L-lysine-alpha-oxidase caused the decrease of L-lysine concentration in blood. It promotes increased intracellular oxidative stress partially due to H_2O_2 synthesis that provides antitumour effect (O. Zhukova, N. Gogichaeva, E. Lukashova, T. Berezov, 2001; H. Kusakabe, K. Kodama, A. Kuninaka et al., 1980). The most expressed antiproliferative L-lysine-alpha-oxidase effects were obtained in relation to the primary tumours in animals. In vivo experiments it was shown that this enzyme possessed a more wide spectrum of antitumor actions in comparison with L-asparaginase and was suggested to be perspective in chemical tumor therapy (I. Smirnova, Ya. Didzhyapetrene, S. Alexejev et al., 2001).

The effective decrease of the facial herpes and herpetic neuralgia symptoms was observed due to the application of lysine containing ointment in combination with zinc oxide, lithium carbonate and some plant extracts (B.B. Singh, J. Udani, S.P. Vinjamury et al., 2005). It should be stressed that in majority of patients (87%) the disease symptoms disappeared on the 6th day of treatment instead of usual 21-day duration of the diseased period.

As the addition to the standard set of drugs L-lysine in an oral dose 1000 mg is used in genital herpes treatment. On the whole, the obtained data may be estimated as a proof of L-lysine antivirus activity.

Generally, all the above reviews demonstrated some separate direct and indirect L-lysine effects on the immune system reactivity. In particular, the selective L-lysine incorporation into T-lymphocytes and the influence on DNA-synthesis as the most significant universal mechanism of cell activity should be stressed. However, at present the reported separate findings could not be integrated into the concept of the systemic L-lysine action on the defensive organism function. In addition, the problem of L-lysine influence on the nervous and immune systems interactions presents considerable interest.

Based on the above introduction analysis we have determined the aim of our research.

3.2. Investigation of Amino Acid Immunotropic Effects in the Non-Stress and Emotional-Painful Stress Conditions of Animals

The humoral immune response, the delayed-type hypersensitivity development, as well as oxygen-dependent bactericidal neutrophil activity and the phagocytic neutrophil activity were investigated for appraisal of L-lysine influence on the immunological reactivity and the congenital defensive mechanisms.

Amino acid effects were studied in the emotional-neutral situation for animals and in the immunosupression condition, provoked by painful-emotional stress, as the particular phenomenon of the interaction between the nervous and immune systems.

The investigation has been conducted in 321 Wistar rats males and 106 CBA mice males divided in the groups including 10 animals per one. Conditions of their life were described in Chapter 2.

3.2.1. L-Lysine Influence on the Humoral Immune Response Development in the Non-Stress and Emotional-Painful Stress Conditions of Rats

As the antigen for the rats immunization T-dependent antigen was used – ram erythrocytes, which were kept in the Olsver sterile solution at +4 °C. Before using ram erythrocytes were several times washed in isotonic NaCl solution by 1500 rot/min centrifuging until the time of absolute transparention of oversediment liquid during period of 10 minutes. The antigen was intraperitoneally administered as 5-% meal using the calculation 2 x 10^8 cells per 1 kg of body mass. The humoral immune response was valued in number of antibody forming cells (in 10^6 spleen cells and on all the organ) and also in number of lymphocytes in the spleen. It was determined by the direct local hemolysis method in the Channigham chamber (C. Malberg, E. Siegel, 1987). L-lysine in the doses 0.15, 0.5, 1.5, 5, 15, 50, 150 & 450 mcg/kg 1 time a day during 4 days after ram erythrocytes immunization all the rats was administered intraperitoneally.

Reproduction of the Emotional-Painful Stress

The emotional-painful stress was created by gradually increase irritation of the paired animals in the chamber with electrified floor. Maximum voltage on the lattice was exposure 60 V for rats and for mice – 40 V. We conducted 2 series of foot-shocks one after another with 1 min interval once a day at the same time during the experiment. The duration of the electroshock was 2 min (I. Bobyntsev, L. Severyanova, 2002).

The stress trials began on the next day after the ram erythrocytes immunization of rats and continued for 4 days. The intraperitoneal administration of L-lysine was performed 15 min before electrical irritation. All control animals were injected with saline. Animals were killed on the 5-th day after the ram erythrocytes immunization.

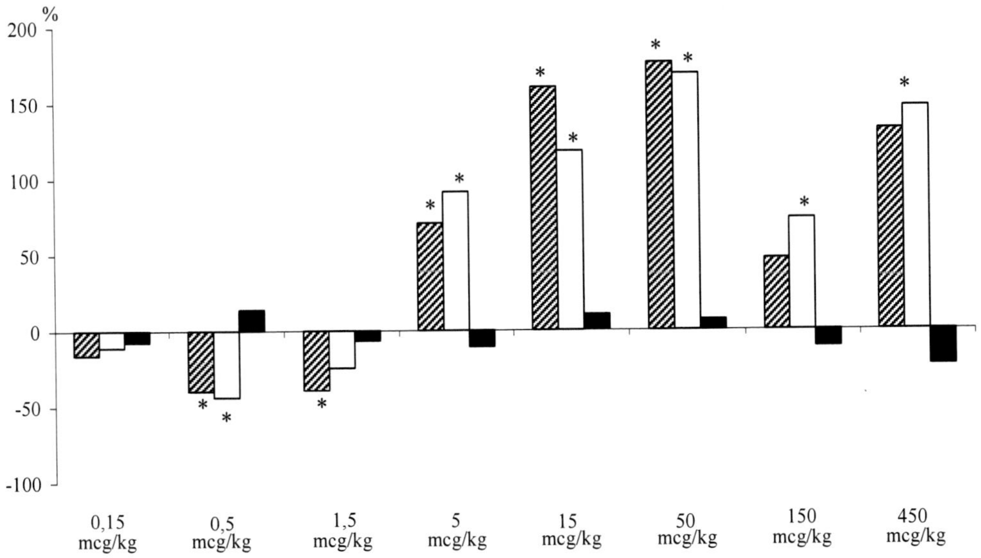

Figure 31. Humoral immune response changes after L-lysine administration. Note: percentage changes of the number of antibody forming cells per the whole spleen (shaded columns), per on 10^6 spleen cells (white columns) & the changes of the number of lymphocytes in the spleen (dark columns), * – $p<0.05$-0.001.

These results of the investigation showed the immunotropic effects of L-lysine and their characteristics. As it is depicted in Table 18 and Figure 31, an amino acid in the majority of used doses influenced essentially the humoral immune response development in the initial condition of rats. Thus, L-lysine administration in the doses 15.0 and 50.0 mcg/kg resulted in the higher and significant increase of the number of antibody forming cells, per 10^6 splenocytes (by 118-169%, $p<0.01$-0.001) and per the whole organ (by 176-182%, $p<0.05$-0.01). Such changes were also reproduced in the L-lysine-treated rat group with the dose 5.0 mcg/kg (by 71-91%, $p<0.05$-0.01) in both per 10^6 spleen cells and per the whole organ. However, no considerable change of the number of lymphocytes in the spleen was found: it was either insignificantly increased in the rat groups which received L-lysine in the doses 15.0 and 50.0 mcg/kg (by 7-10%, $p>0.05$) or decreased in the case of a dose 5.0 mcg/kg (by 11%, $p>0.05$). The significant increase of the number of antibody-forming cells per 10^6 spleen cells was also revealed in the rat groups treated with amino acid in the doses 5.0, 150.0 and 450.0 mcg/kg (by 74-147%, $p<0.01$-0.001). Herewith, no statistically significant changes of the number of antibody-forming cells per the whole spleen and common number of splenocytes were observed.

On the contrary, the administration of L-lysine in the minimum doses 0.5 and 1.5 mcg/kg resulted in the significant decrease of the humoral immune response, namely: the number of antibody-forming cells per 10^6 spleen cells (by 44%, $p<0.01$), as well as per the whole spleen (by 39-40%, $p<0.05$). Such tendency was also determined in the rat group that received L-lysine in the dose 0.15 mcg/kg. No statistically significant changes of the number of lymphocytes in the spleen were revealed.

Table 18. The changes of the number of antibody forming cells (AFC) & the number of lymphocytes (M±m) in the rat spleen after L-lysine administration in non-stress condition

Group \ Index	AFC (cells/organ)	AFC in 10^6 spleen cells	The number of lymphocytes (mill./organ)
Control-1	14936±2062	15.6±1.2	946.6±101.6
L-lysine: 0.15 mcg/kg	12555±2295	13.9±1.6	875.8±74.7
0.5 mcg/kg	8992±1546*	8.7±1.6*	1076.9±135.0
1.5 mcg/kg	9065±1537*	11.8±2.2	883.5±155.1
5.0 mcg/kg	25518±3232*	29.8±3.3*	845.2±46.2
Control-2	20672±3499	20.1±2.0	1002.9±103.4
L-lysine: 15.0 mcg/kg	58376±18048*	43.9±7.2*	1105.6±133.8
50.0 mcg/kg	57116±10840*	54.2±9.8*	1068.7±96.6
150.0 mcg/kg	30377±3969	35.0±2.4*	891.6±129.3
450.0 mcg/kg	48001±15690	49.8±9.5*	770.0±143.2

Note: * – $p<0.05-0.001$ in comparison with the control values.

Thus, the obtained results were testified to the predominant stimulating L-lysine influence on the humoral immune response development in the initial condition of rats.

Analyzing the results of the research we may also conclude that it is rational to evaluate the effects of a matter with the wide spectrum of the biological activity by means of the determination of all the main indices of the humoral immune response in the used stress model. According to the results of our study the number of antibody-forming cells per 10^6 splenocytes may be estimated as the most adequate criterion of the immunotropic action. This criterion does not depend on the amino acid influence on the lymphocytes migration processes, which in particular are not connected with the development of an immune response.

The study of L-lysine influence on the humoral immune response development in the emotional-painful stress conditions of rats showed the following effects of an amino acid (Table 19 & Figure 32). The comparison of the indices between the control intact rats group and the groups which have undergone emotional-painful stress revealed the pronounced decrease of the number of antibody-forming cells per 10^6 splenocytes and per the whole organ (by 65%, $p<0.01-0.001$). Besides, the common number of lymphocytes in the spleen was significantly increased (by 57%, $p<0.01$). It may be suggested that the decrease of the humoral immune response in the stress condition was provided by the possible activation of the T-suppressors and stress-limiting opioid system (E. Kirillina, L. Zakharova, A. Mikhaylova, A. Vasilenko, 1990). The obtained data may be considered as the evidence that the used model provided the sufficient intensity and duration of the emotional-painful stress. So this model was adequate to the aim and tasks of the investigation.

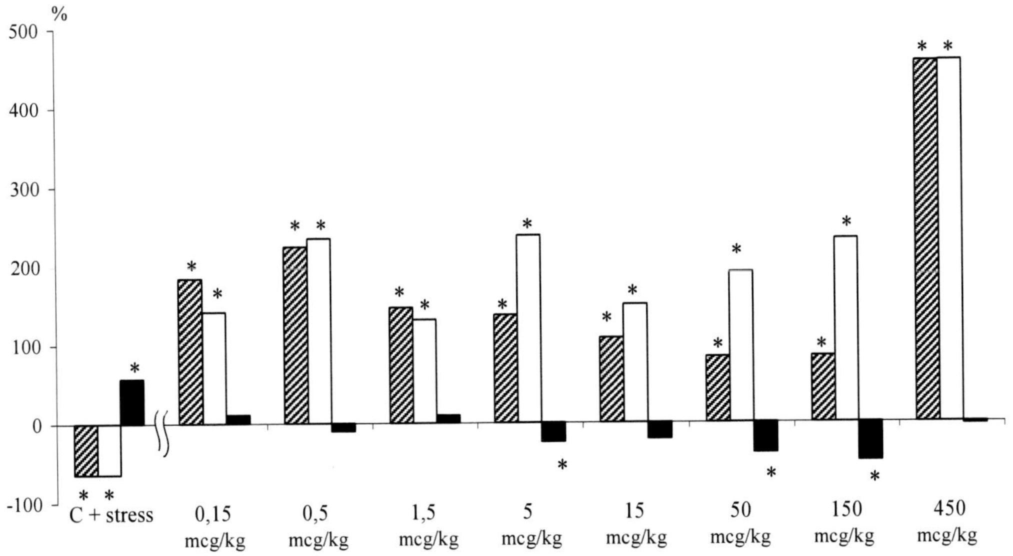

Figure 32. Humoral immune response changes in stress condition and after L-lysine administration. Note: before the broken abscissa-line – percentage changes of the humoral immune response in stress condition; after the broken abscissa-line – percentage changes of the humoral immune response after L-lysine administration in stress condition, * – $p<0.05-0.001$.

In the stress condition the administration of L-lysine in all the doses significantly increased the humoral immune response in comparison with the control values. Thus, the manifestations of the amino acid effects were enhanced. At the same time, the increase of the antibody-forming cells number per 10^6 nuclear splenocytes reached 131-458% ($p<0.01-0.001$) and was associated with the increase of this index per the whole spleen up to 83-457%, ($p<0.05-0.001$). It should be mentioned that the maximum increase of the antibody-forming cells number was found in L-lysine administration in a dose 450.0 mcg/kg. This was a 5.5-fold exceeding of the control values. As for the changes of the nuclear splenocytes number under an amino acid influence in stress condition the index predominantly decreased and was significantly pronounced in L-lysine administration in the doses 5.0, 50.0, and 150.0 mcg/kg (by 25-48%, $p<0.05-0.001$). The opposite effects of the amino acid administration in the doses 0.15, 0.5 and 1.5 mcg/kg in stress condition was found in comparison with the same doses effects in the control (intact) rats.

Table 19. The changes of the number of antibody forming cells (AFC) & the number of lymphocytes (M±m) in the rat spleen after L-lysine administration in stress condition

Index / Group	AFC (cells/organ)	AFC in 10^6 spleen cells	The number of lymphocytes (mill./organ)
Experimental session 1 Non-stressed control	13612±2279	15.6±1.7	870.3±114.7
Control-1	$4822\pm1486^+$	$5.5\pm1.5^+$	881.0±81.2

Table 19. The changes of the number of antibody forming cells (AFC) & the number of lymphocytes (M±m) in the rat spleen after L-lysine administration in stress condition

Index / Group	AFC (cells/organ)	AFC in 10^6 spleen cells	The number of lymphocytes (mill./organ)
L-lysine: 0.15 mcg/kg	13669±2372*	13.3±1.7*	977.8±66.0
0.5 mcg/kg	15630±3981*	18.4±3.2*	791.8±58.5
1.5 mcg/kg	11907±1523*	12.7±1.4*	964.9±98.5
5.0 mcg/kg	11431±1830*	18.6±3.5*	662.9±65.8*
Experimental session 2 Non-stressed control	14513±3625	16.2±4.7	942.2±87.7
Control-2	6986±1406$^+$	4.9±1.1$^+$	1481.6±152.6$^+$
L-lysine: 15.0 mcg/kg	14526±2094*	12.4±1.4*	1170.3±102.3
50.0 mcg/kg	12819±2382*	14.4±2.1*	912.2±87.8*
150.0 mcg/kg	12869±2667*	16.5±2.6*	763.5±96.2*
450.0 mcg/kg	38925±6990*	27.6±3.8*	1433.3±219.8

Note: $^+$ – p<0.05-0.001 in compare with the values of non-stressed animals,
* – p<0.05-0.001 in comparison with the control values.

In summary, in the emotional-painful stress condition L-lysine administration in the wide doses range resulted in the immune stimulating effect that was the analogous one according to the tendency and statistical significance. Since an amino acid stopped the stress-induced immunosuppression it is evident to conclude that L-lysine had a steady stress-limiting effect. In addition, since the immunotropic action of L-lysine was not only maintained but also significantly increased in the stress condition, it was suggested that an amino acid also possessed the immune stimulating action.

3.2.2. L-Lysine Action on the Delayed-Type Hypersensitivity Development in the Non-Stress and Emotional-Painful Stress Conditions of Mice

To study the L-lysine influence on the cell mechanisms of an immune response the delayed hypersensitivity type reaction was reproduced in the CBA males mice (V. Fedosejeva, G. Poryadin, L. Kovalchuk et al., 1993).

To produce the sensibilization condition of mice the ram erythrocytes were intraperitoneally administered as 0.5% suspension in saline (0.1 ml) that contained 10^6 cells. On the 5th day after the sensibilizing injection 10^8 ram erythrocytes in 20.0 mcl saline (the provoking dose) were injected into the foot palm of the right paw (under the aponeurosis), and the control saline dose (20.0 mcl) was injected in the same manner into the left paw.

The delayed-type hypersensitivity development was evaluated in accordance with the inflammation intensity 24 hours later the triggering antigen exposure. Besides, the mice were mortified by means of cervical dislocation, and immediately the both paws were cut at the shank-foot joint level and weighed for the determination of the inflammation index (II) as the percentage difference between the experimental paw (EP) and control paw (CP) masses. The inflammation index was calculated according to the following formula (R. Khaitov, I. Guschin, B. Pinegin, A. Zebrev, 2000):

$$II = \frac{EP_m - CP_m}{CP_m} \times 100\%$$

L-lysine was injected intraperitoneally in the doses 5.0, 15.0, and 50.0 mcg/kg one time per day during 6-days of the experimental session according to the following scheme: on the first day the sensibilizing ram erythrocyte dose and the corresponding L-lysine dose were administered; during the next 4 days the same amino acid dose was injected; and on the last 6th day – the triggering immune response erythrocyte dose and L-lysine injections.

To study delayed hypersensitivity reaction in an emotional-painful stress condition stressful stimulation was performed one time per day during 5 days after mice sensibilization with ram erythrocytes. L-lysine in the doses 5.0, 15.0, and 50.0 mcg/kg was administered 15 min before the beginning of stressful condition. The control animals were injected with the equivalent doses of saline.

The results of the study of L-lysine effects on the inflammation indices of the delayed hypersensitivity development in non-stress condition are presented in Table 20 and Figure 33. As it was shown, the amino acid administration in the maximum used dose 50.0 mcg/kg suppressed the delayed hypersensitivity development and, consequently, the cell immune response. The degree of inflammation index decrease reached 28% ($p<0.05$). Meanwhile in the rat group which was injected with L-lysine in a dose 15.0 mcg/kg the inflammation index values did not differ from the control ones. The administration of a low amino acid dose 5.0 mcg/kg resulted in insignificant decrease of the index (by 3%, $p>0.05$).

Table 20. The changes in the inflammation indices (M±m) of the delayed hypersensitivity development after L-lysine administration in non-stress & stress conditions

Group / Series	Control (n=10;14)	L-lysine		
		5.0 mcg/kg (n=10;12)	15.0 mcg/kg (n=11;12)	50.0 mcg/kg (n=10;13)
Non-stress condition	15.4±1.4	14.9±0.9	15.3±1.0	11.1±1.4*
Stress condition	12.2±1.9	10.8±1.0	15.7±2.0	17.2±2.0*

Note: * – $p<0.05$ in comparison with the control values.

In summary, it may be concluded that L-lysine influenced both the development of humoral immune response and cell immune mechanisms. Thus, the character of the amino acid effects in used doses seemed to have an opposite action in relation to humoral and cellular immune mechanisms, namely: the humoral immune response was activated and the cellular one was suppressed.

In the emotional-painful stress condition L-lysine also influenced the delayed hypersensitivity development (Table 20, Figure 34). The comparison of the inflammation indices in non-stress and stressful conditions showed the immune suppression development (by 21%, $p>0.05$). The obtained data confirmed that the proposed experimental model provided sufficient stress intensity and duration. Concerning the mechanisms of immune suppression the release of glucocorticoids is well known as one of the main anti-inflammatory effect in stress.

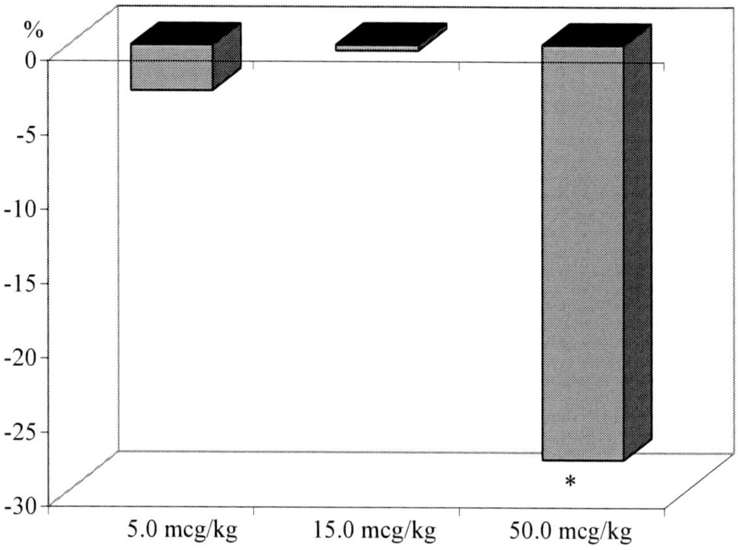

Figure 33. The changes of the inflammation indices in the delayed hypersensitivity reaction after L-lysine administration. Note: percentage changes of the inflammation indices, * – p<0.05.

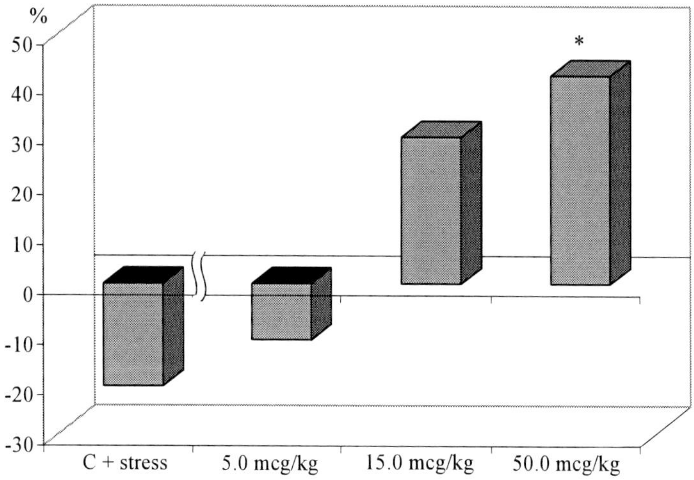

Figure 34. The changes of the inflammation indices in the delayed hypersensitivity reaction in stress condition after L-lysine administration. Note: before the broken abscissa-line – percentage changes of the inflammation indices in stress; after the broken abscissa-line – percentage changes of the inflammation indices after L-lysine administration in stress, * – p<0.05.

In an emotional-painful stress condition the character and expression of L-lysine effects on the cellular immune response essentially differed from the changes in the non-stress case. These concerned the greater amino acid doses administration. In mice injected with L-lysine in a dose 15.0 mcg/kg the inflammation index increased (by 29%), but did not reach the significant values in stress condition. On the contrary, the same amino acid dose administration weakened an inflammatory reaction.

The most expressed increase of an inflammation index (by 42%, $p<0.05$) was found after L-lysine administration in a dose 50.0 mcg/kg. The absolute index values in this group even partially exceeded the inflammation index in non-stressed mice.These data proved that L-lysine in a dose 50.0 mcg/kg essentially stimulated the delayed hypersensitivity development, and, consequently, prevented stressful action, since stress condition resulted in weakened inflammation in the control mice. Based on the obtained findings it may be concluded that L-lysine induced the considerable stress-limiting action on the cellular immune mechanisms.

3.2.3. L-lysine Influence on the Neutrophils Activity in Rats in the Non-Stress and Emotional-Painful Stress Condition

To study L-lysine influence on the indices of neutrophil activity the experiments were performed on Wistar rat males. The amino acid was injected intraperitoneally one time per day during 4 days till the end of experiment. In the experiment set with the reproduction of stress condition the amino acid was administered in the doses 0.5, 5.0, 50.0, and 500.0 mcg/kg 15 min before the beginning of stressful action. The control rats received the equivalent saline volumes.

To get enough neutrophil number 2 ml blood volume was layered on the special mixture surface and centrifugalized during 15 min at 1,500 rotations per 1 min. The oxygen-dependent defensive activity was estimated by means of the spontaneous and stimulated test of nitro-blue tetrazole reduction (NBT-test) (V. Zinkin, M. Godkov, 2004). Both the intact zymosan and opsonized by the 4th group human serum one were used as stimulating agents. The levels of neutrophil activity kinds were evaluated as follows: 1) a spontaneous activity (sNBT); 2) an activity, stimulated by opsonized zymosan (o/zNBT); 3) an activity, stimulated by unopsonized zymosan (u/zNBT). The photometric indices were recorded with Multiskan MCC/340 apparatus ("Labsystems", Finland). The indices were measured in mOD (mille optical density). Based on the above indices the functional activity reserves were evaluated due to the corresponding coefficients as follows: CAo = o/zNBT / sNBT; and CAu = u/zNBT / sNBT.

The different neutrophil activity in response to different stimulation was determined due to the opsonization coefficient: OC = o/zNBT / u/zNBT.

The phagocytic neutrophil activity was studied after 30 min incubation with latex at 37 °C (A. Shtelzner, 1987) and evaluated with the following parameters: the phagocytic number (this is a number of latex particles, phagocytized by neutrophil, PN); the phagocytic index (the percentage of neutrophils, taking part in phagocytosis phenomenon, PI). Both characteristics of phagocytic neutrophil activity were determined due to the viewing of painted blood slides.

The obtained data demonstrated the L-lysine influence on the neutrophil activity indices in the intact rats (Table 21, Figures 35, 36).

The activity of oxygen-dependent bactericidal mechanisms showed the dose-dependent changes. Thus, the amino acid administration in a dose 0.5 mcg/kg caused the decrease of most of indices in spontaneous and stimulated NBT-test. In comparison with control values the oxygen-dependent defensive mechanisms of neutrophils were significantly inhibited (by

15%, p<0.05). The spontaneous neutrophil activity revealed the same tendency in changes, but these were insignificant ones (by 23%). The index o/zNBT in this group of rats was supported on the control value. The coefficients CAo and CAu seemed to be slightly increased (by 34% and 15% respectively, p>0.05) that reflected the slight increase of neutrophil functional reserve. As for the phagocytic activity of polynuclear leukocytes it significantly decreased since the values of phagocytic index and phagocytic number were diminished (by 14%, p<0.05 and by 25%, p<0.01 respectively).

Table 21. Neutrophil activity indices (M±m) after L-lysine administration in non-stress condition

Group / Index	Control	0.5 mcg/kg	5.0 mcg/kg	50.0 mcg/kg	500.0 mcg/kg
Spontaneous NBT-test (mOD)	1.19±0.13	0.92±0.09	1.14±0.09	1.76±0.08*	1.80±0.15*
NBT-test, stimulated by opsonized zymosan (mOD)	1.20±0.07	1.20±0.06	1.45±0.16	1.75±0.14*	1.75±0.16*
NBT-test, stimulated by unopsonized zymosan (mOD)	1.47±0.09	1.25±0.04*	1.38±0.10	1.72±0.12	1.97±0.12*
Coefficient of the activation by opsonized zymosan	1.10±0.11	1.47±0.21	1.35±0.17	1.04±0.11	1.01±0.09
Coefficient of the activation by unopsonized zymosan	1.32±0.11	1.52±0.18	1.26±0.11	0.99±0.08*	1.15±0.09
Opsonization coefficient	0.83±0.05	0.98±0.07	1.08±0.13	1.10±0.15	0.92±0.11
Phagocytic index (%)	59.27±2.75	51.20±2.91*	57.60±2.0	57.80±2.22	66.0±4.47
Phagocytic number	1.23±0.08	0.92±0.08*	0.97±0.08*	1.15±0.08	1.38±0.11

Note: * – p<0.05-0.01 in comparison with the control values.

L-lysine administration in a dose 5.0 mcg/kg did not cause significant changes of spontaneous and stimulated neutrophils activity and functional reserve. It should be stressed that the phagocytic number was significantly decreased (by 22%, p<0.05).

When the amino acid was administered in the greater dose the neutrophil activity increased including the indices sNBT (by 47%, p<0.01) and o/zNBT (by 47%, p<0.01) in the rats, which received L-lysine in a dose 50.0 mcg/kg. The index CAu significantly decreased (by 25%, p<0.05) that testified to functional reserve diminution. No significant changes of a phagocytic index or phagocytic number were shown.

L-lysine injection in a maximum dose 500.0 mcg/kg also caused the significant stimulation of oxygen-dependent bactericidal neutrophil activity that was reflected in the essential increase of the indices sNBT-index (by 51%, p<0.05), o/zNBT (to 46%, p<0.05) and u/zNBT (by 34%, p<0.05). The changes of functional reserve, phagocytic index and phagocytic number seemed to be insignificant.

Thus, L-lysine administration exerted the marked dose-dependent influence on the oxygen-dependent bactericidal and phagocytic neutrophil activities. This influence manifested itself in the decrease of corresponding indices in the case of low amino acid doses administration and essential enhancement after the greater doses administration. It should be noted that increased bactericidal activity arised against the background of the decreased functional reserve.

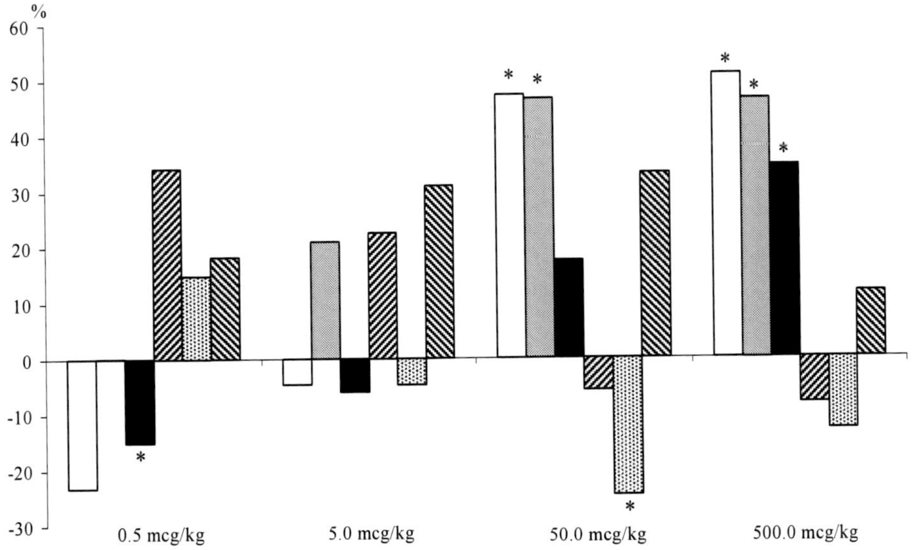

Figure 35. The changes of the neutrophil activity indices in NBT-test after L-lysine administration. Note: percentage changes in mOD of the sNBT (white), o/zNBT (grey), u/zNBT (black), percentage changes of the CAo, CAu & CO (shaded), * – p<0.05-0.01.

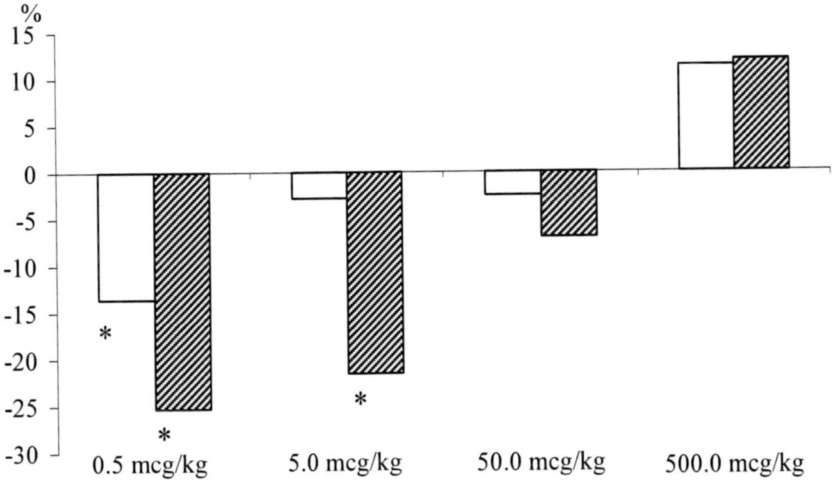

Figure 36. The changes of the phagocytic neutrophil activity indices after L-lysine administration. Note: columns – the percentage changes of the phagocytic index (white) & phagocytic number (shaded), * – p<0.05-0.01.

The nonspecific resistance was also investigated in the emotional-painful stress condition of rats (Table 22, Figures 37, 38). The indices of spontaneous cell activity and unopsonized zymosan-stimulated test insignificantly decreased (by 10% and 9% respectively). The activity of neutrophils in response to opsonized zymosan and functional reserve did not differ virtually from the values of control animals.

Table 22. Neutrophil activity indices (M±m) after L-lysine administration in stress condition

Index \ Group	Control without stress	Control	0.5 mcg/kg	5.0 mcg/kg	50.0 mcg/kg	500.0 mcg/kg
Spontaneous NBT-test (mOD)	1.19±0.13	1.07±0.08	0.96±0.11	0.72±0.05*	1.18±0.09	1.28±0.06*
NBT-test, stimulated by opsonized zymosan (mOD)	1.20±0.08	1.21±0.10	1.06±0.06	1.04±0.08	1.28±0.10	1.19±0.05
NBT-test, stimulated by unopsonized zymosan (mOD)	1.47±0.09	1.34±0.09	1.32±0.07	1.31±0.06	1.40±0.14	1.34±0.06
Coefficient of the activation by opsonized zymosan	1.10±0.11	1.15±0.08	1.25±0.17	1.49±0.13*	1.12±0.08	0.95±0.07
Coefficient of the activation by unopsonized zymosan	1.32±0.11	1.30±0.12	1.55±0.20	1.89±0.14*	1.23±0.12	1.07±0.07
Opsonization coefficient	0.83±0.05	0.90±0.04	0.82±0.06	0.80±0.05	0.96±0.09	0.90±0.04
Phagocytic index (%)	59.27±2.75	71.60±3.51 +	60.40±2.60*	55.20±2.98*	53.0±1.69*	59.55±3.18*
Phagocytic number	1.23±0.08	1.51±0.20	0.99±0.06*	0.89±0.07*	0.77±0.04*	0.96±0.06*

Note: + – $p<0.05$ 001 in comparison with the values of non-stressed animals, * – p<0.05-0.001 in comparison with the control values.

In contrast to this, the phagocytic activity was stimulated under influence of stress that was manifested by the decrease of phagocytic index (by 21%, $p<0.05$) and phagocytic number (by 22%, $p<0.05$). The above described criteria showed the sufficient intensity of stress condition induced in the elaborated model since the polynuclear and macrophages activity could be enhanced in low intensive stress condition (R. Khaitov, V. Leskov, 2001; T. Kirillicheva, N. Goncharov, V. Mitkin et al., 1993).

It should be emphasized that the intensity of stress condition was equal in both the experimental studies of humoral and cellular immune responses and neutrophil activity. In the light of this fact the insignificant changes of neutrophil activity in spontaneous and stimulated NBT-test and increase of phagocytic index may be the evidence of the low sensitivity of these mechanisms to stressful action.

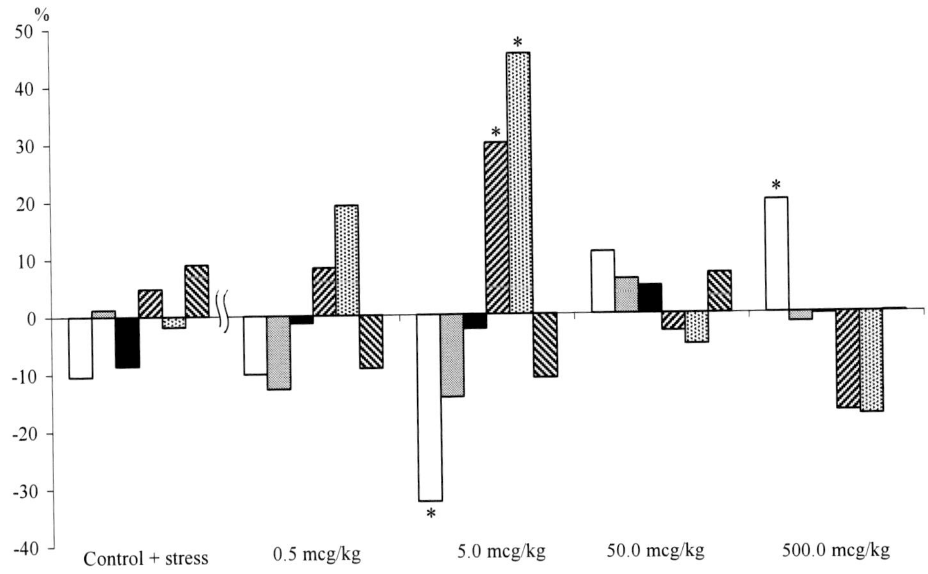

Figure 37. The changes of the neutrophil activity indices in NBT-test in stress condition & after L-lysine administration. Note: the same that on Figore 35.

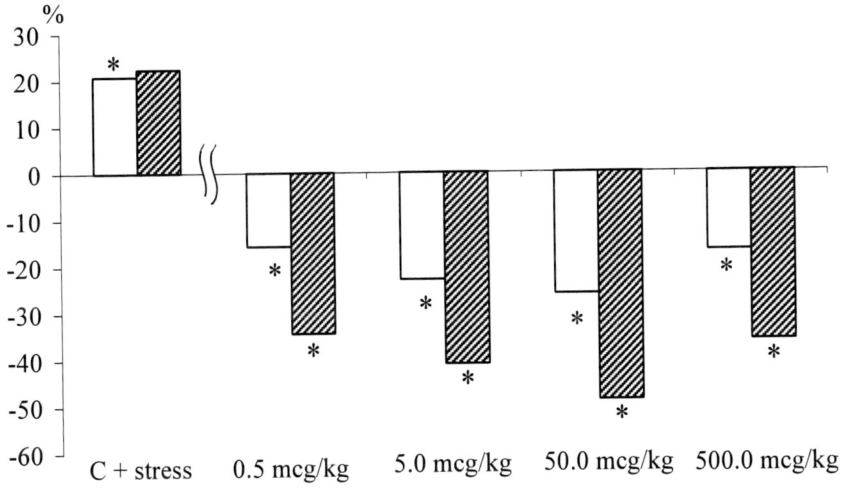

Figure 38. The changes of the phagocytic neutrophil activity indices in stress condition & after L-lysine administration. Note: before the broken abscissa-line – the percentage changes of the phagocytic index (white) & phagocytic number (shaded) in stress; after the broken abscissa-line – percentage changes of these indices after L-lysine administration in stress condition, * – p<0.001-0.05.

No significant changes of neutrophil oxygen-dependent bactericidal activity were shown in stress condition after L-lysine administration in a dose 0.5 mcg/kg. However, the phagocytic index and number were essentially decreased (by 16%, p<0.05 and 34%, p<0.05 respectively).

When the greater amino acid dose 5.0 mcg/kg was administered the neutrophil activity in the spontaneous NBT-test decreased (by 32%, p<0.01). The increased values of CAo and CAu (by 30%, p<0.05 and 45%, p<0.01) serve the manifestations of noticeable functional reserve augmentation. The phagocytic index and number this rat group decreased by 23% and 41% respectively ($p<0.01$).

At last, L-lysine administration in the maximum used dose 500.0 mcg/kg resulted in the increased oxygen-dependent bactericidal neutrophil activity in both stress and non-stress conditions. Thus, this mechanism essentially increased in the spontaneous NBT-test (by 20%, $p<0.05$). Simultaneously, the functional reserve (by 17-18%, $p>0.05$), phagocytic index (by 17%, $p<0.05$) and phagocytic number (by 36%, $p<0.05$) decreased in the group.

Thus, L-lysine administration within the used dose range exerted less expressed influence on the oxygen-dependent bactericidal mechanisms. The stress-limiting amino acid effect was found only after administration in the maximum dose (500.0 mcg/kg). Meanwhile, such congenital defensive mechanisms as the phagocytic polynuclear leukocytes activity were inhibited by L-lysine in stress condition.

In summary, it should be stressed that L-lysine exerted the dose-dependent effects on the oxygen-dependent bactericidal neutrophil activity in rats in stress and non-stress conditions and, consequently, demonstrated stress-limiting action.

3.3. The Analysis of the Mechanisms of L-lysine Immunotropic Effects

Based on the above described data we may conclude that L-lysine influences the humoral and cellular mechanisms of the immune response and congenital defensive mechanisms. Indeed, the amino acid administration in a rather wide range of doses resulted in the immune response activation in non-stressed rats. Only on using the minimum doses (0.5 and 1.5 mcg/kg) the tendency to decrease the antibody-forming cells number per 10^6 nuclear splenocytes and per a whole spleen was revealed. Since the administration of the greater L-lysine doses caused the steady stimulation of a humoral immune response, it may be concluded that the above mentioned amino acid effect is evident.

It should be stressed that L-lysine was injected only 1 time per day during 4 days after the animal immunization with ram erythrocytes, that provided for the physiological but not pharmacological character of the effects. In some rat groups the antibody-forming cells number per 10^6 nuclear splenocytes was three-four fold higher versus the control values. Numerous obvious changes of the humoral immune response confirm the conclusion about a high immunotropic L-lysine activity.

The data obtained in our research correspond to the reviewed findings according to which L-lysine administration could elevate an antibody titre in the animals received BCG-antigen or HBsAg administration (S. Dasgupta, V. Chandran, A. Bhinge et al., 2004; S. Dasgupta,
A. Bhinge, V. Chandran et al., 2003). No concrete facts about the possible mechanisms of these effects realization have been presented. Our own obtained data showed the essential and prolonged changes of the immune response and made it possible to suggest that L-lysine effects may be observed due to both direct action on immune system cells and indirect influence through the regulatory mechanisms. In this connection the interaction of L-lysine with the brain neurotransmitter systems can promote the changes of modulatory nervous system effects.

The elaborated in our research model of an emotional-painful stress condition in rats seemed o be adequate to the aim since its reproduction was followed by the typical stress manifestations including an immune response suppression the possible mechanisms of which are known as the activation of T-suppressors and brain opioid stress-limiting system (E. Kirillina, L. Zakharova, A. Mikhailova, A. Vasilenko, 1990).

In the emotional-painful stress condition the activating L-lysine effect on an humoral immune response was considerably expressed and prevented the development of stress-induced immune suppression in all dose ranges. The absolute numbers of the antibody-forming cells per 10^6 nuclear splenocytes in all the rat groups reached the control values (in intact animals) or suppressed them since the administration of L-lysine prevented stress-induced immune suppression the conclusion is well-grounded concerning the steady stress-limiting amino acid effect. It may be suggested that one of the mechanisms of this effect includes L-lysine anxiolytic action due to inhibition of serotonin binding to $5HT_4$-receptors (M. Smriga, K. Torii, 2003).

L-lysine influence on the cellular mechanisms of the immune response was studied by us with use of delayed hypersensitivity model. The effector link of this immune reaction kind was estimated according to the intensity of an inflammation development at the site of a provoking antigen dose injection. This integrative index is called the index of inflammation. L-lysine administration in a maximum used dose 50.0 mcg/kg resulted in the decrease of inflammatory reaction. On the whole, amino acid effects on an immune response depended on the mechanisms of its development and manifested by the predominant stimulating influence on the humoral immune response and suppressing one on the cellular reaction.

It was reported that L-lysine enhanced the inflammatory reaction at different times after a traumatic injury (D. Vologzhanin, N. Kalinina, A. Sosyukin et al., 2005). It should be emphasized that L-lysine effect was also found if the doses were rather low and administration was rather rare – 1 time per 24 hours during 5-days of the experimental session.

The expressed tendency to the decrease of an inflammation index under stress influence was shown due to the comparison of indices between the control and stressed rats. Such inhibition of the inflammation development at the site of the provoking dose of an antigen injection might be the consequence of the immune suppression and anti-inflammatory glucocorticoids action because of their intensive release in stress condition.

Meanwhile L-lysine administration (50.0 mcg/kg) in an emotional-painful stress condition of rats promoted the development of a delayed hypersensitivity response that was associated with the maximum increase of an inflammation index up to its level in the control non-stressed animals. It may be suggested that the noticeable stress-limiting L-lysine effect promoted the immune inflammation development due to the maintenance of an immune cells activity.

In this connection it is important to emphasize the fact that the stress-limiting amino acid effect was revealed in relation to both effector mechanisms of an immune response.

Concerning the study of L-lysine influence on the congenital non-specific defensive mechanisms only isolated reports were found (G. Belokrylov, I. Molchanova, E. Sorochinskaya, 1986; G. Belokrylov, O. Popova, I. Molchanova et al., 1991). In our study the oxygen-dependent bactericidal and phagocytic neutrophil activities were investigated in the intact (non-stress) and emotional-painful stress conditions of rats. L-lysine exerted the dose-dependent action on these mechanisms of non-specific resistance. The amino acid administration in a minimum dose (0.5 mcg/kg) resulted in the significantly decreased phagocytic neutrophil activity and oxygen-dependent mechanisms which were measured by means of u/zNBT-test. However, when an amino acid dose was increased the effect changed and manifested itself in the essential enhance both the spontaneous and stimulated neutrophil activities with the simultaneous increase of the phagocytic number. It should be noted that the enhance of the oxygen-dependent defensive neutrophil activity was performed against the background of the decreased functional reserve.

The emotional-painful stress in rats was followed by dissimilar changes of non-specific resistance. The indices of the spontaneous and stimulated NBT-test were characterized by the tendency to be decreased while the neutrophil functional reserve did not essentially change. On the contrary the phagocytic index significantly increased in a stress condition.

Such character of the revealed changes can be provided by different sensitivity of the investigated non-specific resistance mechanisms to a stressful action. It is known that a stress of low intensity is capable not only to inhibit but also stimulate the defensive immune response of an organism (E. Korneva, 1993; V. Chereshnev, B. Yushkov, V. Klimin, E. Lebedeva, 2002; E. Korneva, E. Shkhinek, 1988; E. Korneva, 1994; R. Khaitov, V. Leskov, 2001). Based on these data the conclusion may be made that the spontaneous and stimulated activity of the oxygen-dependent bactericidal mechanisms is less resistant to a stressful action than the phagocytic activity.

The delayed hypersensitivity response was also studied with an application of the same emotional-painful stress model. The comparison of the corresponding indices in the non-stress and stress conditions showed that the humoral and cellular immune mechanisms seemed to be more reactive to a stressful action than such non-specific resistance kinds as the phagocytic and oxygen-dependent bactericidal activities of neutrophils.

L-lysine effects on the indices of the oxygen-dependent bactericidal neutrophil activity were more expressed in a stress condition in comparison with their values in the initial (non-stress) condition. The significant stress-limiting amino acid effect was demonstrated when the administered dose reached 500.0 mcg/kg. Simultaneously L-lysine depressed a phagocytic neutrophil activity.

Thus, based on the obtained data we might conclude that L-lysine exerted the dose-dependent effect on the neutrophil oxygen-dependent bactericidal activity in both non-stress and stress conditions of rats when the stress-limiting effect was revealed. On the contrary, L-lysine action on a phagocytic granulocyte activity seemed to be inhibitory and independent of an animal condition.

On the whole, the obtained results suggest, that the activatory L-lysine influence on the development of the antigen-specific humoral and cellular immune response seems to be more expressed and resistant to a stressful action than an amino acid effect on the congenital defensive mechanisms which arised earlier in the evolution process.

As for the mechanisms of the interactions between the nervous, endocrine and immune systems this problem is recognized as very actual and needs to be studied in the future (J. Sandkühler, 2000; A.M. Aloisi, R.M. Craft, S. Marchand, 2006; K.D. Davis, 2006;

W. Jänig, C.R. Chapman, P.G. Green, 2006; D.E. Bentley, U. Baumgärtner, A. Watson et al., 2006).

At present it is determined that the interrelationships of the regulatory systems including the development of a stress condition are realized through the secretion or binding of the signaling molecules as follows: neurotransmitters, regulatory peptides and cytokines which play a role of the intersystem messengers. Thus, the nervous system influence on the immune system activity can be exerted through the sympathetic innervation of the lymphoid organs as well as norepinephrine and neuropeptide Y which are released by the nerve terminals (F. Mignini, V. Streccioni, F. Amenta, 2003; K.C. Torres, L.R. Antonelli, A.L. Souza et al., 2005). The expression of the immune cells receptors was also found in relation to the following neuropeptides: substance P, somatostatin, vasoactive intestinal peptide which are released by C-terminals (M. Delgado, D. Rozo, D. Ganea, 2004; A.M. ten Bokum, L.J. Hofland, P.M. van Hagen, 2000). On the other hand, the immune system controls the nervous system responses through the cytokines in particular interlukine-1, which are produced by the immune cells (A.J. Dunn, J. Wang, T. Ando, 1999). As for the involvement of the endocrine glands into the stress condition development it is performed through the well-known activatory hypothalamic action on the pituitary-adrenal and sympatho-adrenal axes and, consequently, through the cortisol and catecholamines effects (E.R. De Kloet, 2004).

In conclusion, it seems likely that the L-lysine stress-limiting effect is due to its modulatory influence on the regulatory systems interactions that is performed through the changes of the functional activity of cellular proteins and neurotransmitter systems.

CONCLUSION

According to the main concepts of modern neurochemistry, brain proteins and certain amino acids (L-asparagine, L-glutamate, L-arginine, glicine, GABA) are of exclusively significant importance for the brain activity. The concept of their dual role in brain functions was developed. On the one hand amino acids take part in proteins synthesis and influence their active properties, on the other hand, – they have an independent significance due to formation of neurotransmitter systems (K. Rajevsky, 1989). Presented in our study review of modern scientific literature and our own research showed that the above mentioned amino acids group may be added by L-lysine.

L-lysine is a diaminomonocarbonic essential amino acid included into almost all proteins of animal and plant origin. The highest L-lysine content (up to 85%) was found in neuronal nuclear proteins (O. Terpilovskaya, V. Ivanov, 1990; H.-D. Yacubke, H. Yeshkait, 1985). The particular significance of the amino acid for the CNS activity was confirmed by the fact that the more upper brain nuclei the greater L-lysine neuronal content. The highest amino acid content was shown in the highly developed brain area – cerebral cortex (R. Khudoerkov, 2001).

According to modern reports review two main aspects of L-lysine functional effects were summarized, namely: cellular and intercellular effects. L-lysine seemed to control the cellular processes as follows: proliferation, differentiation and apoptosis. The idea was considered that these effects were based on the close connection of L-lysine containing proteins with chromatin structures. As for the intercellular mechanisms the possible amino acid influence on neurotransmitter systems was demonstrated. In particular, it was determined that anticonvulsant and anxiolytic L-lysine effects mechanisms is the action on the brain GABA-ergic system (Y.F. Chang, X.M. Gao, J.S. Chen, 1991; X.M. Gao, Y.F. Chang, 1989).

On the ground of the mentioned data we thought that the perspective idea was to study L-lysine influence on the integrative brain activity, immune system reactivity and interaction between the nervous and immune regulatory systems.

To investigate the neurotropic L-lysine effects different models of pain-induced behavior in rats (the electrical stimulation of tail and paws, "hot plate" test) as well as rat behavior in unknown environment ("glass arena" and "open field" tests) and digestive conditioning in T-maze were used.

It was determined that the action of the intraperitoneal L-lysine administration on the thermal pain sensitivity was characterized by the parabolic curve of dose-dependence. According to this the algesic amino acid effect was found if the administered doses were minimum and maximum (0.15 and 50.0 mcg/1 kg of body weight) but the administration of intermediate doses (1.5 and 5.0 mcg/kg) resulted in analgesic effect. L-lysine increased the frequency of the avoidance reaction in relation to thermoalgesic stimulation independent of the dose.

The systemic L-lysine administration in the used dose range increased pain sensitivity to electrical tail stimulation in rats. The expression of affective-defensive reactions also enhanced on dosage increase. Meanwhile, the intraperitoneal amino acid administration was the characterized by U-form dose-dependent influence on the aggressive-defensive behavior development induced by gradually increasing unescapable painful stimulation of paired rats in an electrical chamber. According to the curve character the algesic and aggressogenic L-lysine action was reproduced if the doses were 1.5 and 5.0 mcg/kg and the opposite effects arised in amino acid administration in the minimum (0.5 mcg/kg) and maximum (50.0 mcg/kg) doses.

The following obtained data should be accentuated, namely: the biological significance of the concrete situation of stimulation played the greater role in the pain-induced behavior development than the intensity of stimulation. We arrived at a conclusion that L-lysine effect on the supraspinal structures seemed to be more significant than on the spinal cord level.

On the systemic administration before the learning session L-lysine improved the indices of the food conditioned reflex elaboration in rats as well as enhanced the emotional reactions to unknown environment in the "open field" and "glass arena" tests but decreased the exploratory activity.

The determined neurotropic L-lysine effects are schematically summarized in Figure 39.

In order to estimate L-lysine influence on the immune reactivity and congenital resistance the humoral immune response and delayed hypersensitivity type as well as oxygen-dependent bactericidal and phagocytic neutropfil activity.

At last, L-lysine effects were studied in emotional-painful stress and stress-induced immunosupression as the special phenomenon of the interaction between nervous and immune systems.

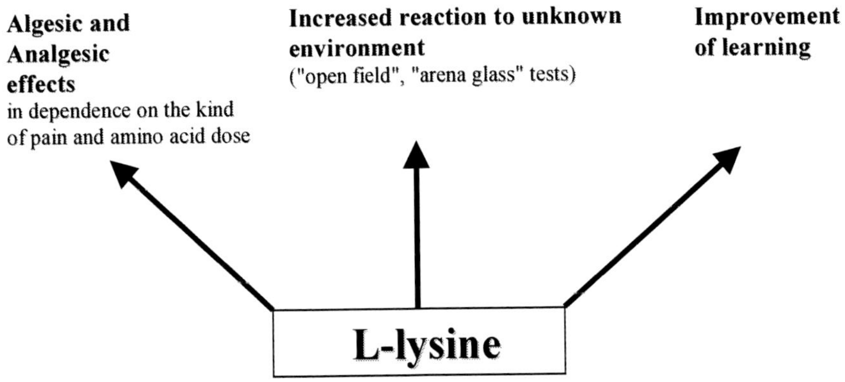

Figure 39. Neurotropic effects of L-lysine.

It was shown that the systemic L-lysine administration in the doses 5-450 mcg/kg of body weight activated the development of immune response to the ram erythrocyte injection and suppressed it after the small doses administration (0.5 and 1.5 mcg/kg). The indices of delayed hypersensitivity type decreased in the most degree after the dose 50 mcg/kg administration.

L-lysine administration in the equivalent doses prevented the immunosupression development in the model of painful stimulation in the electrical chamber. In this condition an amino acid enhanced both humoral and cellular immune response and consequently accomplished immune system stimulating and stress-limiting effects.

In the intact and stressful conditions of rats L-lysine changed oxygen-dependent bactericidal neutrophil activity that might be the manifestation of stress-limiting effect while amino acid administration inhibited the phagocytic granulocyte activity independent of an animal condition.

The obtained in our study immunotropic and stress-limiting L-lysine effects are schematically illustrated in Figure 40.

In summary, the results from our study permitted to postulate L-lysine influence on the nervous and immune regulatory systems activity and their interaction. It is possible to suggest that the neurotropic and immunotropic effects may include the action on the proliferation, differentiation and apoptosis processes in the regulatory systems (in particular, in the immune system) as well as on the intercellular interaction by means of the changes of neurotransmitter systems activity.

The results from our study promoted the development of the modern conception of L-lysine physiological role in an animal and human and determination of perspectives for future investigation. In addition the obtained data may be used to propose practical recommendations in relation to L-lysine-containing diets. Since the amino acid accomplished stress-limiting action it is possible to elaborate special diets to decrease psychoemotional overstrain. The special method of L-lysine administration may be also elaborated to correct the immune system activity.

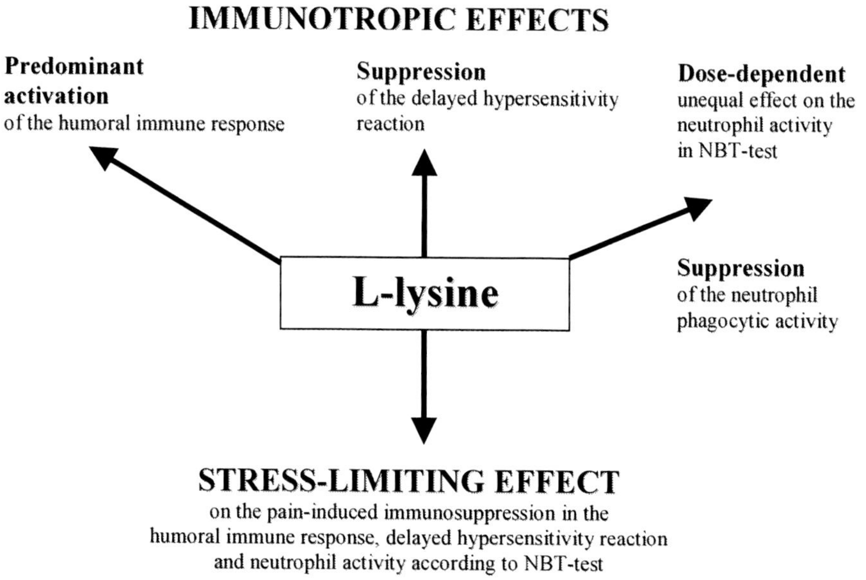

Figure 40. L-lysine influence on the immune response and the congenital defensive mechanisms.

References

Aloisi, AM; Craft, RM; Marchand, S. The effects of gonadal hormones on pain. *Proceedings of the 11th World Congress on Pain* (Seattle: IASP Press), 2006, 301-309.

Armstrong, DW; Gasper, MP; Lee, SH et al. Factors controlling the level and determination of D-amino acids in the urine and plasma of laboratory rodents. *Amino Acids*, 1993 Vol. 5, 299-315.

Ashmarin, IP; Kamensky, AA; Melekhov SL. Influence of the adrenocorticotrope hormone fragment $ACTH_{4-10}$ on the learning session by positive confirmation in white rats. *Lectures of USSR Academy of Sciences*, 1978 Vol. 240 N 5, 1245-1247.

Ashmarin, IP; Stukalov, PV; Eschenko, ND. Biochemistry of the brain. St.-Petersburg: Publisher SPbSU; 1999. 328 p.

Ayala, E; Krikorian, D. Effect of L-lysine monohydrochloride on cutaneous herpes simplex virus in the guinea pig. *J. Med. Virol*, 1989 Vol. 28, 16-20.

Beauman, JG. Genital herpes: a review. *American Family Physician*, 2005 Vol. 72 N 8, 1527-1534.

Bentley, DE; Baumgärtner, U; Watson, A et al. The time-course of sensory and affective pain processing: evidence from laser-evoked potentials. *Proceedings of the 11th World Congress on Pain* (Seattle: IASP Press), 2006, 397-405.

Belokrylov, GA; Molchanova, IV; Sorochinskaya, EI. The ability of some protein amino acids to stimulate thymus-dependent immune response. *Bulletin of the experimental biology and medicine*, 1986 Vol. 102 N 7, 51-53.

Belokrylov, GA; Popova, OYa; Molchanova, IV et al. The different effects of peptides and amino acids included in them on the antibody production and activity of phagocytes in mice. *Bulletin of the experimental biology and medicine*, 1991 Vol. 111 N 1, 53-55.

Belokrylov, GA; Derevnina, ON; Popova, OYa et al. The differences in the immune response, fagocytosis and detoxic properties under peptides and amino acids preparations influence. *Bulletin of the experimental biology and medicine*, 1996 Vol. 121 N 5, 509-512.

Belokrylov, GA; Popova, OYa; Sorochinskaya, EI. The likeness of the immuno-, fagocytosis-modulate and antitoxic characters of the dipeptides and amino acids composed them. *Bulletin of the experimental biology and medicine*, 1999 Vol. 127 N 6, 674-676.

Bobyntsev, II; Severyanova, LA. The immunotropic effects of the gonadotropin-releasing hormone analogue in rats in the emotional-painful stress condition. *Bulletin of the experimental biology and medicine*, 2002 Vol. 133 N 5, 504-506.

Bobyntsev, II; Dolgintsev, ME; Kryukov, AA; Severyanova, LA. The influence of L-lysine on the humoral immune response development and neutrophil activity in the non-stress and emotional-painful stress conditions. *Herald of the new medical technologies*, 2007 Vol. 14 N 1, 135-137.

Borodkin, YuS; Shabanov, PD. The neurochemical mechanisms of the extraction of the memory sign. Leningrad: Science; 1986. 150 p.

Broquist, HP. Lysine-pipecolic acid metabolic relationships in microbes and mammals. *Annu. Rev. Nutr*, 1991 Vol. 11, 435-448.

Cederbaum, SD; Shaw, KN; Dancis, J et al. Hyperlysinemia with saccharopinuria due to combined lysine-ketoglutarate reductase and saccharopine dehydrogenase deficiencies presenting as cystinuria. *J. Pediatr*, 1979 Vol. 95 N 2, 234-238.

Chalisova, NI; Penniyaynen, VA; Haze, G. Regulatory role of the some amino acids on the apoptosis development in the organotypic culture of the nervous and lymphoid tissues. *IM Sechenov Russian physiological journal*, 2002 Vol. 88 N 5, 627-633.

Chalisova, NI; Penniyaynen, VA. Modulatory role of the irreplaceable and replaceable amino acids in the organotypic culture of the tissues in different ages rats. *IM Sechenov Russian physiological journal*, 2003 Vol. 89 N 5, 591-597.

Chang, YF. Lysine metabolism in the human and the monkey: demonstration of pipecolic acid formation in the brain and other organs. *Neurochem. Res*, 1982 Vol. 7 N 5, 577-588.

Chang, YF; Myslinski, NR. Effects of L-lysine and its metabolites on pentylenetetrazol-induced seizures. *Neurosci. Lett*, 1985 Vol. 59 N 1, 79-84.

Chang, YF; Hargest, V; Chen, JS. Modulation of benzodiazepine by lysine and pipecolic acid on pentylenetetrazol-induced seizures. *Life Sci*, 1988 Vol. 43 N 15, 1177-1188.

Chang, YF; Gao, XM; Chen, JS. Correlation between enhancement of [3H]flu-nitrazepam binding and suppression of pentylenetetrazol-induced seizures by L-lysine. *Eur. J. Pharmacol*, 1991 Vol. 193 N 2, 239-247.

Chang, YF; Wang, Y; Cauley, RK; Gao, XM. Chronic L-lysine develops anti-pentylenetetrazol tolerance and reduces synaptic GABAergic sensitivity. *Eur. J. Pharmacol*, 1993 Vol. 233 N 2-3, 209-217.

Chang, YF; Gao, XM. L-lysine is a barbiturate-like anticonvulsant and modulator of the benzodiazepine receptor. *Neurochem. Res*, 1995 Vol. 20 N 8, 931-937.

Charles, AK; Chang, YF. Uptake, release, and metabolism of D- and L-alpha-aminoadipate by rat cerebral cortex. *J. Neurochem*, 1981 Vol. 36 N 3, 1127-1136.

Charles, AK; Chang, YF; Myslinski, NR. Blood-brain barrier transport of L-pipecolic acid in various rat brain regions. *Neurochem. Res*, 1983 Vol. 8 N 9, 1087-1096.

Charles, AK. Pipecolic acid receptors in rat cerebral cortex. *Neurochem. Res*, 1986 Vol. 11 N 4, 521-525.

Chereshnev, VA; Yushkov, BG; Klimin, VG; Lebedeva, EV. Immunophysiology. Yekaterinburg: UrO RAS; 2002. 259 p.

Colpaert, FC; Hamon, M; Wiesenfeld-Hallin, Z. 5-HT_{1A} receptors in chronic pain processing and control. *Proceedings of the 11th World Congress on Pain* (Seattle: IASP Press), 2006, 147-154.

Crawford, DH; Chen, S; Boyd, CAR. Cationic amino acid transport in human T lymphocytes is markedly increased in the CD45RA, CD8+ population after activation. *Immunology*, 1994 Vol. 82, 357-360.

Dancis, J; Hutzler, J; Woody, NC; Cox, RP. Multiple enzyme defects in familial hyperlysinemia. *Pediatr. Res*, 1976 Vol. 10 N 7, 686-691.

Dancis, J; Hutzler, J; Cox, RP. Familial hyperlysinemia: enzyme studies, diagnostic methods, comments on terminology. *Am. J. Hum. Genet*, 1979 Vol. 31 N 3, 290-299.

Dancis, J; Hutzler, J. The metabolism of D- and L-pipecolic acid in the rabbit and rat. *Biochim. Biophys. Acta*, 1981 Vol. 675 N 3-4, 411-415.

Dancis, J; Hutzler, J. Comparative rates of metabolism of pipecolic acid in several animal species. *Comp. Biochem. Physiol. B*, 1982 Vol. 73 N 4, 1011-1012.

Dasgupta, S; Bhinge, A; Chandran, V et al. Role of L-lysine HCl in immunopotentiation towards development of suitable tuberculosis vaccination. *Vaccine*, 2003 Vol. 21 N 32, 4722-4727.

Dasgupta, S; Chandran, V; Bhinge, A et al. Role of L-lysine HCl in adoptive immune therapy towards development of suitable tuberculosis vaccination. *Indian J. Exp. Biol*, 2004 Vol. 42 N 8, 758–765.

Davis, KD. Recent advances in imaging pain. *Proceedings of the 11th World Congress on Pain* (Seattle: IASP Press), 2006, 387-396.

De Kloet, ER. Hormones and the stressed brain. *Ann. NY Acad. Sci*, 2004 Vol. 1018 N 1, 1-15.

Delgado, M; Rozo, D; Ganea, D. The significance of vasoactive intestinal peptide in immunomodulation. *Pharmacol. Rev*, 2004 Vol. 56 N 2, 249-290.

Dolgintsev, ME. The neurotropic and immunotropic effects of L-lysine: dissertation ... PhD medical sciences: 03.00.13: defended 24.05.07. Kursk; 2007. 127 p.

Dunn, AJ; Rees, HD; Iuvone, PM. ACTH and the stress-induced changes of lysine incorporation into brain and liver proteins. *Pharmacol. Biochem. Behav*, 1978 Vol. 8 N 4, 455-465.

Dunn, AJ; Wang, J; Ando, T. Effects of cytokines on cerebral neurotransmission. Comparison with the effects of stress. *Adv. Exp. Med. Biol*, 1999 Vol. 461 N 1, 117-127.

Fedosejeva, VN; Poryadin, GV; Kovalchuk, LV et al. Direction on the immunological and allergological methods in the hygiene investigations. Moscow; 1993. 319 p.

Feigenbaum, P; Chang, YF. Pipecolic acid antagonizes barbiturate-enhanced GABA binding to bovine brain membranes. *Brain Res*, 1986 Vol. 372 N 1, 176-179.

Flodin, NW. The metabolic roles, pharmacology, and toxicology of lysine. *J. Am. Coll. Nutr*, 1997 Vol. 16 N 1, 7-21.

Fountain, SJ; North, RA. A C-terminal lysine that controls human P2X4 receptor desensitization. *J. Biol. Chem*, 2006 Vol. 281 N 22, 15044–15049.

Fujita, T; Hada, T; Higashino, K. Origin of D- and L-pipecolic acid in human physiological fluids: a study of the catabolic mechanism to pipecolic acid using the lysine loading test. *Clin. Chim. Acta*, 1999 Vol. 287 N 1-2, 145-156.

Fujita, T; Fujita, M; Kodama, T et al. Determination of D- and L-pipecolic acid in food samples including processed foods. *Ann. Nutr. Metab*, 2003 Vol. 47, 165-169.

Gao, XM; Chang, YF. Enhancement of benzodiazepine receptor binding by L-lysine is chloride-dependent and due to increase in binding affinity. *Eur. J. Pharmacol*, 1989 Vol. 173 N 2-3, 197-200.

Gogichaeva, NV; Lukasheva, EV; Gavrilova, EM et al. Conjugation of L-lysine α-oxidase with antibodies. *Questions of the medical chemistry*, 2000 Vol. 46 N 4, 410-418.

Griffith, RS; Norins, AL; Kagan, C. A multicentered study of lysine therapy in Herpes simplex infection. *Dermatologica*, 1978 Vol. 156, 257-267.

Griffith, RS; De Long, DC; Nelson, JD. Relation of arginine-lysine antagonism to Herpes simplex growth in tissue culture. *Chemotherapy*, 1981 Vol. 27, 209-213.

Griffith, RS; Walsh, DE; Myrmel, KH et al. Success of L-lysine therapy in frequently recurrent herpes simplex infection. Treatment and prophylaxis. *Dermatologica*, 1987 Vol. 175, 183-190.

Gubler, EV; Genkin, AA. The employment of the nonparametric statistically criterion in the medico-biological investigations. Leningrad: Medicine; 1973. 141 p.

Gureeva, TA; Golubeva, NV; Lubkova, ON. Aminopeptidases in human leukemic lymphoid cells. *Questions of the medical chemistry*, 1999 Vol. 45 N 4, 309-313.

He, M. Pipecolic acid in microbes: biosynthetic routes and enzymes. *J. Ind. Microbiol. Biotechnol*, 2006 Vol. 33 N 6, 401-407.

Hess, A; Sergejeva, M; Budinsky, L et al. Imaging of hyperalgesia in rats by functional MRI // Eur. J. Pain. – 2007. – Vol. 11, N 1. – P. 109–119.

Ilyuchenok, RYu. Pharmacology of the behavior and memory. Novosibirsk: Science; 1972. 223 p.

Ishikawa, E; Imagawa, M; Hashida, S et al. Enzyme-labeling of antibodies and their fragments for enzyme immunoassay and immunohistochemical staining. *J. Immunoassay*, 1983 Vol. 4 N 3, 209-227.

Jezova, D; Makatsori, A; Duncko, R et al. High trait anxiety in healthy subjects: association with low neuroendocrine responses during psychosocial stress. *Prog. Neuro-psychopharmacol. Biol. Psychiatry*, 2004 N 28, 1331-1336.

Jezova, D; Makatsori, A; Smriga, M et al. Subchronic treatment with amino acid mixture of L-lysine and L-arginine modifies neuroendocrine activation during psychosocial stress in subjects with high trait anxiety. *Nutr. Neurosci*, 2005 Vol. 8 N 3, 155-160.

Jones, CK; Peters, SC; Shannon, HE. Synergistic interaction between the dual serotonergic, noradrenergic reuptake inhibitor duloxetine and the non-steroidal anti-inflammatory drug ibuprofen in inflammatory pain in rodents. *Eur. J. Pain*, 2007 Vol. 11 N 2, 208-215.

Jänig, W; Chapman, CR; Green, PG. Pain and body protection: sensory, autonomic, neuroendocrine, and behavioral mechanisms in the control of inflammation and hyperalgesia. *Proceedings of the 11th World Congress on Pain* (Seattle: IASP Press), 2006, 331-347.

Kalueff, AV. Role of central GABA in anxiety and depression pathogenesis: neurogenetics, neurochemistry and brain circuitry. *The Russian journal of neuroscience*, 2006 Vol. 4 N 2, 29-41.

Kase, Y; Takahama, K; Hashimoto, T et al. Electrophoretic study of pipecolic acid, a biogenic imino acid, in the mammalian brain. *Brain Res*, 1980 Vol. 193, 608-613.

Khaitov, RM; Guschin, IS; Pinegin, BV; Zebrev, AI. Methodical instructions by study of the pharmacological substances immunotropic activity. Moscow: Minzdrav RF; 2000. P. 257-258.

Khaitov, RM; Leskov, VP. Immunity and stress. *IM Sechenov Russian physiological journal*, 2001 Vol. 87 N 8, 1060-1072.

Khudoerkov, RM. Ammonia-argental method in the alkali diapason pH is as a manner of the morpho-functional peculiarities of the nervous elements revealed. *Bulletin of the experimental biology and medicine*, 1992 Vol. 113 N 6, 660-663.

Khudoerkov, RM. Cytochemistry of the peptides in the disclose of the conformity to natural laws of the brain structural and functional organization. *Herald of the Russian Academy of the medical sciences*, 2001 N 4, 43-48.

Kirillicheva, TB; Goncharov, NP; Mitkin, VV et al. The peculiarities of influence of the immobilizing stress on the 5'-nucleotidase macrophages action and endogen cortisol level in CBA & C57Bl/6 mice. *Bulletin of the experimental biology and medicine*, 1993 Vol. 115 N 3, 280-282.

Kirillina, EA; Zakharova, LA; Mikhaylova; Vasilenko, AM. T-suppressor and opiate-ergic mechanisms of the immune response depression in stress. *Immunology*, 1990 N 3, 68-70.

Kiryanova, NA. The neurotropic and immunotropic effects of L-arginine: dissertation ... PhD medical sciences: 03.00.13: defended 20.04.06. Kursk; 2006. 123 p.

Kontos, HA; Wei, EP. Cerebral arteriolar dilations by KATP channel activators need L-lysine or L-arginine. *Am. J. Physiol*, 1998 Vol. 274 N 2, 974-981.

Korneva, EA; Shkhinek, EK. Hormones and immune system. Leningrad: Science; 1988. 251 p.

Korneva, EA. Immunophysiology is as a new science discipline. *I.M. Sechenov Physiological journal*, 1994 Vol. 80 N 6, 116-124.

Korneva, EA. Immunophysiology. St.-Petersburg: Science; 1993. 684 p.

Kruglikov, RI. The neurochemical mechanisms of the training and memory. Moscow: Science; 1981. 211 p.

Kusakabe, H; Kodama, K; Kuninaka, A et al. A new antitumor enzyme, L-lysine alpha-oxidase from Trichoderma viride. Purification and enzymological properties. *J. Biol. Chem*, 1980 Vol. 255 N 3, 976-981.

Magrum, LJ; The, PS; Kreiter, MR et al. Transfer ribonucleic acid charging in rat brain after consumption of amino acid-imbalanced diets. *Nutr. Neurosci*, 2002 Vol. 5 N 2, 125-130.

Malberg, K; Siegel, E. Local hemolysis method. Immunological methods. / edition Frimel, G. Moscow: World; 1987. P. 57-72.

Martin, AM; Pouchnik, DJ; Walker, JL; Wyrick, JJ. Redundant roles for histone H3 N-terminal lysine residues in subtelomeric gene repression in Saccharomyces cerevisiae. *Genetics*, 2004 Vol. 167, 1123-1132.

Mignini, F; Streccioni, V; Amenta, F. Autonomic innervation of immune organs and neuroimmune modulation. *Auton. Autacoid. Pharmacol*, 2003 Vol. 23 N 1, 1-25.

Monconduit, L; Bourgeais, L; Bernard, J-F et al. The ventromedial thalamus: a link for noxious cutaneous inputs from the whole body surface. *Proceedings of the 9th World Congress on Pain* (Seattle: IASP Press), 2000, 411-417.

Nagato, I; Akino, T; Ohno, K et al. Free D-amino acids in human plasma in relation to senescence and renal disease, *Clin. Sci*, 1987 Vol. 73, 105-108.

Nagy, I., Reeh, P; Tomassy, GS; Santha, P. Control of activity: relationship between excitatory and inhibitory receptors expressed by nociceptors. *Proceedings of the 11th World Congress on Pain* (Seattle: IASP Press), 2006, 109-118.

Nurmagambetov, TZh; Amirov, BB; Kuanysheva, TK, Sharmanov, TSh. Induction of monooxygenases and incorporation of radioactivity from 2-^{14}C-lysine into hepatic microsomes of phenobarbital-treated rats fed a diet deficient in lysine, methionine, threonine and vitamins A, C and E. *Bulletin of the experimental biology and medicine*, 1991 Vol. 111 N 3, 256-259.

Plokhinsky, NA. Biometry. Moscow: Publisher MSU; 1970. 367 p.

Pokrovsky, VM; Shulzhenko, LV. Chronobiological parameters of functional activity of tissue basophils. *Kuban sciences medical herald*, 2000 Vol. 50 N 2, 88-90.

Poshivalov, VP. Experimental psychopharmacology of the aggression behavior. Leningrad: Science; 1986. 175 p.

Rajevsky, KS. The exciting amino acids are as an neuromediators. The results of the science and technique. Series "Human and animals physiology". Moscow; 1989. Vol. 36. 184 p.

Rakityanskaya, IA; Kucher, AG; Abramova, TV. Assessment of the effect of certain amino acids isolated from soy-beans on the functional state of blood cells, lympho- and granulocytopoiesis *in vitro*. *Nephrology*, 2000 Vol. 4 N 1, 59-62.

Rao, VV; Chang, YF. L-pipecolic acid metabolism in human liver: detection of L-pipecolate oxidase and identification of its reaction product. *Biochim. Biophys. Acta*, 1990 Vol. 1038 N 3, 295-299.

Rao, VV; Chang, YF. Assay for L-pipecolate oxidase activity in human liver: detection of enzyme deficiency in hyperpipecolic acidaemia. *Biochim. Biophys. Acta*, 1992 Vol. 1139 N 3, 189-195.

Rao, VV; Tsai, MJ; Pan, X; Chang, YF. L-pipecolic acid oxidation in rat: subcellular localization and developmental study. *Biochim. Biophys. Acta*, 1993 Vol. 1164 N 1, 29-35.

Rees, HD; Dunn, AJ. The role of the pituitary-adrenal system in the footshock-induced increase of [3H]lysine incorporation into mouse brain and liver proteins. *Brain Res*, 1977 Vol. 120 N 2, 317-325.

Ren, K; Zhuo, M; Willis, WD. Multiplicity and plasticity of descending modulation of nociception: implications for persistent pain. *Proceedings of the 9th World Congress on Pain* (Seattle: IASP Press), 2000, 387-400.

Rodgers, RJ; Semple, JM. Pituitary-adrenocortical axis and shock-induced fighting in rats. *Physiol. Behav*, 1978 Vol. 20 N 5, 533-537.

Sandkühler, J. Long-lasting analgesia following TENS and acupuncture: spinal mechanisms beyond gate control. *Proceedings of the 9th World Congress on Pain* (Seattle: IASP Press), 2000, 359-369.

Schmidt, R; Schmelz, M; Forster, C et al. Novel classes of responsive and unresponsive C nociceptors in human skin. *J. Neurosci*, 1995 Vol. 1 N 1, 333-341.

Severyanova, LA. Adoptive hormones influence on the integrative brain activity. Moscow: Science; 1988. 124 p.

Severyanova, LA; Bobyntsev, II. The neurotropic effects luliberin analogue in intraventricular injection in rats with different ethanol sensitiveness. *Bulletin of the experimental biology and medicine*, 1995 Vol. 119 N 2, 129-132.

Severyanova, L; Bobyntsev, I; Kryukov, A et al. Effects of L-lysine and L-argi-nine on the different kinds of pain sensitivity and nociceptive behavior. *11th World Congress on Pain* (August 21-26, 2005, Sydney, Australia). *Abstracts*, 2005, 353-354.

Severyanova, LA; Dolgintsev, ME. The modern concept of L-lysine action on the nervous and immune regulator systems. *Kursk science-practical herald "Human and his health"*, 2007 N 2, 67-79.

Selischeva, AA; Alexejev, SB; Smirnova, IP; Podboronov, VM. Effectiveness of the anti-herpetic action of the different L-lysine α-oxidase drug form. *Antibiotics and chemotherapy*, 2003 Vol. 48 N 1, 9-12.

Sergejev, PV; Shimanovsky, NL. Physiological active substances receptors. Moscow: Medicine; 1987. 396 p.

Sernov, LN; Gatsura, VV. Elements of the experimental pharmacology. Moscow; 2000. 352 p.

Shtelzner, A. Local hemolysis method. Immunological methods. / edition Frimel, G. Moscow: World; 1987. P. 378-389.

Singh, BB; Udani, J; Vinjamury, SP et al. Safety and effectiveness of an L-lysine, zinc, and herbal-based product on the treatment of facial and circumoral herpes. *Altern. Med. Rev*, 2005 Vol. 10 N 2, 123-127.

Smirnova, IP; Alexejev, SB; Berezov, TT. To the study question of the connection mechanism HIV-infection with autoimmunity. *Questions of the medical chemistry*, 1996 Vol. 42 N 3, 211-216.

Smirnova, IP; Diorditsa, SV; Alexejev SB; Zaitsev IZ. Influence of L-lysin-alfa-oxidase on the reproduction of herpes simplex virus type 1 *in vitro*. *Questions of the medical chemistry*, 1998 Vol. 44 N 4, 384-387.

Smirnova, IP; Alexejev SB; Diorditsa, SV et al. Influence of L-lysin-alfa-oxidase on the herpetic genital infection development in guinea pigs. *Bulletin of the experimental biology and medicine*, 1999 Vol. 128 N 12, 654-656.

Smirnova, IP; Didzhyapetrene Ya; Alexejev SB et al. Influence of L-lysin-alfa-oxidase on the skin carcinoma mice induced methylcholantren. *Antibiotics and chemotherapy*, 2003 N 4, 13-15.

Smriga, M; Torii, K. Release of hypothalamic norepinephrine during MSG intake in rats fed normal and nonprotein diet. *Physiol. Behav*, 2000 Vol. 70 N 3-4, 413-415.

Smriga, M; Mori, M; Torii, K. Circadian release of hypothalamic norepinephrine in rats in vivo is depressed during early L-lysine deficiency. *J. Nutr*, 2000 Vol. 130 N 6, 1641-1643.

Smriga, M; Murakami, H; Mori, M; Torii, K. Effects of L-lysine deficient diet on the hypothalamic interstitial norepinephrine and diet-induced thermogenesis in rats *in vivo*. *BioFactors*, 2000 N 12, 137-142.

Smriga, M; Kameishi, M; Uneyama, H; Torii, K. Dietary L-lysine deficiency increases stress-induced anxiety and fecal excretion in rats. *J. Nutr*, 2002 Vol. 132 N 12, 3744-3746.

Smriga, M; Torii, K. Metabolic interactions between restraint stress and L-lysine: the effect on urea cycle components. *Amino Acids*, 2003 Vol. 24 N 4, 435-437.

Smriga, M; Torii, K. L-lysine acts like a partial serotonin receptor 4 antagonist and inhibits serotonin-mediated intestinal pathologies and anxiety in rats. *Proc. Natl. Acad. Sci. USA*, 2003 Vol. 100 N 26, 15370-15375.

Smriga, M; Torii, K. Prolonged treatment with L-lysine and L-arginine reduces stress-induced anxiety in an elevated plus maze. *Nutr. Neurosci*, 2003 Vol. 6 N 2, 125-128.

Smriga, M; Ghosh, S; Mouneimne, Y et al. Lysine fortification reduces anxiety and lessens stress in family members in economically weak communities in Northwest Syria. *Proc. Natl. Acad. Sci. USA*, 2004 Vol. 101 N 22, 8285-8288.

Spidel, JL; Craven, RC; Wilson, CB et al. Lysines close to the Rous sarcoma virus late domain critical for budding. *J. Virol*, 2004 Vol. 78 N 19, 10606-10616.

Srinongkote, S; Smriga, M; Nakagawa, K; Toride, Y. A diet fortified with L-lysine and L-arginine reduces plasma cortisol and blocks anxiogenic response to transportation in pigs. *Nutr. Neurosci*, 2003 Vol. 6 N 5, 283-289.

Tanphaichitr, V; Broquist, HP. Role of lysine and ε-N-trimethyllysine in carnitine biosynthesis. II. Studies in the rat. *J. Biol. Chem*, 1973 Vol. 248 N 6, 2176-2181.

ten Bokum, AM; Hofland, LJ; van Hagen, PM. Somatostatin and somatostatin receptors in the immune system: a review. *Eur. Cytokine Netw*, 2000 Vol. 11 N 1, 161-176.

Terpilovskaya, ON; Ivanov, VA. Structure-functional organization of the mammals nervous cells chromatin. *Successes of the modern biology*, 1990 Vol. 110 N 1 (4), 118-133.

Tikhomirov, SM; Nikolajev, SV. Diapason of the possible practical luliberin employment is as a neurotropic peptide. *Problems of an endocrinology*, 1988 N 4, 54-56.

Torebjörk, E. Subpopulations of human C nociceptors and their sensory correlates. *Proceedings of the 9th World Congress on Pain* (Seattle: IASP Press), 2000, 199-206.

Torres, KC; Antonelli, LR; Souza, AL et al. Norepinephrine, dopamine and dexamethasone modulate discrete leukocyte subpopulations and cytokine profiles from human PBMC. *J. Immunol*, 2005 Vol. 166 N 2, 144-157.

Torii, K; Yokawa, T; Tabuchi, E et al. Recognition of deficient nutrient intake in the brain of rats with the L-lysine deficiency monitored by functional magnetic resonance imaging, electrophysiologically and behaviorally. *Amino Acids*, 1996 N 10, 73-81.

Treede, RD; Margel, W. Modern concepts of pain and hyperalgesia beyond the polymodal C-nociceptor. *News in Physiological Sciences*, 1995 Vol. 10, 216-228.

Uhe, AM; Collier, GR; O'Dea, K. A comparison of the effects of beef, chicken and fish protein on satiety and amino acid profiles in lean male subjects. *J. Nutr*, 1992 Vol. 122 N 3, 467-472.

Vianey-Liaud, C; Divry, P; Poinas, C; Mathieu, M. Lysine metabolism in man. *Ann. Biol. Clin.* (Paris), 1991 Vol. 49 N 1, 18-26.

Villanueva, L; Nathan, PW. Multiple pain pathways. *Proceedings of the 9th World Congress on Pain* (Seattle: IASP Press), 2000, 371-386.

Vogt, BA; Porro, CA; Faymonville, M-E. Pain processing and modulation in the cingulate gyrus // *Proceedings of the 11th World Congress on Pain* (Seattle: IASP Press), 2006, 415-430.

Vologzhanin (DA); Kalinina (NM); Sosyukin (AE) et al. The metabolic bases of the formation of immune insufficiency during trauma. Russian biomedical journal Medline [online], 2005 Vol. 168 N 6 (597-625). http://www.medline.ru

Weidner, C; Schmelz, M; Schmidt, R et al. Functional attributes discriminating mechano-insensitive and mechano-responsive C nociceptors in human skin. *J. Neurosci*, 1999 Vol. 19 N 22, 10184-10190.

Wickwire, BM; Harris, CM; Harris, TM et al. Pipecolic acid biosynthesis in Rhixoctonia leguminicola. I. The lysine, saccharopine, Δ^1-piperideine-6-carboxylic acid pathway. *J. Biol. Chem*, 1990 Vol. 265 N 25, 14742-14747.

Wickwire, BM; Wagner, C; Broquist, HP. Pipecolic acid biosynthesis in Rhixoctonia leguminicola. II. Saccharopine oxidase: a unique flavin enzyme involved in pipecolic acid biosynthesis. *J. Biol. Chem*, 1990 Vol. 265 N 25, 14748-14753.

Wood, JN; Akopian, AN; Cesare, P. et al. The primary nociceptor: special functions, special receptors. *Proceedings of the 9th World Congress on Pain* (Seattle: IASP Press), 2000, 47-62.

Yacubke, H-D; Yeshkait, H. Amino acids, peptides, albumens. Moscow: World; 1985. 456 p.

Zhukova, OS; Gogichaeva, NV; Lukasheva, EV; Berezov, TT. The investigation of the conjugates L-lysine α-oxidase with the monoclonal antibodies cytotoxic effect on the human tumor cells *in vitro*. *Questions of the medical chemistry*, 2001 Vol. 47 N 6, 588-592.

Zinkin, VYu; Godkov, MA. Method of the quantity appraisal of the human neutrophils granulocytes oxygen-dependent metabolism. *Clinical laboratories diagnostics*, 2004 N 8, 26-29.

Reviewed by Kubatiev A.A., a Member of the Russian Academy of Medical Sciences, the Head of the Scientific Research Institute of General Pathology and Pathophysiology (The Russian Academy of Medical Sciences, Baltiyskaya st. 8, Moscow, 125315, Russia, tel. 007(495)151-17-56, e-mail: niiopp@mail.ru).

Reviewed by Porjadin G.V., a Corresponding Member of the Russian Academy of Medical Sciences, the Head of the Pathophysiology Department of the Russian State Medical

University (Ostrovityanova st. 1, Moscow, 117997, Russia, tel. 007(495)434-45-74, e-mail: gen@rsmu.ru).

In: Diamino Amino Acids
Editors: H. A. Schäfer, L. M. Wohlbier, pp. 103-122
ISBN: 978-1-60456-892-9

Chapter III

D-SERINE AND SCHIZOPHRENIA

Kenji Hashimoto*

Division of Clinical Neuroscience, Chiba University Center for Forensic Mental Health, Chiba, Japan.

ABSTRACT

Several lines of evidence suggest that a dysfunction in glutamatergic neurotransmission via N-methyl-D-aspartate (NMDA) receptors is involved in the pathophysiology of schizophrenia, and that D-serine, an endogenous agonist at the glycine sites of NMDA receptors, plays a role in neurotransmission via these receptors. Previously, we reported that serum levels of D-serine were significantly decreased in patients with schizophrenia, and that the ratio of D-serine to total serine in the cerebrospinal fluid (CSF) of first-episode and drug-naive patients with schizophrenia was also decreased. D-Serine is synthesized from L-serine by the enzyme serine racemase and is degraded by D-amino acid oxidase (DAO). Extracellular D-serine is removed by alanine-serine-cysteine transporter-1 (Asc-1) in neurons. These findings suggest that synthetic and/or metabolic pathways for D-serine may be altered in the brain of schizophrenia patients. Furthermore, D-serine has been shown to be effective for treating several symptoms of this disorder. Accumulating evidence implicates DAO and its interacting gene G72 in the pathogenesis of schizophrenia. In this chapter, the author reviews the role of D-serine, an endogenous co-agonist of glycine modulatory sites on NMDA receptors, in the pathophysiology of schizophrenia. The author also discusses the glycine modulatory sites on NMDA receptors as potential therapeutic targets for schizophrenia.

---

* Correspondence concerning this article should be addressed to: Kenji Hashimoto, Ph.D., Division of Clinical Neuroscience, Chiba University Center for Forensic Mental Health, 1-8-1 Inohana, Chiba 260-8670, Japan. Phone. +81-43-226-2147; Fax: +81-43-226-2150; E-mail: hashimoto@faculty.chiba-u.jp.

1. Introduction

Schizophrenia is a chronic, severe, and disabling brain disorder that affects approximately 1 percent of the population during their lifetime. Although schizophrenia affects men and women with equal frequency, the disorder often appears earlier in men, usually in the late teens or early twenties, than in women, who are generally affected in the twenties to early thirties. People with schizophrenia often suffer terrifying symptoms such as hearing internal voices not heard by others, or believing that other people are reading their minds, controlling their thoughts, or plotting to harm them. These symptoms may leave them fearful and withdrawn. Their speech and behavior can be so disorganized that they may be incomprehensible or frightening to others. Available treatments can relieve many symptoms, but most people with schizophrenia continue to suffer some symptoms throughout their lives; it has been estimated that no more than one in five individuals recovers completely [1].

The causes of schizophrenia are unknown, but multiple lines of evidence suggest that genetic and environmental factors contribute to its pathophysiology [2-5]. Furthermore, schizophrenia has been implicated in multiple neurotransmission abnormalities involving the dopamine (DA), serotonin (5-hydroxytryptamine: 5-HT), acetylcholine (ACh), glutamate, and γ-aminobutyric acid (GABA) systems [5-16]. This chapter focuses on the role of glutamatergic neurotransmission in the pathophysiology of schizophrenia.

2. NMDA Receptors and Schizophrenia

L-Glutamic acid (glutamate) is accepted as the major excitatory neurotransmitter in the nervous system. It plays a major role in brain development, affecting neuronal migration, neuronal differentiation, axon genesis, and neuronal survival [17]. Multiple lines of evidence suggest that a dysfunction in glutamatergic neurotransmission via N-methyl-D-aspartate (NMDA) receptors (Figure 1) might be involved in the pathophysiology of schizophrenia [5-14]. The most impressive data supporting this hypothesis are derived from observations that NMDA receptor antagonists, including phencyclidine (PCP) and ketamine (Figure 2), can produce schizophrenia-like symptoms (positive and negative symptoms as well as cognitive impairments) in humans. It has been reported that subanesthetic doses of PCP (0.1 mg/kg) transiently produce a wide range of schizophrenia-like symptoms, including estrangement, loss of body boundaries, formal thought disorder, hallucinations, and psychosis [18]. In addition, PCP dramatically exacerbates schizophrenic symptoms [7,18,19]. Based on a randomized, double-blind placebo-controlled study, Krystal et al. [20] reported that ketamine produces positive and negative symptoms in a dose-dependent manner. Higher doses of ketamine elicit significant perceptual effects, including alterations of body, environment, and time perceptions. In another double-blind, placebo-controlled study, Malhotra et al. [21] found that ketamine produces a brief psychosis marked by thought disorder and withdrawal-retardation in healthy volunteers. Furthermore, a double-blind crossover design study indicated that ketamine alters mood in healthy volunteers [22]. Also, in a double-blind, placebo-controlled study by Lahti et al. [23], ketamine induced a brief (less than 30 minutes), dose-related worsening of positive symptoms in schizophrenia patients maintained on haloperidol. Krystal et al. [24] reported a placebo-controlled, randomized, double-blind psychopharmacological trial of ketamine and amphetamine in humans. At the doses studied, ketamine and amphetamine each produced positive symptoms and euphoria. However, only ketamine produced perceptual changes, while only amphetamine produced hostility, grandiosity, and somatic concern. Amphetamine and ketamine produced conceptual disorganization, but only ketamine produced concrete ideation and unusual mannerisms. Ketamine produced negative symptoms and disrupted delayed recall. When administered together, ketamine and amphetamine showed three types of interactive effects: (1) amphetamine attenuated the ketamine-produced impairment of working memory; (2) amphetamine and ketamine had additive effects on thought disorder, arousal, and euphoria; and (3) amphetamine and ketamine had less-than-additive effects on psychosis. These findings implicate NMDA receptors and dopamine systems in psychosis. However, glutamate and dopamine may differentially contribute to psychosis, thought disorder, and euphoria. Regarding medication development for cognitive dysfunction, the pattern of the interactive effects of ketamine and amphetamine is consistent with the hypothesis that facilitation of prefrontal cortical dopamine levels would attenuate some cognitive impairments associated with deficits in NMDA receptor function [24].

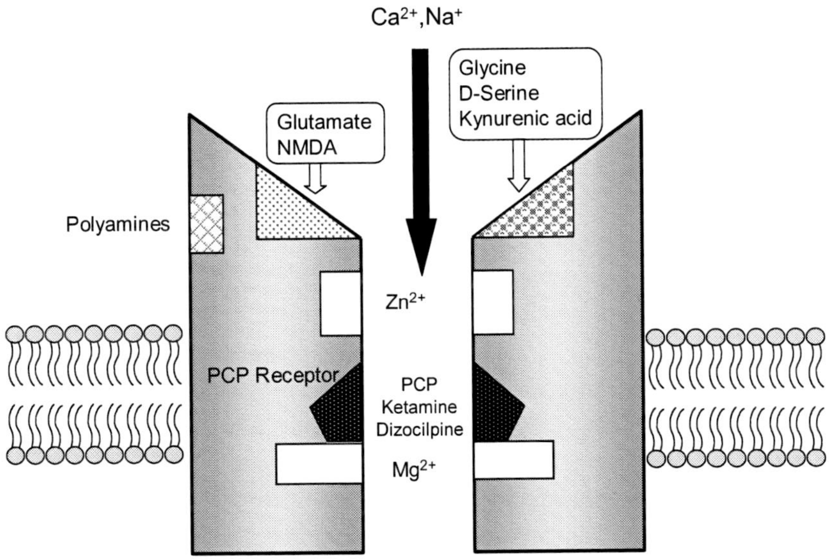

Figure 1. The NMDA receptor complex Glutamate and NMDA bind to the agonist site on the NMDA receptors Phencyclidine (PCP), ketamine, and dizocilpine bind PCP-receptor in the inside of the NMDA receptors. Glycine and D-serine bind to a co-agonist site (glycine modulatory site) on the NMDA receptors.

In addition to glutamate, NMDA receptors are modulated by glycine, D-serine, kynurenic acid, polyamines, and specific divalent cations, including magnesium and zinc (Figure 1). Therefore, it seems that abnormal levels of these endogenous substances might therefore potentially lead to decreased activation of NMDA receptors, and that the endogenous dysfunction of NMDA receptor-mediated neurotransmission might contribute to the pathophysiology of schizophrenia [5,14].

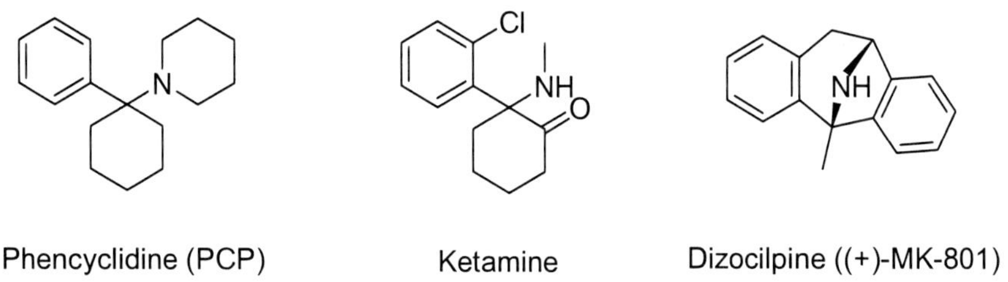

Figure 2. Chemical structures of phencyclidine (PCP), ketamine, and dizocilpine ((+)-MK-801).

Kim et al. [25] first reported that the cerebrospinal fluid (CSF) levels of glutamate in schizophrenia patients were significantly lower than those of normal controls. However, several studies did not find changes in the glutamate levels in the CSF of patients with schizophrenia [26,27] or in the CSF of first-episode and drug-naive patients with schizophrenia [28]. In postmortem studies of schizophrenia, glutamate levels were significantly abnormal in the hippocampus and frontal cortex [29]. Other studies, however, found no differences in brain glutamate levels between patients with schizophrenia and normal controls [26,27,30]. Because the postmortem interval (PMI) has a pronounced effect on glutamate and other amino acids, it will be important to examine PMI's effects on these amino acids in postmortem brains; the analysis of intergroup differences should then be adjusted accordingly. Previously we determined the levels of amino acids (glutamate, glutamine, glycine, D-serine, L-serine) in postmortem brains from patients with schizophrenia by normalizing the effects of the postmortem interval with equations derived from control studies using rodent postmortem brains [30]. The frontal cortex of those brains showed no changes in the levels of any of those five amino acids [30].

In the CNS, glutamine synthesis occurs exclusively in the astrocytes from glutamate and ammonia. Glutamine plays major roles in nitrogen and carbon homeostasis, in the detoxification of ammonia, and as a precursor for the synthesis of neurotransmitters glutamate and GABA in specialized excitatory and inhibitory neurons [31]. Glutamate is released from presynaptic neurons, and this amino acid can interact with postsynaptic glutamate receptors, including kainite, α-amino-3-hydroxy-5-methyl-4-isoxazole propionic acid (AMPA), and NMDA receptors. Released glutamate is taken up by surrounding astrocytes via the glutamate transporter, where it is converted to glutamine, transported back to the presynaptic neurons, and reconverted to glutamate [31]. In a magnetic resonance spectroscopy (MRS) study, glutamine levels were significantly increased in the medial prefrontal cortex of never-treated schizophrenia patients, whereas the levels of N-acetylaspartate (NAA) and glutamate were not altered in those patients [32]. A recent 4.0 T MRS study found that glutamine levels were significantly increased in the left anterior cingulate cortex and thalamus of never-treated schizophrenia patients, whereas the levels of other substances (NAA, glutamate, taurine, choline, creatine, myo-inositol) were not altered in those patients [33]. These findings of increased glutamine in schizophrenia patients may indicate an abnormality in the cycle of conversion of glutamine to glutamate in the human brain. Recently, we found that the ratio of glutamine to glutamate in the CSF was significantly higher in first-episode and drug-naive patients with schizophrenia than in controls, although the levels of both were not significantly different from those of the controls [28]. These findings suggest that the increased ratio of glutamine to glutamate in the CSF sample may reflect impairment of the glutamate-glutamine cycle in the brains of schizophrenia patients (Figure 3) [5,14,28].

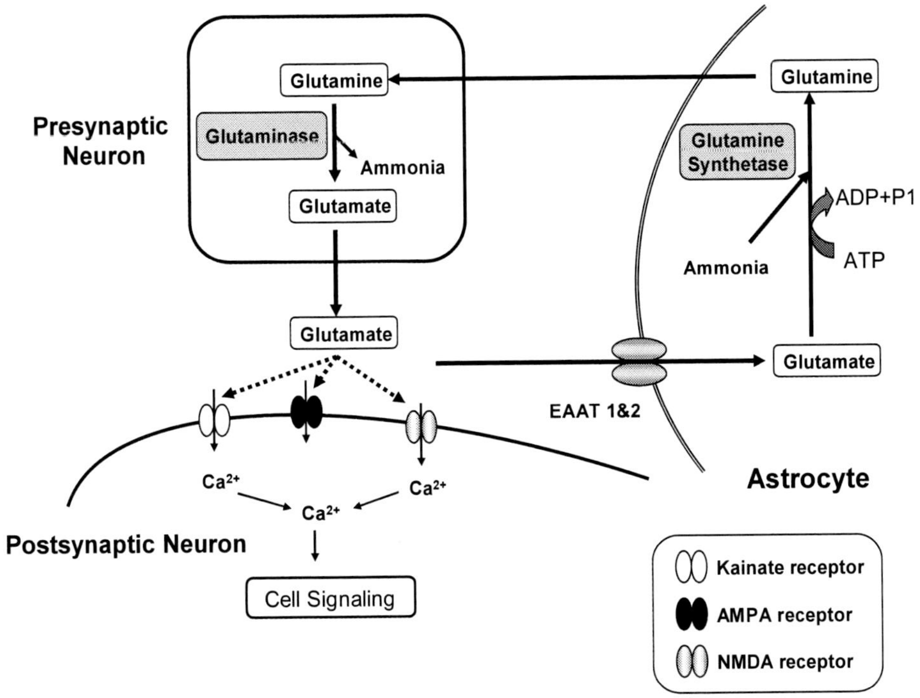

Figure 3. Glutamate-glutamine cycle in brain Glutamate released from the neurons binds glutamate receptors (e.g., kainite, AMPA, NMDA) on postsynaptic neurons, and released glutamate is also taken up mostly by the astrocytes via the glutamate transporter (EAAT 1 and 2). In the astrocytes, glutamine synthetase converted glutamate to glutamine using ammonia and ATP. The action of glutaminase, which hydrolyses glutamine to glutamate and ammonia within neurons, completes the glutamate-glutamine cycle in the brain.

3. D-Serine and NMDA Receptors

Research in the past decade has consistently revealed significant levels of D-serine in mammalian brains. In the rat brain, the distribution of D-serine was similar to that of NMDA receptors, suggesting that D-serine might be an endogenous agonist on NMDA receptors [34-39]. Immunohistochemical studies revealed that endogenous D-serine is localized in the astrocytes of the forebrain grey matter, near or ensheathing the NMDA receptor synapses [40]. Agonists of the non-NMDA subtypes of the glutamate receptors promote the release of D-serine from cultured astrocytes [41]. Thus, D-serine was first detected in glial cells, which led to the idea that astrocytes are the only source of D-serine. The origin of D-serine in mammals was unclear until serine racemase, which catalyzes the direct conversion of L-serine to D-serine, was isolated from the brain [42-44]. However, recent studies demonstrated that serine racemase and D-serine exist in neurons [45-50]. Expression of the mRNA and protein of serine racemase was detected in primary cultures of rat neurons [49]. Furthermore, an *in situ* hybridization study detected serine racemase mRNA in a large number of neurons throughout the brain, especially in forebrain regions such as the cerebral cortex, striatum, and hippocampus [50]. Karvelishvily et al. [48] reported that neuronal D-serine was released upon activation of kainite, AMPA, and NMDA receptors. These findings suggest that neuronal D-serine release provides a mechanism for activating NMDA receptors by an autocrine or paracrine pathway [38,39,51]. Very recently, it was demonstrated that knockout mice for the serine racemase gene show decreased (>80 %) levels of D-serine in the brain [52,53], suggesting the major role of serine racemase in the production of D-serine in the mouse brain. Therefore, knockout mice for the serine racemase gene will contribute to the analysis of the physiological and pathophysiological roles of D-serine *in vivo*.

D-Amino acid oxidase (DAO) (EC 1.4.3.3) was first identified by Hans Krebs in 1935 [54,55]. DAO is a flavoenzyme that catalyzes the oxidation of D-amino acids, including D-serine [35,56-59]. In mammals, the highest concentrations of DAO are in the kidney, liver, and brain. The contrasting maps of D-serine immunoreactivity and DAO activity in adult rat brain show the influence of DAO on D-serine levels [35]. DAO in the brainstem, medulla, and spinal cord keeps D-serine levels low. In addition, forebrain DAO activity is undetected and D-serine levels are very high, strongly implying that DAO is the major enzyme capable of destroying endogenous D-serine in the brain [35]. In the brain, gene expression of DAAO is detected in astrocytes [60-62].

While serine racemase and DAO, which are respectively responsible for the specific synthesis and degradation of D-serine, have been characterized, the transport system for the removal of D-serine from the extracellular space has been less well defined. The most likely candidate for D-serine transport is the Na^+-independent alanine-serine-cysteine 1 (Asc-1) transporter [63,64]. Areas with high levels of Asc-1-immunoreactivity included the hypothalamus, the medial septal area, the globus pallidus, the entopeduncular nucleus, and the cingulate and retrosplenial cortices. Moderate Asc-1-immunoraectivity was observed in several areas, including layers III and V of the neocortex, thalamus, nucleus accumbens, caudate putamen, bed nucleus of stria terminalis, all amygdaloid nuclei, hippocampus (CA1-CA3 and hilus of the dentate gyrus), as well as in several brainstem nuclei. At the ultrastructural level, Asc-1 immunoreactivity was confined mainly to presynaptic terminals. Immunostaining in either glial cell bodies or perivascular sites was not observed, and white matter was completely devoid of Asc-1-immunoreactivity. These findings suggest that Asc-1-immunoreactivity is widespread in the brain and limited to neuronal structures and that Asc-1 may contribute to synaptic clearance of D-serine in the brain [65]. Behavioral phenotyping combined with electroencephalogram recordings revealed that Asc-1 gene knockout mice developed tremors, ataxia, and seizures that resulted in early postnatal death. Both tremors and seizures were reduced by the NMDA receptor antagonist dozocilpine [66]. These findings suggest that the Asc-1 transporter plays a critical role in regulating neuronal excitability, and that the transporter is vital for normal CNS function and essential to the postnatal survival of mice [66]. Total [^{3}H]D-serine uptake in the forebrain and cerebellar synaptosomes from Asc-1 knockout mice was reduced to 34% and 22%, respectively, of that observed in wild-type mice [67]. When the Na^+-dependent transport components were removed by omission of Na^+ in the assay buffer, D-serine uptake in knockout mice was reduced to 8% and 3% of that measured in the forebrain and cerebellum, respectively, of wild-type mice. These results suggest that Asc-1 is the only high-affinity D-serine transporter in the mouse brain and is the predominant mechanism for D-serine reuptake [67].

4. Susceptibility Genes

A number of family, twin, and adoption studies demonstrated the involvement of genetic factors in the transmission of vulnerability to schizophrenia. Two genes, G72 and G30, were discovered by Chumakov et al. [68] in an association study using single nucleotide polymorphisms in chromosome region 13q22-q34, an area that has been linked to schizophrenia [69]. Common DNA variants at the G72/G30 locus were associated with schizophrenia in both French-Canadian and Russian case-control samples. The G72 protein was shown to physically interact with DAO and to regulate DAO activity *in vitro*. Hence G72 has been named a DAO activator. The DAO gene was also associated with schizophrenia in a Chinese population [70]. Since DAO degrades D-serine, it has been hypothesized that abnormal hyperactivity of G72 leads to attenuated NMDA receptor-mediated glutamatergic transmission. Interestingly, Korostishevsky et al. [71] reported the increased G72 expression in the dorsolateral prefrontal cortex (DLPFC) of schizophrenia patients as compared to healthy controls. A German group detected an association of DAO with schizophrenia but no statistical interaction between the G72/G30 and DAO genotypes [72]. In a study of a Japanese cohort, no evidence of association was detected for either DAO or serine racemase, and serum concentrations of D-serine did not correlate with the DAO or serine racemase genotypes [73]. Furthermore, no association was found between the serine racemase gene and schizophrenia in a German population [74]. In contrast, an association between variants in the serine racemase gene and schizophrenia was demonstrated [75,76].

5. D-Serine and Schizophrenia

Treatment with D-serine revealed significant improvements in positive, negative, and cognitive symptoms in schizophrenia patients treated with antipsychotic drugs [77], suggesting that levels of D-serine may be decreased in the brains of schizophrenia patients. We reported that the serum levels of L-serine were significantly higher in patients with schizophrenia than in normal controls, and that the serum levels of D-serine and the ratio of D-serine to total serine were markedly lower in patients with schizophrenia than in normal controls [73,78]. At the ultrastructual level, D-serine staining appears patchy, and is most abundant around the blood vessels in the forebrain [40]. Furthermore, no changes in D-serine levels in the forebrain of mice lacking D-amino acid oxidase (DAO) were demonstrated [79,80], suggesting that DAO does not regulate D-serine levels in the forebrain, where NMDA receptors are abundant. Some lines of evidence suggest that D-serine may be swept away by the blood vessels, and that the resulting D-serine in the blood may originate from the brain [5,14].

We found that the ratio of D-serine to total serine in the CSF was significantly lower in first-episode and drug-naive patients with schizophrenia than in controls [81], suggesting that the synthetic and/or metabolic pathway of D-serine may be impaired in the brains of schizophrenia patients. Furthermore, Bendikov et al. [82] reported a 25% decrease in D-serine levels in the CSF of schizophrenia patients, although D-serine levels in the parietal cortices of postmortem brains with schizophrenia were not altered. In a well-characterized set of autopsied brains (Brodmann area 6) obtained from the Stanley Foundation, we also reported that D-serine levels in samples of postmortem schizophrenic brain were not different from those of controls [30]. Another possibility may be due to different effects of D-serine at development. It has been suggested that D-serine may play different roles in the adult brain than in the developing brain, which may be mirrored in distinctive localizations at different ages [83]. A recent study has demonstrated that D-serine plays an important role in neuronal migration [84], suggesting that D-serine serves as a co-agonist for NMDA receptor-dependent cell migration at the developmental stage. Therefore, it is possible that D-serine levels in the brain may be altered at the developmental stage or at the onset of schizophrenia.

Postmortem brain studies demonstrated decreased levels of serine racemase protein in the frontal cortex (39% decrease) and hippocampus (21% decrease) of schizophrenia patients compared to controls [82], whereas increased levels of serine racemase were found in the dorsolateral cortex [85] and hippocampus [86] of schizophrenia patients. In schizophrenia patients, DAO mRNA was increased in the cerebellum [85]. In contrast, DAO activity in postmortem brain samples was twice as high in the schizophrenia group than in a control group [87], suggesting that D-serine availability may be altered in the brains of schizophrenia patients because of increased D-serine degradation.

6. Potential Therapeutic Approach

The NMDA receptor hypofunction hypothesis of schizophrenia suggests that increasing NMDA receptor function via pharmacological manipulation could provide a potential novel strategy for the management of schizophrenia. Currently, the glycine modulatory sites on NMDA receptors present the most attractive therapeutic targets for the treatment of schizophrenia [88-98]. The most likely approach is to increase the availability of the obligatory co-agonists (D-serine and glycine) at modulatory sites on the NMDA receptors (Figure 4).

One way to increase D-serine levels is to inhibit DAO, the enzyme responsible for D-serine degradation. Recently, Adage et al. [99] reported that a selective DAO inhibitor AS057278 (5-methoxypyrazole-3-carboxylic acid) (Figure 5) has the potential for antipsychotic action toward both cognitive and positive symptoms of schizophrenia. Administration of AS057278 (10 mg/kg, i.v.) caused a slight increase of D-serine levels in rat cortex and midbrain. Furthermore, AS057278 was able to normalize PCP-induced prepulse inhibition after acute (80 mg/kg) and chronic (20 mg/kg, b.i.d.) oral administration in mice [99]. This weak effect of AS057278 may be a result of the increased endogenous D-serine in the systemic circulation due to DAO inhibition in kidney and liver, since D-serine is capable of penetrating the blood-brain barrier. Furthermore, a brain-penetrate DAO inhibitor may not be able to significantly enhance NMDA receptor-mediated neurotransmission by itself, since there is little overlap in the distributions of DAO and NMDA receptors [34,35].

As described above, D-serine is effective for several symptoms of schizophrenia, but a large amount of D-serine (30 mg/kg/day) is necessary. D-serine administered orally is degraded substantially by DAO in the peripheral organs, diminishing its oral bioavailability. These findings prompted us to design small-molecule DAO inhibitors that can be co-administered with D-serine to minimize its degradation by DAO. DAO inhibition should not only improve the bioavailability of D-serine but also reduce its nephrotoxic effects, since D-serine is nephrotoxic in rats [100,101]. The mechanisms of D-serine-induced nephrotoxicity are believed to be associated with oxidative stress caused by hydrogen peroxide, a product of DAO-mediated metabolism of D-serine [102]. Recently, Ferraris et al. [103] reported a series of small-molecule DAO inhibitors based on a benzo[d]isoxazol-3-ol core structure and *in vivo* effects of DAO inhibition on D-serine pharmacokinetics. Among benzo[*d*]isoxazol-3-ol derivatives, 5-chloro-benzo[*d*]isoxazol-3-ol (CBIO) (Figure 5) potently inhibited DAO with an IC_{50} value of 200 nM. Oral administration of CBIO (30 mg/kg) in conjunction with D-serine (30 mg/kg) significantly enhanced the plasma and brain levels of D-serine in rats compared to the oral administration of D-serine (30 mg/kg) alone. Furthermore, we found that co-administration of D-serine and CBIO significantly enhanced D-serine's effect on NMDA receptor antagonist-induced PPI deficits in mice (Hashimoto et al., in preparation). Therefore, co-administration of D-serine and a DAO inhibitor, CBIO, may provide a novel therapeutic approach to cognitive disorders associated with NMDA receptor hypofunction ([103]; Hashimoto et al., in preparation).

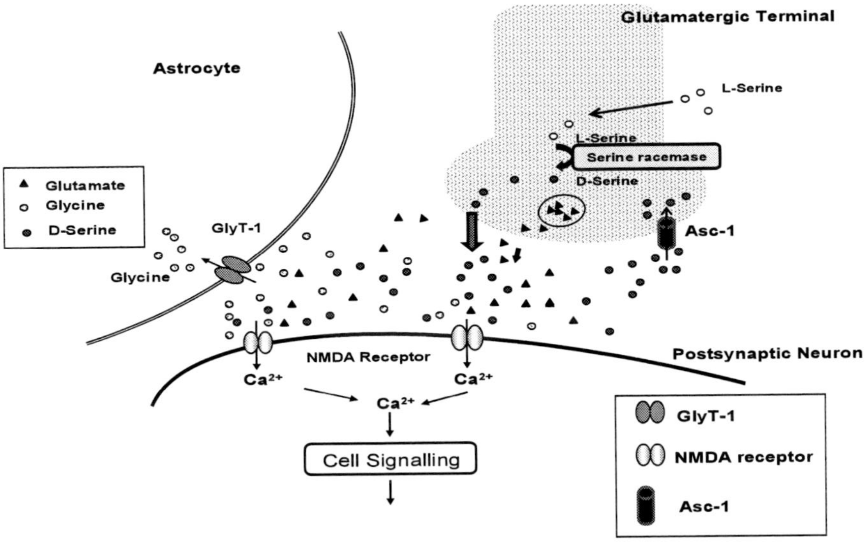

Figure 4. Role of NMDA receptors, glycine transporter (GlyT-1), Asc-1 and D-serine in the brain D-Serine is synthesized from L-serine via serine racemase in neurons. Extracellular glycine is removed by glycine transporter-1 (GlyT-1) in astrocytes. Extracellular D-serine is removed by alanine-serine-cysteine transporter-1 (Asc-1) in neurons. Thus, serine racemase, GlyT-1 and Asc-1 are implicated in the function of the NMDA receptors in the brain.

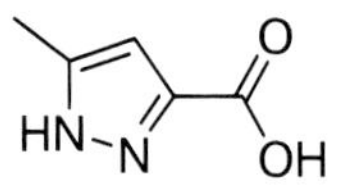

AS057278
(5-methylpyrazole-3-carboxylic acid)

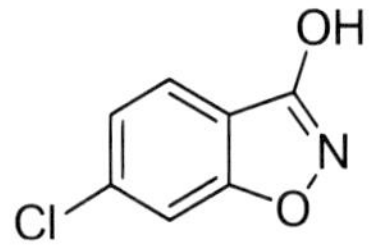

CBIO
(6-Chloro-benzo[d]isoxazol-3-ol)

Figure 5. Chemical structures of DAO inhibitors AS057278 and CBIO.

A limitation in the use of D-serine for the treatment of neuropsychiatric diseases is that large doses must be administered in order for sufficient D-serine to pass the blood-brain barrier. Asc-1 inhibition is an alternative approach [104]. (S)-Methyl-L-cysteine was a potent and selective inhibitor of [^{3}H]D-serine uptake in rat cortical synaptosomes, rat cerebellar synaptosomes, and HEK293 cells expressing the human Asc-1 transporter [104]. With these findings taken together, it seems that DAO inhibitors, Asc-1 inhibitors, or agents that promote the conversion of L-serine to D-serine by enhancing serine racemase activity would be potential therapeutic drugs for schizophrenia [5,90,104]. In addition, agents that modulate the serine racemase-interacting proteins, including the glutamate receptor interacting protein (GRIP) [84,105] and a protein interacting with C kinase 1 (PICK1) [105-108], may also be potential therapeutic drugs for schizophrenia.

Another approach that can enhance NMDA receptor neurotransmission is to increase the availability of the obligatory co-agonist glycine at modulatory sites by inhibiting glycine transporter-1 (GlyT-1) on glial cells [88-98]. NMDA receptor antagonists such as PCP have been used as animal models of cognitive deficits, since cognitive deficits are a core feature of schizophrenia, which predicts vocational and social disabilities for those who have it [109,110]. We reported that repeated administration of PCP (10 mg/kg/day for 10 days) caused cognitive deficits in mice for a long time (more than 6 weeks after the final administration), and that PCP-induced cognitive deficits could be improved by subsequent subchronic (2 weeks) administration of clozapine, but not haloperidol [111]. Therefore, reversal of PCP-induced cognitive deficits may be a potential animal model of atypical antipsychotic activity in relation to the amelioration of cognitive deficits in schizophrenia [111-117]. Recently, we found that treatment with (R)-(N-[3-(4'-fluorophenyl)-3-(4'-phenylphenoxy)propyl])sarcosine (NFPS), a selective glycine transporter-1 (GlyT-1) inhibitor, or D-serine could attenuate PCP-induced cognitive deficits in mice [117]. Furthermore, we reported that D-serine and NFPS improved dizocilpine-induced cognitive deficits in rats [118]. These findings suggest that enhancing the NMDA receptor function by GlyT-1 inhibitors is an effective treatment for cognitive deficits in schizophrenia patients.

Interestingly, in a recent 6-week double-blind, placebo-controlled trial of the GlyT-1 inhibitor N-methylglycine (sarcosine: 2 g/day), schizophrenia patients who received sarcosine treatment showed significant improvements in their positive, negative, cognitive, and general psychiatric symptoms [119]. Sarcosine was also well-tolerated, and no significant side effects were noted [119]. A recent double-blind, placebo-controlled study demonstrated that sarcosine, more than D-serine, can benefit not only patients with long-term stable disease but also persons with acute schizophrenia, suggesting that a GlyT-1 inhibitor may be more efficacious than NMDA-glycine site agonists for adjuvant treatment of schizophrenia, at least during the acute phase [120]. A recent study demonstrated that sarcosine treatment (2 g/day) was effective for acutely symptomatic drug-free schizophrenia patients, although placebo- or active-controlled, larger studies will be needed to fully assess sarcosine's effects [121,122]. Taken together, these findings suggest that pharmacological modulation of glycine modulatory sites on NMDA receptors by GlyT-1 inhibitors might be beneficial in the treatment of cognitive deficits as well as psychosis in several psychiatric diseases, including schizophrenia [88-98].

7. Conclusion

As discussed above, D-serine plays a role as an endogenous co-agonist at modulatory sites on NMDA receptors, and abnormality in the synthetic and metabolic pathways of D-serine may be implicated in the pathophysiology of schizophrenia. Based on the NMDA receptor hypofunction hypothesis, drugs to increase the availability of the obligatory co-agonists (D-serine and glycine) at modulatory sites on NMDA receptors would be potential therapeutic drugs for schizophrenia. To date, D-serine, DAO inhibitors plus D-serine, GlyT-1 inhibitors, and Asc-1 inhibitors are the most attractive potential therapeutic drugs for this disorder.

Recent studies using knockout mice for the serine racemase gene showed decreased levels (>80%) of D-serine in the mouse brain, suggesting the major role of serine racemase in the production of D-serine in the brain. Thus, small amounts (approximately 20%) of D-serine in the mouse brain may be synthesized by an unidentified process. Further detailed studies using these knockout mice will contribute to the understanding of the physiological and pathophysiological roles of D-serine *in vivo*. Finally, gaining a further understanding of the roles of D-serine and NMDA receptors in the pathophysiology of schizophrenia would reveal new perspectives for treating this disorder.

8. Acknowledgements

I would like to thank my collaborators who are listed as the co-authors of our articles in the reference list. This study was supported in part by grants from the Minister of Education, Culture, Sports, Science, and Technology of Japan (to K.H.), and the Program for Promotion of Fundamental Studies in Health Sciences of the National Institute of Biomedical Innovation of Japan (to K.H.).

REFERENCES

[1] Schizophrenia - National Institute of Mental Health, NIH Publication, 2007.
[2] Harrison PJ, Owen MJ. Genes for schizophrenia? Recent findings and their pathophysiological implications. *Lancet* 2003; 361(9355): 417-419.
[3] Craddock N, O'Donovan MC, Owen MJ. The genetics of schizophrenia and bipolar disorders: dissecting psychosis. *J Med Genet* 2005; 42(3): 193-204.
[4] Harrison PJ, Weinberger DR. Schizophrenia genes, gene expression, and neuropathology: on the matter of their convergence. *Mol Psychiatry* 2005; 10(1): 40-68.
[5] Hashimoto K, Hattori E. Candidate genes and models. 4. Neurotrasmission. In: *Neurogenetics of Psychiatric Disorders*, Edited by Sawa A, McInnis MG, Informa HealthcareUSA, Inc., 2007; pp.81-100.
[6] Carlsson A, Waters N, Holm-Waters S, Tedroff J, Nilsson M, Carlsson ML. Interactions between monoamines, glutamate, and GABA in schizophrenia: new evidence. *Annu Rev Pharmacol Toxicol* 2001; 41: 237-260.
[7] Javitt DC, Zukin SR. Recent advances in the phencyclidine model of schizophrenia. *Am J Psychiatry* 1991; 148(10): 1301-1308.
[8] Olney JW, Farber NB. Glutamate receptor dysfunction and schizophrenia. *Arch Gen Psychiatry* 1995; 52(12): 998-1007.
[9] Goff DC, Coyle JT, The emerging role of glutamate in the pathophysiology and treatment of schizophrenia. *Am J Psychiatry* 2001; 158(9): 1367-1377.
[10] Tsai G, Coyle JT. Glutamatergic mechanisms in schizophrenia. *Annu Rev Pharmacol Toxicol* 2002; 42: 165-179.
[11] Krystal JH, D'Souza DC, Petrakis IL, Belger A, Berman RM, Charney DS, Abi-Saab W, Madonick S. NMDA agonists and antagonists as probes of glutamatergic dysfunction and pharmacotherapies in neuropsychiatric disorders. *Harv Rev Psychiatry* 1999; 7(3): 125-143.
[12] Sawa A, Snyder SH. Schizophrenia: diverse approaches to a complex disease. *Science* 2002; 296(5568): 692-695.
[13] Hashimoto K, Okamura N, Shimizu E, Iyo M. Glutamate hypothesis of schizophrenia and approach for possible therapeutic drugs. *Curr Med Chem - CNS Agents* 2004; 4(2): 147-154.
[14] Hashimoto K, Shimizu E, Iyo M. Dysfunction of glia-neuron communication in pathophysiology of schizophrenia. *Curr Psychiatry Rev* 2005; 1(2): 151-163.
[15] Lewis DA. GABAergic local circuit neurons and prefrontal cortical dysfunction in schizophrenia. *Brain Res Brain Res Rev* 2000; 31(2-3): 270-276.
[16] Benes FM, Berretta S. GABAergic interneurons: implications for understanding schizophrenia and bipolar disorder. *Neuropsychopharmacology* 2001; 25(1): 1-27.
[17] Coyle JT, Leski M, Morrison JH. The diverse roles of L-glutamic acid in brain signal transduction. In: *Neuropsychopharmacology – The fifth Generation of Progress*, Edited by Davis KL, Charnney D, Coyle JT, Nemeroff C, Lippincott Williams & Wilkins, Philadelphia, PA, 2002, pp.70-90.

[18] Luby ED, Cohen BD, Rosenbaum G, Gottlieb JS, Kelley R. Study of a new schizophrenomimetic drug; sernyl. *AMA Arch Neurol Psychiatry* 1959; 81(3): 363-369.

[19] Allen RM, Young SJ. Phencyclidine-induced psychosis. *Am J Psychiatry* 1978; 135(9): 1081-1084.

[20] Krystal JH, Karper LP, Seibyl JP, Freeman GK, Delaney R, Bremner JD, Heninger GR, Bowers MB Jr, Charney DS. Subanesthetic effects of the noncompetitive NMDA antagonist, ketamine, in humans. Psychotomimetic, perceptual, cognitive, and neuroendocrine responses. *Arch Gen Psychiatry* 1994; 51(3): 199-214.

[21] Malhotra AK, Pinals DA, Weingartner H, Sirocco K, Missar CD, Pickar D, Breier A. NMDA receptor function and human cognition: the effects of ketamine in healthy volunteers. *Neuropsychopharmacology* 1996; 14(5): 301-307.

[22] Hartvig P, Valtysson J, Lindner KJ, Kristensen J, Karlsten R, Gustafsson LL, Persson J, Svensson JO, Oye I, Antoni G. Central nervous system effects of subdissociative doses of (S)-ketamine are related to plasma and brain concentrations measured with positron emission tomography in healthy volunteers. *Clin Pharmacol Ther* 1995; 58(2): 165-173.

[23] Lahti AC, Koffel B, LaPorte D, Tamminga CA: Subanesthetic doses of ketamine stimulate psychosis in schizophrenia. *Neuropsychopharmacology* 1995; 13(1): 9-19.

[24] Krystal JH, Perry EB Jr, Gueorguieva R, Belger A, Madonick SH, Abi-Dargham A, Cooper TB, Macdougall L, Abi-Saab W, D'Souza DC. Comparative and interactive human psychopharmacologic effects of ketamine and amphetamine: implications for glutamatergic and dopaminergic model psychoses and cognitive function. *Arch Gen Psychiatry* 2005; 62(9): 985-994.

[25] Kim JS, Kornhuber HH, Schmid-Burgk W, Holzmüller B. Low cerebrospinal fluid glutamate in schizophrenic patients and a new hypothesis on schizophrenia. *Neurosci Lett* 1980; 20(3): 379-382.

[26] Perry TL. Normal cerebrospinal fluid and brain glutamate levels in schizophrenia do not support the hypothesis of glutamatergic neuronal dysfunction. *Neurosci Lett* 1982; 28(1): 81-85.

[27] Korpi ER, Kaufmann CA, Marnela KM, Weinberger DR. Cerebrospinal fluid amino acid concentrations in chronic schizophrenia. *Psychiatry Res* 1987; 20(4): 337-345.

[28] Hashimoto K, Engberg G, Shimizu E, Nordin C, Lindstrom LH, Iyo M. Elevated glutamine/glutamate ratio in cerebrospinal fluid of first episode and drug naive schizophrenic patients. *BMC Psychiatry* 2005; 5(1): 6.

[29] Tsai G, Passani LA, Slusher BS, Carter R, Baer L, Kleinman JE, Coyle JT. Abnormal excitatory neurotransmitter metabolism in schizophrenic brains. *Arch Gen Psychiatry* 1995; 52(10): 829-836.

[30] Hashimoto K, Sawa A, Iyo M. Increased levels of glutamate in brains from patients with mood disorders. *Biol Psychiatry* 2007; 25(11): 1310-1316.

[31] Rothman DL, Behar KL, Hyder F, Shulman RG. *In vivo* NMR studies of the glutamate neurotransmitter flux and neuroenergetics: implications for brain function. *Annu Rev Physiol* 2003; 65: 401-427.

[32] Bartha R, Williamson PC, Drost DJ, Malla A, Carr TJ, Cortese L, Canaran G, Rylett RJ, Neufeld RW. Measurement of glutamate and glutamine in the medial prefrontal

cortex of never-treated schizophrenic patients and healthy controls by proton magnetic resonance spectroscopy. *Arch Gen Psychiatry* 1997; 54(10): 959-965.

[33] Theberge J, Bartha R, Drost DJ, Menon RS, Malla A, Takhar J, Neufeld RW, Rogers J, Pavlosky W, Schaefer B, Densmore M, Al-Semaan Y, Williamson PC. Glutamate and glutamine measured with 4.0 T proton MRS in never-treated patients with schizophrenia and healthy volunteers. *Am J Psychiatry* 2002; 159(11): 1944-1946.

[34] Hashimoto A, Oka T. Free D-aspartate and D-serine in the mammalian brain and periphery. *Prog Neurobiol* 1997; 52(4): 325-353.

[35] Schell MJ. The N-methyl D-aspartate receptor glycine site and D-serine metabolism: an evolutionary perspective. *Philos Trans R Soc Lond B Biol Sci* 2004; 359(1446): 943-964.

[36] Martineau M, Baux G, Mothet JP. D-serine signalling in the brain: friend and foe. *Trends Neurosci* 2006; 29(8): 481-491

[37] Oliet SH, Mothet JP. Molecular determinants of D-serine-mediated gliotransmission: from release to function. *Glia* 2006; 54(7): 726-737.

[38] Wolosker H. D-serine regulation of NMDA receptor activity. *Sci STKE* 2006; 2006(356): pe41.

[39] Wolosker H. NMDA receptor regulation by D-serine: new findings and perspectives. *Mol Neurobiol* 2007; 36(2): 152-164.

[40] Schell MJ, Brady RO Jr, Molliver ME, Snyder SH. D-serine as a neuromodulator: regional and developmental localizations in rat brain glia resemble NMDA receptors. *J Neurosci* 1997; 17(5): 1604-1615.

[41] Schell MJ, Molliver ME, Snyder SH. D-serine, an endogenous synaptic modulator: localization to astrocytes and glutamate-stimulated release. *Proc Natl Acad Sci USA* 1995; 92(9): 3948-3952.

[42] Wolosker H, Sheth KN, Takahashi M, Mothet JP, Brady RO Jr, Ferris CD, Snyder SH. Purification of serine racemase: biosynthesis of the neuromodulator D-serine. *Proc Natl Acad Sci USA* 1999; 96(2): 721-725.

[43] Wolosker H, Blackshaw S, Snyder SH. Serine racemase: a glial enzyme synthesizing D-serine to regulate glutamate-N-methyl-D-aspartate neurotransmission. *Proc Natl Acad Sci USA* 1999; 96(23): 13409-13414.

[44] De Miranda J, Santoro A, Engelender S, Wolosker H. Human serine racemase: moleular cloning, genomic organization and functional analysis. *Gene* 2000; 256(1-2): 183-188.

[45] Yasuda E, Ma N, Semba R. Immunohistochemical evidences for localization and production of D-serine in some neurons in the rat brain. *Neurosci Lett* 2001; 299(1-2): 162-164.

[46] Williams SM, Diaz CM, Macnab LT, Sullivan RK, Pow DV. Immunocytochemical analysis of D-serine distribution in the mammalian brain reveals novel anatomical compartmentalizations in glia and neurons. *Glia* 2006; 53(4): 401-411.

[47] Shleper M, Kartvelishvily E, Wolosker H. D-serine is the dominant endogenous coagonist for NMDA receptor neurotoxicity in organotypic hippocampal slices. *J Neurosci* 2005; 25(41): 9413-9417.

[48] Kartvelishvily E, Shleper M, Balan L, Dumin E, Wolosker H. Neuron-derived D-serine release provides a novel means to activate N-methyl-D-aspartate receptors. *J Biol Chem* 2006; 281(20): 14151-14162.

[49] Yoshikawa M, Nakajima K, Takayasu N, Noda S, Sato Y, Kawaguchi M, Oka T, Kobayashi H, Hashimoto A. Expression of the mRNA and protein of serine racemase in primary cultures of rat neurons. *Eur J Pharmacol* 2006; 548(1-3): 74-76.

[50] Yoshikawa M, Takayasu N, Hashimoto A, Sato Y, Tamaki R, Tsukamoto H, Kobayashi H, Noda S. The serine racemase mRNA is predominantly expressed in rat brain neurons. *Arch Histol Cytol* 2007; 70(2): 127-134.

[51] Boks MP, Rietkerk T, van de Beek MH, Sommer IE, de Koning TJ, Kahn RS. Reviewing the role of the genes G72 and DAAO in glutamate neurotransmission in schizophrenia. *Eur Neuropsychopharmacol* 2007; 17(9): 567-572.

[52] Basu AC, Tsai GE, Han L, Jiang ZI, Benneyworth M, Ehmsen JT, Mustafa AK, Dore S, Snyder SH, Coyle JT. Gene knockout study of serine racemase. 37th Annual Meeting of Society for Neuroscience, San Diego, CA, Nov. 06, 2007. Washington DC: Society for Neuroscience, 2007; 576.7.

[53] Ying-Luan Z, Zhao YL, Mori H. Role of D-serine in the mammalian brain. *Brain Nerve* 2007; 59(7): 725-730.

[54] Krebs HA. Metabolism of amino-acids: Deamination of amino-acids. *Biochem J* 1935; 29(7): 1620-1644.

[55] Krebs HA. Metabolism of amino-acids: The synthesis of glutamine from glutamic acid and ammonia, and the enzymic hydrolysis of glutamine in animal tissues. *Biochem J* 1935; 29(8): 1951-1969.

[56] Fukui K, Miyake Y. Molecular cloning and chromosomal localization of a human gene encoding D-amino-acid oxidase. *J Biol Chem* 1992; 267(26): 18631-18638.

[57] Kawazoe T, Park HK, Iwana S, Tsuge H, Fukui K. Human D-amino acid oxidase: an update and review. *Chem Rec* 2007; 7(5): 305-315.

[58] Pollegioni L, Piubelli L, Sacchi S, Pilone MS, Molla G. Physiological functions of D-amino acid oxidases: from yeast to humans. *Cell Mol Life Sci* 2007; 64(11): 1373-1394.

[59] Park HK, Shishido Y, Ichise-Shishido S, Kawazoe T, Ono K, Iwana S, Tomita Y, Yorita K, Sakai T, Fukui K. Potential role for astroglial D-amino acid oxidase in extracellular D-serine metabolism and cytotoxicity. *J Biochem* 2006; 139(2): 295-304.

[60] Horiike K, Tojo H, Arai R, Yamano T, Nozaki M, Maeda T. Localization of D-amino acid oxidase in Bergmann glial cells and astrocytes of rat cerebellum. *Brain Res Bull* 1987; 19(5): 587-596.

[61] Horiike K, Tojo H, Arai R, Nozaki M, Maeda T. D-amino-acid oxidase is confined to the lower brain stem and cerebellum in rat brain: regional differentiation of astrocytes. *Brain Res* 1994; 652(2): 297-303.

[62] Urai Y, Jinnouchi O, Kwak KT, Suzue A, Nagahiro S, Fukui K. Gene expression of D-amino acid oxidase in cultured rat astrocytes: regional and cell type specific expression. *Neurosci Lett* 2002; 324(2): 101-104.

[63] Fukasawa Y, Segawa H, Kim JY, Chairoungdua A, Kim DK, Matsuo H, Cha SH, Endou H, Kanai Y. Identification and characterization of a Na^+-independent neutral amino acid transporter that associates with the 4F2 heavy chain and exhibits substrate

selectivity for small neutral D- and L-amino acids. *J Biol Chem* 2000; 275(13): 9690-9698.

[64] Nakauchi J, Matsuo H, Kim DK, Goto A, Chairoungdua A, Cha SH, Inatomi J, Shiokawa Y, Yamaguchi K, Saito I, Endou H, Kanai Y. Cloning and characterization of a human brain Na^+-independent transporter for small neutral amino acids that transports D-serine with high affinity. *Neurosci Lett* 2000; 287(3): 231-235.

[65] Helboe L, Egebjerg J, Møller M, Thomsen C. Distribution and pharmacology of alanine-serine-cysteine transporter 1 (asc-1) in rodent brain. *Eur J Neurosci* 2003; 18(8): 2227-2238.

[66] Xie X, Dumas T, Tang L, Brennan T, Reeder T, Thomas W, Klein RD, Flores J, O'Hara BF, Heller HC, Franken P. Lack of the alanine-serine-cysteine transporter 1 causes tremors, seizures, and early postnatal death in mice. *Brain Res* 2005; 1052(2): 212-221.

[67] Rutter AR, Fradley RL, Garrett EM, Chapman KL, Lawrence JM, Rosahl TW, Patel S. Evidence from gene knockout studies implicates Asc-1 as the primary transporter mediating D-serine reuptake in the mouse CNS. *Eur J Neurosci* 2007; 25(6): 1757-1766.

[68] Chumakov I, Blumenfeld M, Guerassimenko O, Cavarec L, Palicio M, Abderrahim H, Bougueleret L, Barry C, Tanaka H, La Rosa P, Puech A, Tahri N, Cohen-Akenine A, Delabrosse S, Lissarrague S, Picard FP, Maurice K, Essioux L, Millasseau P, Grel P, Debailleul V, Simon AM, Caterina D, Dufaure I, Malekzadeh K, Belova M, Luan JJ, Bouillot M, Sambucy JL, Primas G, Saumier M, Boubkiri N, Martin-Saumier S, Nasroune M, Peixoto H, Delaye A, Pinchot V, Bastucci M, Guillou S, Chevillon M, Sainz-Fuertes R, Meguenni S, Aurich-Costa J, Cherif D, Gimalac A, Van Duijn C, Gauvreau D, Ouellette G, Fortier I, Raelson J, Sherbatich T, Riazanskaia N, Rogaev E, Raeymaekers P, Aerssens J, Konings F, Luyten W, Macciardi F, Sham PC, Straub RE, Weinberger DR, Cohen N, Cohen D. Genetic and physiological data implicating the new human gene G72 and the gene for D-amino acid oxidase in schizophrenia. *Proc Natl Acad Sci USA* 2002; 99(21): 13675-13680.

[69] Owen MJ, Williams NM, O'Donovan MC. The molecular genetics of schizophrenia: new findings promise new insights. *Mol Psychiatry* 2004; 9(1): 14-27.

[70] Liu X, He G, Wang X, Chen Q, Qian X, Lin W, Li D, Gu N, Feng G, He L. Association of DAAO with schizophrenia in the Chinese population. *Neurosci Lett* 2004; 369(3): 228-233.

[71] Korostishevsky M, Kaganovich M, Cholostoy A, Ashkenazi M, Ratner Y, Dahary D, Bernstein J, Bening-Abu-Shach U, Ben-Asher E, Lancet D, Ritsner M, Navon R. Is the G72/G30 locus associated with schizophrenia? single nucleotide polymorphisms, haplotypes, and gene expression analysis. *Biol Psychiatry* 2004; 56(3): 169-176.

[72] Schumacher J, Jamra RA, Freudenberg J, Becker T, Ohlraun S, Otte AC, Tullius M, Kovalenko S, Bogaert AV, Maier W, Rietschel M, Propping P, Nöthen MM, Cichon S. Examination of G72 and D-amino-acid oxidase as genetic risk factors for schizophrenia and bipolar affective disorder. *Mol Psychiatry* 2004; 9(2): 203-207.

[73] Yamada K, Ohnishi T, Hashimoto K, Ohba H, Iwayama-Shigeno Y, Toyoshima M, Okuno A, Takao H, Toyota T, Minabe Y, Nakamura K, Shimizu E, Itokawa M, Mori N, Iyo M, Yoshikawa T. Identification of multiple serine racemase (SRR) mRNA

isoforms and genetic analyses of SRR and DAO in schizophrenia and D-serine levels. *Biol Psychiatry* 2005; 57(12): 1493-1503.

[74] Strohmaier J, Georgi A, Schirmbeck F, Schmael C, Jamra RA, Schumacher J, Becker T, Höfels S, Klopp N, Illig T, Propping P, Cichon S, Nöthen MM, Rietschel M, Schulze TG. No association between the serine racemase gene (SRR) and schizophrenia in a German case-control sample. *Psychiatr Genet* 2007; 17(2): 125.

[75] Goltsov AY, Loseva JG, Andreeva TV, Grigorenko AP, Abramova LI, Kaleda VG, Orlova VA, Moliaka YK, Rogaev EI. Polymorphism in the 5'-promoter region of serine racemase gene in schizophrenia. *Mol Psychiatry* 2006; 11(4): 325-326.

[76] Morita Y, Ujike H, Tanaka Y, Otani K, Kishimoto M, Morio A, Kotaka T, Okahisa Y, Matsushita M, Morikawa A, Hamase K, Zaitsu K, Kuroda S. A genetic variant of the serine racemase gene is associated with schizophrenia. *Biol Psychiatry* 2007; 61(10): 1200-1203.

[77] Tsai G, Yang P, Chung LC, Lange N, Coyle JT. D-serine added to antipsychotics for the treatment of schizophrenia. *Biol Psychiatry* 1998; 44(11): 1081-1089.

[78] Hashimoto K, Fukushima T, Shimizu E, Komatsu N, Watanabe H, Shinoda N, Nakazato M, Kumakiri C, Okada S, Hasegawa H, Imai K, Iyo M. Decreased serum levels of D-serine in patients with schizophrenia: evidence in support of the N-methyl-D-aspartate receptor hypofunction hypothesis of schizophrenia. *Arch Gen Psychiatry* 2003; 60(6): 572-576.

[79] Nagata Y. Involvement of D-amino acid oxidase in elimination of D-serine in mouse brain. *Experientia* 1992; 48(8): 753-455.

[80] Hashimoto A, Nishikawa T, Konno R, Niwa A, Yasumura Y, Oka T, Takahashi K. Free D-serine, D-aspartate and D-alanine in central nervous system and serum in mutant mice lacking D-amino acid oxidase. *Neurosci Lett* 1993; 152(1-2): 33-36.

[81] Hashimoto K, Engberg G, Shimizu E, Nordin C, Lindström LH, Iyo M. Reduced D-serine to total serine ratio in the cerebrospinal fluid of drug naive schizophrenic patients. *Prog Neuropharmacol Biol Psychiatry* 2005; 29(5): 767-769.

[82] Bendikov I, Nadri C, Amar S, Panizzutti R, De Miranda J, Wolosker H, Agam G. A CSF and postmortem brain study of D-serine metabolic parameters in schizophrenia. *Schizophr Res* 2007; 90(1-3): 41-51.

[83] Wang LZ, Zhu XZ. Spatiotemporal relationships among D-serine, serine racemase, and D-amino acid oxidase during mouse postnatal development. *Acta Pharmacol Sin* 2003; 24(10): 965-974.

[84] Kim PM, Aizawa H, Kim PS, Huang AS, Wickramasinghe SR, Kashani AH, Barrow RK, Huganir RL, Ghosh A, Snyder SH. Serine racemase: activation by glutamate neurotransmission via glutamate receptor interacting protein and mediation of neuronal migration. *Proc Natl Acad Sci USA* 2005; 102(6): 2105-2110.

[85] Verrall L, Walker M, Rawlings N, Benzel I, Kew JN, Harrison PJ, Burnet PW. D-Amino acid oxidase and serine racemase in human brain: normal distribution and altered expression in schizophrenia. *Eur J Neurosci* 2007; 26(6): 1657-1669.

[86] Steffek AE, Haroutunian V, Meador-Woodruff JH. Serine racemase protein expression in cortex and hippocampus in schizophrenia. *Neuroreport* 2006; 17(11): 1181-1185.

[87] Madeira C, Freitas ME, Panizzutti R. D-Serine degradation in human brain: correlation with schizophrenia. Unpublished data.
[88] Sur C, Kinney GG. The therapeutic potential of glycine transporter-1 inhibitors. *Expert Opin Investig Drugs* 2004; 13(5): 515-521.
[89] Coyle JT, Tsai G. The NMDA receptor glycine modulatory site: a therapeutic target for improving cognition and reducing negative symptoms in schizophrenia. *Psychopharmacology (Berl)* 2004; 174(1): 32-38.
[90] Kinney GG, Sur C. Glycine site modulators and glycine transporter-1 inhibitors as novel therapeutic targets for the treatment of schizophrenia. *Curr Neuropharmacol* 2005; 3(1): 35-43.
[91] Lindsley CW, Wolkenberg SE, Kinney GG. Progress in the preparation and testing of glycine transporter type-1 (GlyT-1) inhibitors. *Curr Top Med Chem* 2006; 6(17): 1883-1896.
[92] Thomsen C. Glycine transporter inhibitors as novel antipsychotics. *Drug Discovery Today* 2006; 3(4): 539-545.
[93] Hashimoto K. Glycine transporter inhibitors as therapeutic agents for schizophrenia. *Recent Patents on CNS Drug Discovery* 2006; 1(1): 43-54.
[94] Hashimoto K. The NMDA receptor hypofunction hypothesis for schizophrenia and glycine modulatory sites on the NMDA receptors as potential therapeutic drugs. *Clin Psychopharmacol Neurosci* 2006; 4(1): 3-10.
[95] Hashimoto K. Glycine transporter-1 inhibitors as novel therapeutic drugs for schizophrenia. *CNS Agents Med Chem* 2007; 7(3): 177-182.
[96] Javitt DC. Is the glycine site half saturated or half unsaturated? Effects of glutamatergic drugs in schizophrenia patients. *Curr Opin Psychiatry* 2006; 19(2): 151-157.
[97] Javitt DC. Glutamate and Schizophrenia: Phencyclidine, N-Methyl-D-Aspartate Receptors, and Dopamine-Glutamate Interactions. *Int Rev Neurobiol* 2007; 78: 69-108.
[98] Sur C, Kinney CG. Glycine transporter I inhibitors and modulation of NMDA receptor-mediated exciatory neurotransmission. *Curr Drug Targets* 2007; 8(5): 643-649.
[99] Adage T, Trillat AC, Quattropani A, Perrin D, Cavarec L, Shaw J, Guerassimenko O, Giachetti C, Gréco B, Chumakov I, Halazy S, Roach A, Zaratin P. *In vitro* and *in vivo* pharmacological profile of AS057278, a selective D-amino acid oxidase inhibitor with potential anti-psychotic properties. *Eur Neuropsychopharmacol* 2008; 18(3): 200-214.
[100] Ganote CE, Peterson DR, Carone FA. The nature of D-serine-induced nephrotoxicity. *Am J Pathol* 1974; 77(2): 269-282.
[101] Carone FA, Ganote CE. D-serine nephrotoxicity. The nature of proteinuria, glucosuria, and aminoaciduria in acute tubular necrosis. D-serine nephrotoxicity. *Arch Pathol* 1975; 99(12): 658-662.
[102] Williams RE, Lock EA. Sodium benzoate attenuates D-serine induced nephrotoxicity in the rat. *Toxicology* 2005; 207(1): 35-48.
[103] Duvall B, Ferraris D, Ko YS, Thomas A, Majer P, Rojas C, Tsukamoto T, Hashimoto K. Synthesis of D-amino acid oxidase inhibitors and their effects on the plasma and brain levels of D-serine. 234th ACS National Meeting & Exposition, Boston, MA, USA. August 19-23, 2007; Abstract #287.

[104] Thomsen C, Helboe L, Egebjerg JJ. Use of Asc-1 inhibitors to treat neurological and psychiatric disorders. WO/2003/077998.

[105] Mustafa AK, Kim PM, Snyder SH. D-Serine as a putative glial neurotransmitter. *Neuron Glia Biology* 2004; 1(3): 275-281.

[106] Fujii K, Maeda K, Hikida T, Mustafa AK, Balkissoon R, Xia J, Yamada T, Ozeki Y, Kawahara R, Okawa M, Huganir RL, Ujike H, Snyder SH, Sawa A. Serine racemase binds to PICK1: potential relevance to schizophrenia. *Mol Psychiatry* 2006; 11(2): 150-157.

[107] Matsuzawa D, Hashimoto K, Miyatake R, Shirayama Y, Shimizu E, Maeda K, Suzuki Y, Mashimo Y, Sekine Y, Inada T, Ozaki N, Iwata N, Harano M, Komiyama T, Yamada M, Sora I, Ujike H, Hata A, Sawa A, Iyo M. Identification of functional polymorphisms in the promoter region of the human PICK1 gene and their association with methamphetamine psychosis. *Am J Psychiatry* 2007; 164(7): 1105-1114.

[108] Hikida T, Mustafa AK, Maeda K, Fujii K, Barrow RK, Saleh M, Huganir RL, Snyder SH, Hashimoto K, Sawa A. Modulation of D-serine levels in brains of mice lacking PICK1. *Biol Psychiatry* 2008; in press.

[109] Green MF, Nuechterlein KH, Gold JM, Barch DM, Cohen J, Essock S, Fenton WS, Frese F, Goldberg TE, Heaton RK, Keefe RS, Kern RS, Kraemer H, Stover E, Weinberger DR, Zalcman S, Marder SR. Approaching a consensus cognitive battery for clinical trials in schizophrenia: the NIMH-MATRICS conference to select cognitive domains and test criteria. *Biol Psychiatry* 2004; 56: 301-307.

[110] Green MF. Stimulating the development of drug treatments to improve cognition in schizophrenia. *Annu Rev Clin Psychol* 2007; 3: 159-180.

[111] Hashimoto K, Fujita Y, Shimizu E, Iyo M. Phencyclidine-induced cognitive deficits in mice are improved by subsequent subchronic administration of clozapine, but not haloperidol. *Eur J Pharmacol* 2005; 519(1-2): 114-117.

[112] Hashimoto K, Fujita Y, Ishima T, Hagiwara H, Iyo M. Phencyclidine-induced cognitive deficits in mice are improved by subsequent subchronic administration of tropisetron: Role of α7 nicotinic receptors. *Eur J Pharmacol* 2006; 553(1-3): 191-195.

[113] Hashimoto K, Fujita Y, Iyo M. Phencyclidine-induced cognitive deficits in mice are improved by subsequent subchronic administration of fluvoxamine: role of sigma-1 receptors. *Neuropsychopharmacology* 2007; 32(3): 514-521.

[114] Hashimoto K, Ishima T, Fujita Y, Matsuo M, Kobashi T, Takahagi M, Tsukada H, Iyo M. Phencyclidine-induced cognitive deficits in mice are improved by subsequent subchronic administration of the novel selective α7 nicotinic receptor agonist SSR180711. *Biol Psychiatry* 2008; 63(1): 92-97.

[115] Fujita Y, Ishima T, Kunitachi S, Hagiwara H, Zhang L, Iyo M, Hashimoto K. Phencyclidine-induced cognitive deficits in mice are improved by subsequent subchronic administration of the antibiotic drug minocycline. *Prog Neuropsychopharmacol Biol Psychiatry* 2008; 32(2): 336-339.

[116] Hagiwara H, Fujita Y, Ishima T, Kunitachi S, Shirayama Y, Iyo M, Hashimoto K. Phencyclidine-induced cognitive deficits in mice are improved by subsequent subchronic administration of the antipsychotic drug perospirone: role of serotonin 5-HT_{1A} receptor. *Eur Neuropsychopharmacology* 2008; in press.

[117] Hashimoto K, Fujita Y, Ishima T, Chaki S, Iyo M. Phencyclidine-induced cognitive deficits in mice are improved by subsequent subchronic administration of glycine transporter-1 inhibitor NFPS and D-serine. *Eur Neuropsychopharmacol* 2008; in press.

[118] Karasawa J, Hashimoto K, Chaki S. D-serine and a glycine transporter inhibitor improve MK-801-induced cognitive deficits in a novel object recognition test in rats. *Behav Brain Res* 2008; 186(1): 78-83.

[119] Tsai G, Lane HY, Yang P, Chong MY, Lange N. Glycine transporter I inhibitor, N-methylglycine (sarcosine), added to antipsychotics for the treatment of schizophrenia. *Biol Psychiatry* 2004; 55(5): 452-456.

[120] Lane HY, Chang YC, Liu YC, Chiu CC, Tsai G. Sarcosine or D-serine add-on treatment for acute exacerbation of schizophrenia. a randomized, double-blind, placebo-controlled study. *Arch Gen Psychiatry* 2005; 62(11): 1196-1204.

[121] Lane HY, Liu YC, Huang CL, Chang YC, Liau CH, Perng CH, Tsai GE. Sarcosine (N-methylglycine) treatment for acute schizophrenia: a randomized, double-blind study. *Biol Psychiatry* 2008; 63(1): 9-12.

In: Diamino Amino Acids
Editors: H. A. Schäfer, L. M. Wohlbier, pp. 123-137
ISBN: 978-1-60456-892-9

Chapter IV

ANALYSIS OF METHYLATED ARGININES BY HIGH-PERFORMANCE LIQUID CHROMATOGRAPHY-FLUORESCENCE DETECTION: A REVIEW

Makoto Tsunoda

Laboratory of Bio-Analytical Chemistry, Graduate School of Pharmaceutical Sciences, The University of Tokyo, Tokyo 113-0033, Japan.

ABSTRACT

N^G-Monomethyl-L-arginine (L-NMMA), N^G,N^G-dimethyl-L-arginine (ADMA), and N^G,$N^{G'}$-dimethyl-L-arginine (SDMA) are emerging cardiovascular risk factors that may impair endothelial function by interfering with nitric oxide synthesis. Especially, ADMA was found to be correlated with several diseases such as hypercholesterolemia, hypertension, type 2 diabetes and hyperhomocysteinemia. Hence, accurate measurement of these methylated arginines to clarify their physiological roles *in vivo* must be established. In recent years, various analytical methods have been developed for the analysis of methylated arginines in bio samples. The most widely used assays are based on solid-phase extraction of basic components in bio samples, followed by derivatization and chromatographic analysis by high-performance liquid chromatography (HPLC) with fluorescence detection. As a fluorescent derivatization reagent, *o*-phthaldialdehyde (OPA), 6-aminoquinolyl-*N*-hydroxysuccinimidyl carbamate (AQC) and 4-fluoro-7-nitro-2,1,3-benzoxadiazole (NBD-F) have been used. This chapter reviews the various HPLC-fluorescence methods described so far for the analysis of methylated arginine in bio samples such as plasma, serum or urine. It discusses the main experimental conditions used for sample pretreatment and for analysis including fluorescence derivatization and chromatographic conditions.

1. Introduction

Nitric oxide (NO) is formed enzymatically from L-arginine in the presence of nitric oxide synthase (NOS), and is important in cellular homeostasis and in the pathophysiology of human diseases [1,2]. Methylated arginines (N^G-monomethyl-L-arginine (L-NMMA), N^G,N^G-dimethyl-L-arginine (ADMA), and N^G,$N^{G'}$-dimethyl-L-arginine (SDMA)) are involved in NOS regulation. Figure 1 shows the chemical structures of methylated arginines. Asymmetrical methylarginines, L-NMMA and ADMA, presented in plasma are inhibitors for NOS [3]. Though SDMA has no biological activity like L-NMMA and ADMA, it can interrupt the transportation of cationic amino acids such as L-arginine into cells. Thus, high concentrations of SDMA in plasma may reduce NO production [4]. The origin and the excretion and metabolism of methylated arginines are briefly described as below.

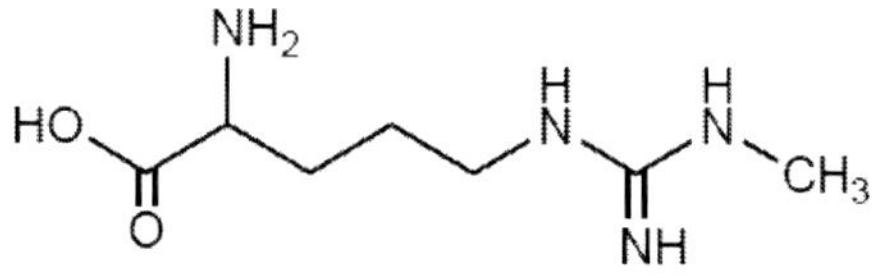

L-NMMA

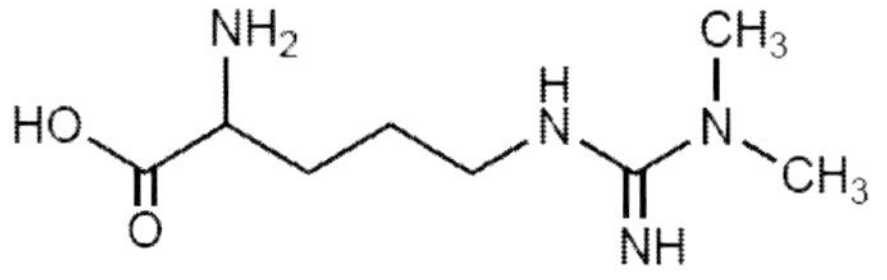

ADMA

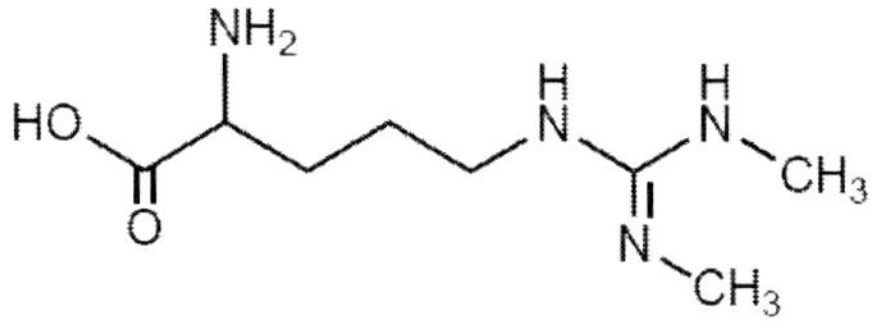

SDMA

Figure 1. Chemical structures of methylated arginines.

Origin of Methylated Arginines

Methylated arginines are the results of the degradation of methylated proteins. Protein arginine methyl transferase (PRMT) enzyme is responsible for the arginine methylation of proteins [5]. The presence of methylated arginine residues has been identified in a wide range of proteins such as myelin basic protein, heat shock proteins, and nuclear proteins. PRMT has two subtypes, PRMT1 and PRMT2. PRMT1 catalyses the formation of L-NMMA and SDMA while PRMT2 catalyses the formation of L-NMMA and ADMA. Proteolysis of the methylated proteins liberates free L-NMMA, ADMA and SDMA.

Excretion and Metabolism of Methylated Arginines

Methylated arginines are eliminated from the body by renal excretion. Also, asymmetrically methylated arginines, L-NMMA and ADMA, but not SDMA, are metabolized by dimethylarginine dimethylaminohydrolase (DDAH). There are two isoforms of DDAH, DDAH-1 and DDAH-2 [6]. DDAH-1 predominates in tissues expressing neural NOS, and DDAH-2 predominates in tissues expressing endothelial NOS. Both DDAH-1 and DDAH-2 are highly expressed in liver and kidneys. It is suggested that changes in DDAH activity may contribute to elevated ADMA levels in various diseases.

These methylated arginines are emerging cardiovascular risk factors that may impair endothelial function by interfering with NOS. Especially, ADMA was found to be elevated and correlated with several diseases such as hypercholesterolemia, hypertension, type 2 diabetes and hyperhomocysteinemia [7].

2. Analytical Methods for the Determination of Methylated Arginines

As mentioned above, the determination of methylated arginines, especially ADMA, is significant not only for clinical diagnosis but clarifying their roles *in vivo*. Hence, various analytical methods have been developed. There are three major analytical methods for the determination of methylated arginines, namely, enzyme-linked immunosorbent assay (ELISA), mass spectrometry (MS), and high-performance liquid chromatography (HPLC)-fluorescence detection method.

In the next section, HPLC with fluorescence detection method will be focused. ELISA was developed for the only determination of ADMA in human serum and plasma, mouse and rat plasma, and cell culture supernatant [8]. MS technique was used coupling with liquid [9-14] or gas chromatography [15,16].

3. HPLC-Fluorescence Detection Methods

HPLC-fluorescence detection method is widely used for the determination of methylated arginines. Using HPLC-fluorescence detection method, we can measure ADMA and SDMA simultaneously, while immunoassay method is applicable for determining ADMA only. Furthermore, the apparatus for HPLC-fluorescence detection method is basically simple and cost-effective especially compared to MS. There are four sections based on fluorescence derivatizing reagent(s), *o*-phthaldialdehyde (OPA), 6-aminoquinolyl-*N*-hydroxysuccinimidyl carbamate (AQC), 4-fluoro-7-nitro-2,1,3-benzoxadiazole (NBD-F), and others. Figure 2 shows the chemical structures of these reagents.

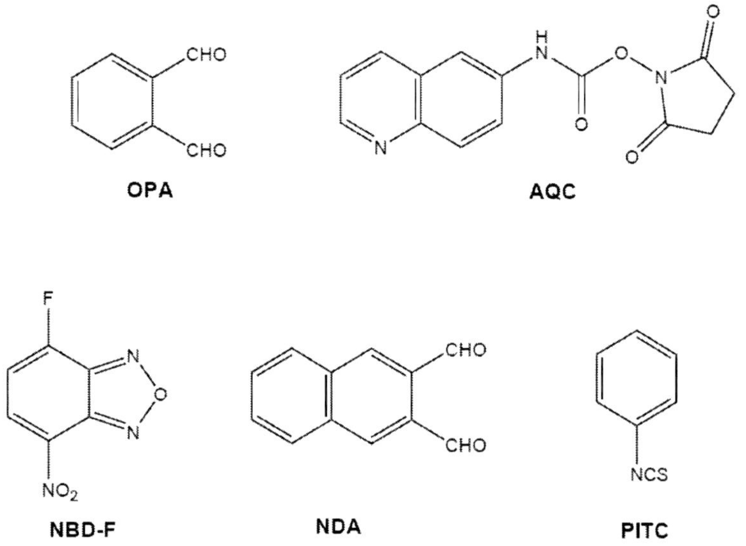

Figure 2. Chemical structures of several fluorescent derivatization reagents, which were used for the determination of methylated arginines.

OPA

OPA is one of the most widely used fluorescent reagents for amines. OPA, in the presence of a thiol, reacts with primary amines at room temperature rapidly forming 1-alkylthio-2-alkyl-substituted isoindoles [17]. The OPA derivatives are highly fluorescent (lex = 340 nm, lem = 455 nm) enough to detect picomole quantities of amino acids.

OPA has several advantages as a fluorescent reagent for diamino amino acids. First, OPA itself is not fluorescent. Secondly, OPA reacts only with primary amines. Third, OPA reacts with amines very quickly even at room temperature.

OPA is usually used as a pre-column derivatization reagent. Typical reaction solution contains 0.4 mM OPA, 0.2 M borate buffer (pH 8.5) and 0.05% 2-mercaptoethanol (v/v) [18]. The reaction proceeds in less than 1 min at room temperature. The reaction mixture is immediately injected onto the HPLC system because some of the derivatives are known to be unstable [17].

Chen *et al.* reported a determination method for ADMA, which provided no data about SDMA [19]. They found that the concentration of ADMA was significantly elevated in plasma from patients with pregnancy induced hypertension in contrast to healthy pregnant individuals. This method was used to investigate the effect of taurine on plasma ADMA levels in experimental endotoxemia [20]. Furthermore, they reported a simultaneous determination method for ADMA, SDMA and L-arginine [21]. Extraction of sample was not required, and the use of mobile phase containing tetrahydrofuran (THF) allowed for separating methylated arginines within 35 min. The detection limit of the method was 1.0 pmol with signal-to-noise ratio of 3 to 1. Petterson *et al.*, using a separation column with particle size of 3 μm, achieved almost baseline-resolution of OPA derivatives of ADMA and SDMA in 70 min [22].

The method reported by Pi *et al.* is more desirable because they utilized a solid-phase extraction before the OPA derivatization of arginine and methylated arginines [18]. After the solid phase extraction using AG50W-X8 resin, L-arginine, L-NMMA, L-homoarginine, ADMA and SDMA were separated within 20 min. The limit of quantification for the metabolites was 4 pmol on column. This method was successfully applied to human plasma, urine and rat kidney cytosol. Paiva *et al.* also employed the solid-phase extraction to determine ADMA concentration in human plasma [23]. They found that the type 2 diabetic patients had lower plasma ADMA levels than the nondiabetic control subjects (0.29 ± 0.15 *vs* 0.34 ± 0.16 μmol/L, $P < 0.03$). Very recently, after the solid-phase extraction, simultaneous determination of L-arginine, L-glutamine, N^G-hydroxyl-arginine, L-citrulline, N^G-monomethyl-L-arginine, L-homoarginine, ADMA, SDMA, L-ornithine, putrescine, agmatine, spermidine and spermine was performed in human urine samples [24]. The limits of detection for L-NMMA, ADMA and SDMA were 0.56, 0.54 and 0.50 pmol/ 20 μl injection, respectively.

There are other reports that determined plasma ADMA levels using the solid phase extraction [25-28]. Bode-Böger *et al.* reported that plasma ADMA concentration increased about two-fold in hypercholesterolemia, but was unaffected by dietary L-arginine [25]. Böger *et al.* also reported that ADMA is elevated in young hypercholesterolemia individuals and that elevation of ADMA is associated with impaired endothelium-dependent vasodilation and reduced urinary nitrate excretion [26]. Yoo and Lee found that patients with ischemic stroke had significantly higher concentrations of plasma ADMA than controls (1.85 ± 1.32 vs. 0.93 ± 0.32 μmol/l, P=0.0001), which suggested that accumulation of endogenous ADMA may account for increased risk of stroke in hyperhomocysteinemia [27]. Feng *et al.* found that ADMA is increased in the circulation of rats with congestive heart failure [28].

Zhang and Kaye, without using soild-phase extraction, determined arginine and its seven metabolites (citrulline, N^G-hydroxy-L-arginine, N^G-monomethyl-L-arginine, ADMA, SDMA, ornithine and agmatine) [29]. By effective gradient elution, this method enabled to separate them within 35 min. Only 20 μl of human plasma samples was analyzed and the results were comparable to other reports.

As mentioned above, OPA derivatives are unstable. To overcome the drawback, Dobashi *et al.*, using OPA as a post-column derivatization reagent, developed an analyzing system for methylated arginines [30]. This system consisted of an on-line extraction, a separation on reversed phase chromatography with ion-pair reagent, a post-column derivarization with OPA, and fluorescence detection. L-Arginine, NMMA, ADMA and SDMA were separated within 40 minutes. The limit of detection for ADMA was 0.08 pmol on column, which enabled the determination of methylated arginines in only 10 μL of plasma. Conte *et al.* also reported a method for the determination of methylated arginines in human plasma and cerebrospinal fluid using post-column derivatization with OPA [31]. They found that mean plasma levels of ADMA were higher in chronic renal failure patients than in healthy control subjects (1.19 ± 0.22 μmol/L vs. 0.63 ± 0.081 μmol/L; p<0.001), and that the cerebrospinal fluid levels of ADMA in patients with multiple sclerosis were higher than those in the control subjects.

Another approach to circumvent the instability of OPA derivatives is the use of 3-mercaptopropionic acid, which provides higher stability than that in mercaptoetanol [32]. They found that plasma from healthy volunteers (n = 53) contained 94 ± 26 μM arginine, 0.42 ± 0.06 μM ADMA, and 0.47 ± 0.08 μM SDMA.

Bulau *et al.* developed a HPLC method to accurately quantify methylated arginines in free and protein-incorporated amino acid pools of cell and tissue extracts, using protein precipitation and hydrolysis, HPLC separation, and fluorescence detection for the simultaneous quantification of L-arginine, L-NMMA, ADMA, and SDMA using OPA as a fluorescene derivatization reagent [33]. The method was suitable for the determination of arginine methylation in mouse kidney and heart, and for the analysis of changes in arginine methylation in cell cultures exposed to stimuli that modify arginine methylation.

AQC

AQC is known as an excellent derivatizing reagent for amino acid analysis [34]. The reaction mixture has no significant interference from the only major fluorescent reagent by-product, 6-aminoquinoline. The excitation and emission wavelengths are 250 nm and 395 nm, respectively.

AQC was used for the determination of ADMA and SDMA in plasma from hemodialysis patients before and after dialysis, from continuous ambulatory peritoneal dialysis patients, and from healthy subjects [35]. The AQC reagent was used as AccQ Fluor reagent kit from Waters (Milford, MA, USA). Twenty μl of deproteinized plasma was added to 60 μl of 200 mM borate buffer (pH 8.8) containing 5 mM disodium ethylenediaminetetraacetate (EDTA) and 20 μl of reconstituted AQC reagent.

Heresztyn *et al.* also used AQC for the determination of L-arginine, ADMA and SDMA in human plasma [36]. They adapted the extraction procedure using strong cation exchange cartridges, followed by derivatization with AQC, separation on a C18 column and detection. The extraction of plasma on strong cation exchange cartridges improved on recovery of methylated arginines. The limit of detection in plasma for ADMA and SDMA was 0.1 μM.

NBD-F

NBD-F is one of fluorogenic reagents which have benzofurazan (2,1,3-benzoxadiazole) skeleton. Benzofurazan reagents have several advantages such as large fluorescence quantum yields values of the resultant derivatives, high reactivity to the analytes and long excitation and emission wavelengths of the derivatives, which avoids from the interference derived from the bio-matrixes. The fluorogenic NBD-F reacts with amines to form strong fluorescent derivatives (lex = 470 nm, lem = 530 nm) [37]. The derivatization using NBD-F as a pre-column fluorogenic reagent was applied to the determination of amino acids [38-45], peptide [46-48], protein [49] and other biological compounds with amines [50].

Recently, we have applied NBD-F to the determination of methylated arginines (L-NMMA, ADMA and SDMA) [51]. Here, detailed experimental conditions and results are described.

First, methylated arginines were extracted from plasma samples with Oasis MCX cation-exchange solid phase extraction columns, which was used for OPA as well. After the extraction, methylated arginines were derivatized with NBD-F. The optimum fluorescent derivatization condition was as follows: 105 μl of 100 mM borate buffer (pH 9.0) and 30μl of 40 mM NBD-F in acetonitrile were added to 30 μl of methylated arginines solution. Then, the reaction mixture was incubated at 40 °C for 3 min. To stop the reaction, 435 μl of 0.5% acetic acid (v/v) was added.

For the separation of L-NMMA, ADMA and SDMA, Unison UK-C18 (150 mm × 4.6 mm I.D., Imtakt, Kyoto, Japan) column was used. The optimum separation condition was as follows: The mobile phase A consisted of 50 mM sodium phosphate buffer (pH 3.2) – acetonitrile (91:9, v/v), and mobile phase B was acetonitrile. The gradient program was as follows: 0–18 min 100% A, 18–28 min linear change to 70% A, 28–32 min 70% A. The flow rate was set at 0.75 ml/min, and the column oven was maintained at 40 °C. A typical chromatogram of the separation of NBD-methylated arginines is shown in Figure 3. It should be noted that ADMA and SDMA were baseline separated each other. Limits of the detection for the method were 12 fmol for L-NMMA, and 20 fmol for ADMA and SDMA at signal-to-noise ratio of 3. The values are much better than those of OPA derivatives, which means that NBD derivatives have higher sensitivity than OPA derivatives.

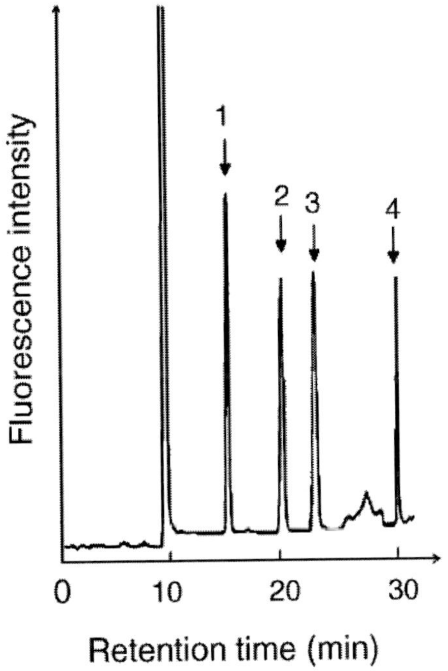

Figure 3. Chromatogram of a standard mixture containing (1) 10 μM L-NMMA; (2) 10 μM ADMA; (3) 10 μM SDMA; and (4) 5 μM N^{ω}-propyl-L-arginine as internal standard. Reproduced with permission from ref. [51].

Another advantage of the use of NBD-F is the stability of NBD derivatives. NBD derivatives of standard sample of methylated arginines were stable for at least 3 days at room temperature. This is useful for the derivatization of a large number samples in one time.

The method mentioned above is ideal in terms of sensitivity and stability for the analysis of methylated arginines. However, solid phase extraction is time-consuming and results in insufficient precision. Hence, to omit the tedious sample treatment, a column-switching system was employed [52]. That is, after the direct derivatization of methylated arginines in plasma samples with NBD-F, derivatized methylated arginines were on-line extracted with a cation exchange column, followed by separation on an ODS column.

To perform direct derivatization of plasma samples with NBD-F, the optimum reaction conditions of methylated arginines in plasma samples with NBD-F were re-examined. The optimum fluorescence derivatization conditions were 100 μl of 200 mM borate buffer (pH 9.0) and 30 μl of 120 mM NBD-F in acetonitrile added to 30 μl of sample. Then, the reaction was carried out at 60 °C for 4 min, and stopped by the addition of 40 μl of 6% acetic acid (v/v).

The column-switching system shown in Figure 4 was used for the determination of NBD-methylated arginines. First, as the pre-column, a cation exchange column (CAPCELLPAK MF-SCX cartridge (20 × 4.0 mm I.D., SHISEIDO, Tokyo, Japan) was used to extract derivatized basic amino acids with NBD-F from plasma samples. After the extraction, the fraction containing the methylated arginines was automatically injected into the separation column by valve switching. Figure 5 illustrates the separation of (A) standard solution and (B) rat plasma sample (30 μl). L-NMMA, ADMA, SDMA and homoarginine were well separated within 42 min.

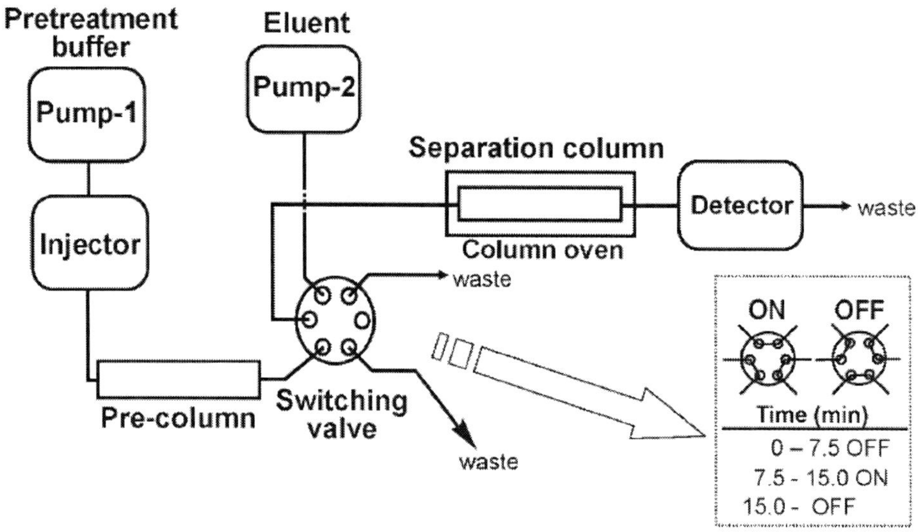

Figure 4. Block diagram of the column-switching HPLC system for the determination of methylated arginines derivatized with NBD-F. Reproduced from ref. [52], with permission of The Royal Society of Chemistry.

Furthermore, using a monolithic silica column (Chromolith™ Performance RP-18e (100mm × 4.6mm I.D., Merck, Darmstadt, Germany)), a fast, simple and sensitive column-switching HPLC method was developed [53]. The separation was done within 15 min, and only ADMA was detectable in rat plasma samples. The developed method was applicable to the determination of DDAH enzyme activity in rat kidney, which was assayed by measuring the amount of ADMA metabolized by the enzyme.

Trapp *et al.* reported a determination method of ADMA, SDMA, L-NMMA, L-arginine, L-homoarginine, L-ornithine, and L-citrulline using capillary electrophoresis [54]. After the derivatization of methylated arginines and structurally related amino acids, the separation of NBD-derivatives was successful by micellar electrokinetic chromatography (MEKC). As the separation buffer, 100 mM sodium deoxycholate in 193 mM borate buffer (pH 9.3) was employed. The limit of quantitation for ADMA in aqueous solution was determined to be 0.1 μM using laser induced fluorescence (LIF) detection in MEKC separation. This method was applicable to quantify ADMA and L-arginine in human plasma samples.

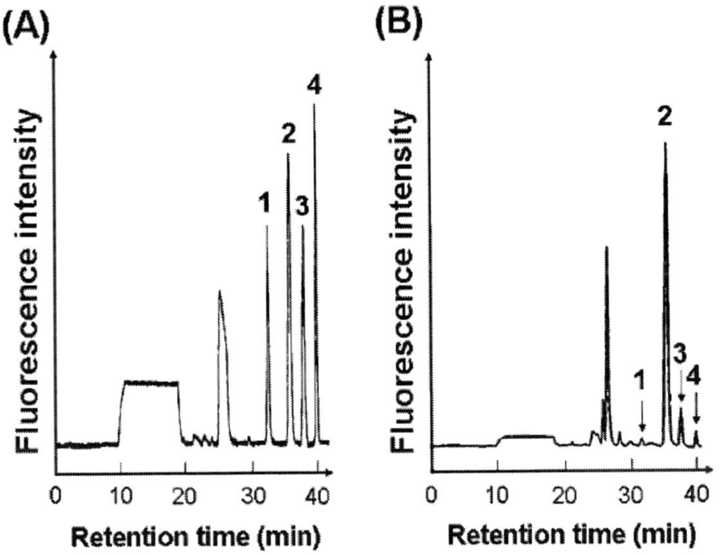

Figure 5. (A) Chromatogram of standards containing 1) 1.0 μM L-NMMA, 2) 1.0 μM homoarginine, 3) 1.0 μM ADMA, and 4) 1.0 μM SDMA. (B) Chromatogram obtained from rat plasma containing 1) 0.17 μM L-NMMA, 2) 10 μM homoarginine as internal standard, 3) 0.76 μM ADMA, and 4) 0.36 μM SDMA. Reproduced from ref. [52],with permission of The Royal Society of Chemistry.

Miscellaneous

Naphthalene-2,3-dicarboxaldehyde (NDA) was used for the determination of L-arginine, L-NMMA, ADMA and SDMA [55]. Marra *et al.* also used NDA as fluorescence derivatizing reagent for the determination of L-arginine, ADMA and SDMA [56]. The method was applicable to human plasma. The ranges obtained (10th and 90th percentile) from 50 healthy subjects (26 males and 24 females, mean age ± SD, 48.1 ± 5.2 years) were 0.38–1.30 μM for ADMA and 1.32–2.70 μM for SDMA. As mentioned above, the fluorescence of OPA derivatives is unstable over time, and they found that the peak areas of NDA derivatives decreased by 5–10% in 20–25 h and by 35% in 72 h. The stability of NDA derivatives is better than OPA derivatives, and worse than NBD derivatives.

Phenylisothiocyanate (PITC) was also used for the determination of methylated arginines. The derivatization of amino acids with PITC forms phenylthiocarbamyl (PTC) amino acids [57]. Ueno *et al.* developed a method to quantitate PTC methylated arginines [58]. They found that the highest concentration of methylated arginines in rat organs was observed in spleen (ADMA: 7.03 nmol/ g wet tissue, SDMA: 0.37 nmol/ g wet tissue).

Fluorescein isothiocyanate (FITC) derivatized methylated arginines were determined with capillary electreophoresis-laser induced fluorescence detection [59]. The methylated arginine derivatives were well separated, with a migration time of around 10 min. The limit of detection for the methylated arginines was 0.05 μM in plasma.

4. Conclusion

This chapter reviewed the analytical methods for the determination of methylated arginines, namely, L-NMMA, ADMA and SDMA with HPLC-fluorescence detection. Compared to other methods (ELISA, LC-MS and GC-MS), HPLC-fluorescence detection methods are useful in terms of sensitivity, reliability, cost and simultaneous determination of methylated arginines. In fact, the roles of methylated arginines in disease and their possibility as a biomarker of cardiovascular disease have been clarified with the use of HPLC-fluorescence detection methods.

Acknowledgement

The author wishes to thank Dr. Kaori Watanabe and Mr. Chiaki Aoyama for their valuable discussions in helping with the preparation of the manuscript.

References

[1] Alderton, W. K., Cooper, C. E. & Knowles, R. G. (2001) Nitric oxide synthases: structure, function and inhibition. *Biochemical Journal, 357*. 593-615.

[2] Achike, F. I. & Kwan, C. Y. (2003) Nitric oxide, human diseases and the herbal products that affect the nitric oxide signalling pathway. *Clinical and Experimental Pharmacology and Physiology, 30*. 605-615.

[3] Leiper, J. & Vallance, P. (1999) Biological significance of endogenous methylarginines that inhibit nitric oxide synthases. *Cardiovascular Research, 43*. 542-548.

[4] Fleck, C., Janz, A., Schweitzer, F., Karge, E., Schwertfeger, M. & Stein, G. (2001) Serum concentrations of asymmetric (ADMA) and symmetric (SDMA) dimethylarginine in renal failure patients. *Kidney International, 59*. S14-S18.

[5] Pahlich, S., Zakaryan, R. P. & Gehring, H. (2006) Protein arginine methylation: Cellular functions and methods of analysis. *Biochimica Et Biophysica Acta-Proteins and Proteomics, 1764*. 1890-1903.

[6] Palm, F., Onozato, M. L., Luo, Z. M. & Wilcox, C. S. (2007) Dimethylarginine dimethylaminohydrolase (DDAH): expression, regulation, and function in the cardiovascular and renal systems. *American Journal of Physiology-Heart and Circulatory Physiology, 293*. H3227-H3245.

[7] Horowitz, J. D. & Heresztyn, T. (2007) An overview of plasma concentrations of asymmetric dimethylarginine (ADMA) in health and disease and in clinical studies: Methodological considerations. *Journal of Chromatography B-Analytical Technologies in the Biomedical and Life Sciences, 851*. 42-50.

[8] Schulze, F., Wesemann, R., Schwedhelm, E., Sydow, K., Albsmeier, J., Cooke, J. P. & Böger, R. H. (2004) Determination of asymmetric dimethylarginine (ADMA) using a novel ELISA assay. *Clinical Chemistry and Laboratory Medicine, 42*. 1377-1383.

[9] Huang, L. F., Guo, F. Q., Liang, Y. Z., Hu, Q. N. & Cheng, B. M. (2003) Rapid simultaneous determination of arginine and methylated arginines in human urine by high-performance liquid chromatography-mass spectrometry. *Analytica Chimica Acta, 487*. 145-153.

[10]Martens-Lobenhoffer, J., Krug, O. & Bode-Böger, S. M. (2004) Determination of arginine and asymmetric dimethylarginine (ADMAL) in human plasma by liquid chromatography/mass spectrometry with the isotope dilution technique. *Journal of Mass Spectrometry, 39*. 1287-1294.

[11]Schwedhelm, E., Tan-Andresen, J., Maas, R., Riederer, U., Schulze, F. & Boger, R. H. (2005) Liquid chromatography-tandem mass spectrometry method for the analysis of asymmetric dimethylarginine in human plasma. *Clinical Chemistry, 51*. 1268-1271.

[12]Huang, L. F., Guo, F. Q., Liang, Y. Z., Li, B. Y. & Cheng, B. M. (2004) Simultaneous determination of L-arginine and its mono- and dimethylated metabolites in human plasma by high-performance liquid chromatography-mass spectrometry. *Analytical and Bioanalytical Chemistry, 380*. 643-649.

[13]Martens-Lobenhoffer, J. & Bode-Böger, S. M. (2006) Fast and efficient determination of arginine, symmetric dimethylarginine, and asymmetric dimethylarginine in biological fluids by hydrophilic-interaction liquid chromatography-electrospray tandem mass Spectrometry. *Clinical Chemistry, 52*. 488-493.

[14]Bishop, M. J., Crow, B., Norton, D., Paliakov, E., George, J. & Bralley, J. A. (2007) Direct analysis of un-derivatized asymmetric dimethylarginine (ADMA) and L-arginine from plasma using mixed-mode ion-exchange liquid chromatography-tandem mass spectrometry. *Journal of Chromatography B-Analytical Technologies in the Biomedical and Life Sciences, 859*. 164-169.

[15]Tsikas, D., Schubert, B., Gutzki, F. M., Sandmann, J. & Frölich, J. C. (2003) Quantitative determination of circulating and urinary asymmetric dimethylarginine (ADMA) in humans by gas chromatography-tandem mass spectrometry as methyl ester tri(N-pentafluoropropionyl) derivative. *Journal of Chromatography B-Analytical Technologies in the Biomedical and Life Sciences, 798*. 87-99.

[16]Albsmeier, J., Schwedhelm, E., Schulze, F., Kastner, M. & Böger, R. H. (2004) Determination of N^G,N^G-dimethyl-L-arginine, an endogenous NO synthase inhibitor, by gas chromatography-mass spectrometry. *Journal of Chromatography B-Analytical Technologies in the Biomedical and Life Sciences, 809*. 59-65.

[17]Alvarezcoque, M. C. G., Hernandez, M. J. M., Camanas, R. M. V. & Fernandez, C. M. (1989) Formation and instability of ortho-phthalaldehyde derivatives of amino-acids. *Analytical Biochemistry, 178*. 1-7.

[18]Pi, J. B., Kumagai, Y., Sun, G. F. & Shimojo, N. (2000) Improved method for simultaneous determination of L-arginine and its mono- and dimethylated metabolites in biological samples by high-performance liquid chromatography. *Journal of Chromatography B, 742*. 199-203.

[19]Chen, B. M., Xia, L. W. & Zhao, R. Q. (1997) Determination of N^G,N^G-dimethylarginine in human plasma by high-performance liquid chromatography. *Journal of Chromatography B, 692*. 467-471.

[20]Balabanli, B., Erdamar, H., Turkozkan, N., Yaman, H. & Kurt, Y. (2007) Effect of taurine on endotoxin-induced alterations in plasma asymmetric dimethylarginine, L-arginine and nitric oxide in guinea pigs. *Journal of Thrombosis and Thrombolysis, 24*. 53-57.

[21]Chen, B. M., Xia, L. W., Liang, S. X., Chen, G. H., Deng, F. L., Zhang, W. R. & Tao, L. J. (2001) Simultaneous determination of L-arginine and dimethylarginines in human urine by high-performance liquid chromatography. *Analytica Chimica Acta, 444*. 223-227.

[22]Pettersson, A., Uggla, L. & Backman, V. (1997) Determination of dimethylated arginines in human plasma by high-performance liquid chromatography. *Journal of Chromatography B, 692*. 257-262.

[23]Paiva, H., Lehtimaki, T., Laakso, J., Ruokonen, I., Rantalaiho, V., Wirta, O., Pasternack, A. & Laaksonen, R. (2003) Plasma concentrations of asymmetric-dimethyl-arginine in type 2 diabetes associate with glycemic control and glomerular filtration rate but not with risk factors of vasculopathy. *Metabolism-Clinical and Experimental, 52*. 303-307.

[24]Markowski, P., Baranowska, I. & Baranowski, J. (2007) Simultaneous determination of L-arginine and 12 molecules participating in its metabolic cycle by gradient RP-HPLC method application to human urine samples. *Analytica Chimica Acta, 605*. 205-217.

[25]Bode-Böger, S. M., Böger, R. H., Kienke, S., Junker, W. & Frölich, J. C. (1996) Elevated L-arginine/dimethylarginine ratio contributes to enhanced systemic NO production by dietary L-arginine in hypercholesterolemic rabbits. *Biochemical and Biophysical Research Communications, 219*. 598-603.

[26]Böger, R. H., Bode-Böger, S. M., Szuba, A., Tsao, P. S., Chan, J. R., Tangphao, O., Blaschke, T. F. & Cooke, J. P. (1998) Asymmetric dimethylarginine (ADMA): A novel risk factor for endothelial dysfunction - Its role in hypercholesterolemia. *Circulation, 98*. 1842-1847.

[27]Yoo, J. H. & Lee, S. C. (2001) Elevated levels of plasma homocyst(e)ine and asymmetric dimethylarginine in elderly patients with stroke. *Atherosclerosis, 158*. 425-430.

[28]Feng, Q. P., Lu, X. G., Fortin, A. J., Pettersson, A., Hedner, T., Kline, R. L. & Arnold, J. M. O. (1998) Elevation of an endogenous inhibitor of nitric oxide synthesis in experimental congestive heart failure. *Cardiovascular Research, 37*. 667-675.

[29]Zhang, W. Z. & Kaye, D. M. (2004) Simultaneous determination of arginine and seven metabolites in plasma by reversed-phase liquid chromatography with a time-controlled ortho-phthaldialdehyde precolumn derivatization. *Analytical Biochemistry, 326*. 87-92.

[30]Dobashi, Y., Santa, T., Nakagomi, K. & Imai, K. (2002) An automated analyzer for methylated arginines in rat plasma by high-performance liquid chromatography with post-column fluorescence reaction. *Analyst, 127*. 54-59.

[31]Conte, C., Floridi, E., Galli, F., Aisa, C. & Floridi, A. (2005) A new method for the liquid chromatography analysis of methylated arginines in biological fluids and tissues by o-phthaldialdehyde postcolumn derivatization. *Analytical Biochemistry, 338*. 347-349.

[32]Teerlink, T., Nijveldt, R. J., de Jong, S. & van Leeuwen, P. A. M. (2002) Determination of arginine, asymmetric dimethylarginine, and symmetric dimethylarginine in human plasma and other biological samples by high-performance liquid chromatography. *Analytical Biochemistry, 303*. 131-137.

[33]Bulau, P., Zakrzewicz, D., Kitowska, K., Wardega, B., Kreuder, J. & Eickelberg, O. (2006) Quantitative assessment of arginine methylation in free versus protein-incorporated amino acids in vitro and in vivo using protein hydrolysis and high-performance liquid chromatography. *Biotechniques, 40*. 305-310.

[34]Cohen, S. A. & Michaud, D. P. (1993) Synthesis of a fluorescent derivatizing reagent, 6-aminoquinolyl-*N*-hydroxysuccinimidyl carbamate, and its application for the analysis of hydrolysate amino-acids via high-performance liquid-chromatography. *Analytical Biochemistry, 211*. 279-287.

[35]Anderstam, B., Katzarski, K. & Bergstrom, J. (1997) Serum levels of N^G,N^G-dimethyl-L-arginine, a potential endogenous nitric oxide inhibitor in dialysis patients. *Journal of the American Society of Nephrology, 8*. 1437-1442.

[36]Heresztyn, T., Worthley, M. I. & Horowitz, J. D. (2004) Determination of L-arginine and N^G,N^G- and N^G,$N^{G'}$-dimethyl-L-arginine in plasma by liquid chromatography as AccQ-Fluor™ fluorescent derivatives. *Journal of Chromatography B-Analytical Technologies in the Biomedical and Life Sciences, 805*. 325-329.

[37]Imai, K. & Watanabe, Y. (1981) Fluorimetric determination of secondary amino-acids by 7-fluoro-4-nitrobenzo-2-oxa-1,3-diazole. *Analytica Chimica Acta, 130*. 377-383.

[38]Watanabe, Y. & Imai, K. (1981) High-performance liquid-chromatography and sensitive detection of amino-acids derivatized with 7-fluoro-4-nitrobenzo-2-oxa-1,3-diazole. *Analytical Biochemistry, 116*. 471-472.

[39]Watanabe, Y. & Imai, K. (1982) Precolumn labeling for high-performance liquid-chromatography of amino-acids with 7-fluoro-4-nitrobenzo-2-oxa-1,3-dDiazole and its application to protein hydrolysates. *Journal of Chromatography, 239*. 723-732.

[40]Klinker, C. C. & Bowser, M. T. (2007) 4-fluoro-7-nitro-2,1,3-benzoxadiazole as a fluorogenic labeling reagent for the in vivo analysis of amino acid neurotransmitters using online microdialysis-capillary electrophoresis. *Analytical Chemistry, 79*. 8747-8754.

[41]Zhang, H. Y., Le Potier, I., Smadja, C., Zhang, J. Y. & Taverna, M. (2006) Fluorescent detection of peptides and amino acids for capillary electrophoresis via on-line derivatization with 4-fluoro-7-nitro-2,1,3-benzoxadiazole. *Analytical and Bioanalytical Chemistry, 386*. 1387-1394.

[42]Aoyama, C., Santa, T., Tsunoda, M., Fukushima, T., Kitada, C. & Imai, K. (2004) A fully automated amino acid analyzer using NBD-F as a fluorescent derivatization reagent. *Biomedical Chromatography, 18*. 630-636.

[43]Kato, M., Dulay, M. T., Bennett, B., Chen, J. R. & Zare, R. N. (2000) Enantiomeric separation of amino acids and nonprotein amino acids using a particle-loaded monolithic column. *Electrophoresis, 21*. 3145-3151.

[44]Tsunoda, M., Kato, M., Fukushima, T., Santa, T., Homma, H., Yanai, H., Soga, T. & Imai, K. (1999) Determination of aspartic acid enantiomers in bio-samples by capillary electrophoresis. *Biomedical Chromatography, 13*. 335-339.

[45]Song, Y. R., Feng, Y. Z., LeBlanc, M. H., Zhao, S. L. & Liu, Y. M. (2006) Assay of trace D-amino acids in neural tissue samples by capillary liquid chromatography/tandem mass spectrometry. *Analytical Chemistry, 78*. 8121-8128.

[46]Kanazawa, H., Nagatsuka, T., Miyazaki, M. & Matsushima, Y. (1997) Determination of peptides by high-performance liquid chromatography with laser-induced fluorescence detection. *Journal of Chromatography A, 763*. 23-29.

[47]Beijersten, I. & Westerlund, D. (1995) Derivatization of dipeptides with 4-fluoro-7-nitro-2,1,3-benzoxadiazole for laser-induced fluorescence and separation by micellar electrokinetic chromatography. *Journal of Chromatography A, 716*. 389-399.

[48]Yoshinaga, K., Kobayashi, N., Nagatani, Y., Tanaka, Y. & Ikeda, Y. (1994) A sensitive detection method for peptide using 4-fluoro-7-nitrobenzo-2-oxa-1,3-diazole and its application to measure prolyl endopeptidase activity. *Biomedical Chromatography, 8*. 297-300.

[49]Wise, E. T., Singh, N. & Hogan, B. L. (1996) Argon-laser-induced fluorescence detection in sodium dodecyl sulfate-capillary gel electrophoretic separations of proteins. *Journal of Chromatography A, 746*. 109-121.

[50]Rellan, S., Osswald, J., Vasconcelos, V. & Gago-Martinez, A. (2007) Analysis of anatoxin-a in biological samples using liquid chromatography with fluorescence detection after solid phase extraction and solid phase microextraction. *Journal of Chromatography A, 1156*. 134-140.

[51]Nonaka, S., Tsunoda, M., Imai, K. & Funatsu, T. (2005) High-performance liquid chromatographic assay of N^G-monomethyl-L-arginine, N^G,N^G-dimethyl-L-arginine, and $N^G$$N^{G'}$-dimethyl-L-arginine using 4-fluoro-7-nitro-2,1,3-benzoxaldiazole as a fluorescent reagent. *Journal of Chromatography A, 1066*. 41-45.

[52]Tsunoda, M., Nonaka, S. & Funatsu, T. (2005) Determination of methylated arginines by column-switching high-performance liquid chromatography-fluorescence detection. *Analyst, 130*. 1410-1413.

[53]Nonaka, S., Tsunoda, M., Aoyama, C. & Funatsu, T. (2006) Determination of N^G,N^G-dimethyl-L-arginine in rat plasma and dimethylarginine dimethylaminohydrolase activity in rat kidney using a monolithic silica column. *Journal of Chromatography B-Analytical Technologies in the Biomedical and Life Sciences, 843*. 170-174.

[54]Trapp, G., Sydow, K., Dulay, M. T., Chou, T., Cooke, J. P. & Zare, R. N. (2004) Capillary electrophoretic and micellar electrokinetic separations of asymmetric dimethyl-L-arginine and structurally related amino acids: Quantitation in human plasma. *Journal of Separation Science, 27*. 1483-1490.

[55]Unceta, N., San Vicente, A., Goicolea, M. A., Salles, J. & Barrio, R. J. (2002) Determination of methylarginines in human plasma by HPLC with pre-column derivatization using naphthalenedicarboxaldehyde as fluorogenic agent. *Journal of Separation Science, 25*. 665-670.

[56]Marra, M., Bonfigli, A. R., Testa, R., Testa, I., Gambini, A. & Coppa, G. (2003) High-performance liquid chromatographic assay of asymmetric dimethylarginine, symmetric dimethylarginine, and arginine in human plasma by derivatization with naphthalene-2,3-dicarboxaldehyde. *Analytical Biochemistry, 318*. 13-17.

[57]Bidlingmeyer, B. A., Cohen, S. A. & Tarvin, T. L. (1984) Rapid analysis of amino-acids using pre-column derivatization. *Journal of Chromatography, 336*. 93-104.

[58]Ueno, S., Sano, A., Kotani, K., Kondoh, K. & Kakimoto, Y. (1992) Distribution of free methylarginines in rat-tissues and in the bovine brain. *Journal of Neurochemistry, 59*. 2012-2016.

[59]Caussé, E., Siri, N., Arnal, J. F., Bayle, C., Malatray, P., Valdiguié, P., Salvayre, R. & Couderc, F. (2000) Determination of asymmetrical dimethylarginine by capillary electrophoresis-laser-induced fluorescence. *Journal of Chromatography B, 741*. 77-83.

In: Diamino Amino Acids
Editors: H. A. Schäfer, L. M. Wohlbier, pp. 139-153
ISBN: 978-1-60456-892-9

Chapter V

YIELD OF COMPATIBLE SOLUTE HYDROXYECTOINE THROUGH CONVERSION FROM ECTOINE BY RECOMBINANT ESCHERICHIA COLI

Shinichi Nagata[1,*], Daochen Zhu[1], Shunyan Cui[1] and Hideaki Sasaki[2]

[1]Environmental Biochemistry Division, Research Center for Inland Seas, Organization of Advanced Science and Technology, Kobe University, Kobe, 658-0022 Japan;

[2]College of Science and Engineering, Iwaki Meisei University, Fukushima, 970-8551 Japan.

ABSTRACT

Halomonas ventosae DL7 synthesized ectoine and hydroxyectoine as major and minor compatible solutes, respectively, when grown in the presence of 3 M NaCl. The amount of hydroxyectoine synthesized increased with an increase of NaCl concentrations and incubation temperatures. To achieve high amount of hydroxyectoine, an attempt was made to examine the conversion of ectoine accumulated in the cells to hydroxyectoine using the recombinant *Escherichia coli* which harbors the ectoine hydroxylase gene (*thpD*) from *Streptomyces chrysomallus*. The presence of carbon source and NaCl was indispensable for the sufficient uptake of ectoine by recombinant *E. coli* BL21-thpD strain, and thus we attempted to examine what kind of carbon sources are required for activity expression of uptake and conversion in M63 medium. As a result, both activities on uptake as well as conversion were maximal under the co-existence of 10 mM glucose and 50 mM xylose. Other factors to affect both activities such as incubation

* Correspondence concerning this article should be addressed to: Shinichi Nagata, e-mail: nagata@maritime.kobe-u.ac.jp, Phone/fax: +81 78 431-6342

temperatures, external pH, salinities and ectoine concentrations supplied were sequentially examined. Efficient and high yield of hydroxyectoine by recombinant *E. coli* BL21 strain was attained through the incubation of the cells in M63 medium containing 1 M NaCl, 1 mM ectoine, 1 mM isopropyl-β-D-thiogalactopyranosid, 10 mM glucose, and 50 mM xylose (pH 7.2) at 30°C, where uptake of ectoine favorably proceeded and at the same time, conversions of ectoine to hydroxyectoine in the cell cytoplasm resulted in 80 and 96 % at 48 and 72 h of incubations, respectively.

Keywords: Hydroxyectoine, Ectoine, Recombinant *E. coli* BL21, Glucose and xylose

Introduction

Halophilic bacteria accumulate various types of compatible solutes intracellularly to cope with external high salinity [1]. Hydroxyectoine, hydroxyl derivative of cyclic amino acid ectoine (1,4,5,6-tetrahydro-2-methyl-4-pyrimidine carboxylic acid), is well known as one of the compatible solutes mainly detected in halophilic and halotolerant bacteria [2,3]. Both ectoine and hydroxyectoine are potential candidates on the enzyme protection against a variety of stress factors such as heating [4], freezing [5] and drying [6]. Further, it has recently been reported that alcohol fermentation in the presence of huge amount of glucose was stimulated by the supplement of ectoine [7]. According to some recent studies, hydroxyectoine seems to be superior to ectoine on the anti-amyloid therapeutics for treating Alzheimer's disease [8], stabilization against proteolysis [9], desiccation tolerance [10] and thermotolerance [11].

Taking the importance of both ectoine and hydroxyectoine into consideration, synthetic devices to obtain them have been developed. To achieve a large scale of supply for ectoines in compliance with the demands, bacterial milking methods in combination with osmotic downshock [12,13] or fed-batch processes [14] have been exploited. The reason why the utilization of bacteria on the production of ectoines has been developed might be ascribed to be easier and higher purity of product of ectoines in comparison with chemical synthesis which generally consists of several different types of reactions. In this connection, efficient and convenient cyclic system to achieve the high yield of ectoine using bacterial functions at the maximum has been recently reported [15].

Synthesis of hydroxyectoine has been indicated by two different pathways such as direct conversion from ectoine and via alternative pathway that converts from biosynthetic intermediate, Nγ-acetyldiaminobutyrate [16]. From the analyses of many ectoine hydroxylase (ectD)-related proteins reported until now, Bursy *et al.* [17] has recently suggested that hydroxyectoine might be directly synthesized from ectoine by EctD. Transgenic *Escherichia coli* possessing evolutionarily conserved gene cluster, *ectABC*, was verified to acquire the halotolerant nature through the synthesis of ectoine [18,19]. Schubert *et al.* observed that the transgenic *E. coli* strain from *Chromohalobacter salexigens* excreted ectoine continuously after induction by anhydrotetracycline as a rate of 40 mg/L·h [19]. Estepa *et al.* [11] demonstrated the enhancement of thermoprotection by *ectD* in *C. salexigens* through mutation. Based on these results shown above, it seems to be of interest to examine the synthesis of hydroxyectoine from ectoine by using the recombinant non-halophilic microorganism, since acquirement of ectoine is much easier than hydroxyectoine at present. In this paper, we attempted to obtain hydroxyectoine by direct synthesis as well as conversion from ectoine by *E. coli* BL21 which harbors the ectoine hydroxylase gene (*thpD*) from *Streptomyces chrysomallus* [20], taking the different incubation conditions such as carbon sources, incubation temperatures, salinities, and pH into consideration.

MATERIALS AND METHODS

Construction of Recombinant *Escherichia coli* with thpD Gene

Genome DNA was purified from *Streptomyces chrysomallus* [20] by DNeasy Tissue Kit (Qiagen, Hilden, Germany). Ectoine hydroxylase gene (*thpD*) was amplified by PCR with Takara Ex Taq Reaction Kit (Takara Bio, Shiga, Japan) and GeneAmp PCR System 9700 (Applied Biosystems, California, USA). Total PCR reaction volume of 30 μl was composed of 3.0 μl 10X Ex Taq Buffer, 3.0 μl dNTP mixture, 2.1 pmol of each primer, 0.8 units Ex Taq and 1.0 μl *S. chrysomallus* DNA solution containing 0.15 μg DNA. Degenerate oligonucleotide primers (thpD-F; 5'-ATGACCACCGAAGTACGCGCCGATCTGTACCCC-TCGCGCG-3', thpD-R; 5'-CTACTTCACCGGGGTGAAGTCCCGCGCCCCGATGAACT-G-3') were used in PCR. The profile of PCR conditions was as follows: initial denaturation at 95°C for 5 min; 35 cycles of denaturation at 95°C for 30 sec, annealing at 54°C for 30 sec, extension at 72°C for 1 min; and a final extension at 72°C for 7 min. The PCR product of *thpD* was ligated into the pCR T7/CT-TOPO vector by the use of TOPO TA Cloning Kit (Invitrogen, California, USA). The sequence of *thpD* was confirmed by DNA sequencing with ABI PRISM™ 3730xl DNA Analyzer (Applied Biosystems). *thpD* expression plasmids were introduced into *E. coli* BL21(DE3)pLysS strain (Invitrogen), where the cells were incubated in LB medium with and without 1 mM IPTG supplemented. As shown in Figure 1, *thpD* was highly transcribed in *E. coli* BL21-thpD strain in the presence of IPTG (isopropyl-β-D-thiogalactopyranoside).

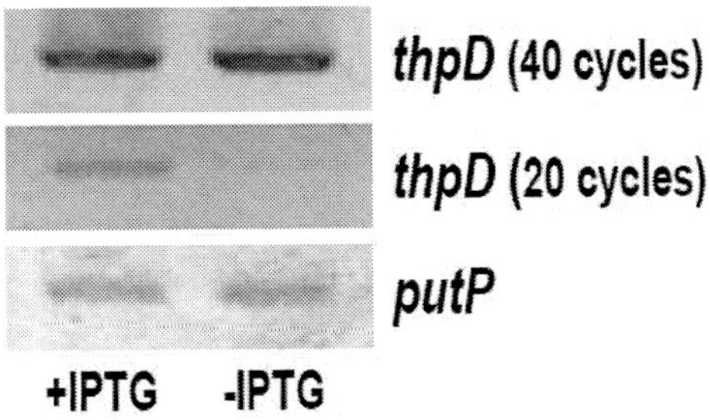

Figure 1. Detection of the transcription of thpD gene in *E. coli* BL21-thpD. The cells were incubated in LB medium with and without 1 mM IPTG supplemented.

Strains and Growth Media

Halomonas ventosae DL7 [21], which was isolated from saltern in Dalian, China, was incubated in a nutrient medium A (per liter) which consists of 5.0 g Bactopeptone (Difco Laboratories, MI, USA) and 3.0 g Bacto yeast extract (Difco). Appropriate amounts of NaCl were added to this medium (pH 7.2) shown in the text.

E. coli BL21-thpD was grown aerobically at 30°C in a medium B which contains (per liter) 5.0 g Polypepton (Nihon Pharmaceutical Co., Ltd., Tokyo, Japan), 3.0 g Dried yeast extract D-3 (Nihon Pharmaceutical) and 30.0 g NaCl, pH 7.2, to which 1 mM IPTG and 20 pM ampicillin were added. After 24 h of incubation the culture was centrifuged (10,000×*g*, 10 min, 4°C) and the pellets obtained were washed once with 50 mM potassium phosphate (KP_i) buffer (PH 7.2) containing the same concentration of NaCl as the medium.

For the conversion experiment from ectoine to hydroxyectoine by recombinant *E. coli* BL21, we used chemically defined medium M63 which contained 100 mM KH_2PO_4, 75 mM KOH, 15 mM $(NH_4)_2SO_4$, 1 mM $MgSO_4$, and 3.9 μM $FeSO_4$ (pH 7.2). Carbon sources tested were added to this medium as described in the text. Cell density was adjusted to $OD_{600} \approx 5.2$-5.3 in M63 medium containing 1 mM IPTG and subsequently incubation was initiated under various conditions such as kinds and concentrations of carbon sources, incubation temperatures, salinity and ectoine concentrations supplied.

Analyses of Intracellular Solutes

Extraction of intracellular compatible solutes was carried out as described previously [22]. Cells grown for designated periods were harvested by centrifugation (10,000×*g*, 4°C, 10 min) and the pellets thus obtained were washed with 50 mM KP_i buffer (pH 7.2) containing isotonic NaCl concentration with the incubation medium and resuspended in 80 % (v/v) ethanol. Cell suspension allowed to stand for 12 h to extract the intracellular solutes and then centrifuged. Supernatant fraction separated was evaporated and suspended in 0.5 ml D_2O for NMR measurements. NMR spectra were recorded on a JEOL JNM-AL300 (JEOL Ltd., Tokyo, Japan) with a probe temperature at 20-22°C as described elsewhere [15].

Extracted fractions by 80 % (v/v) ethanol were also analyzed by HPLC (Hitachi L-5020, Japan) system with 50 mM KP_i buffer (pH 6.0) as eluent and flow rate of 1 mL/min at 35°C in Octyl-80Ts column (Tosoh, Tokyo, Japan). Both ectoine and hydroxyectoine were monitored at 210 nm and quantified by comparison with respective authentic samples [23,24].

Intracellular concentrations of ectoine and hydroxyectoine were expressed as gram per liter culture. Conversion rates of ectoine to hydroxyectoine in the cell cytoplasm were calculated as follows;

Conversion rates (%) = {(Hydroxyectoine)/(Ectoine + Hydroxyectoine)} x 100

Chemicals

Ectoine and hydroxyectoine, both of which were > 97 % purities, were provided by Dainippon Pharmaceutical, Tokyo, Japan and Dr. E. A. Galinski, Universität Bonn, Germany, respectively. Other chemicals used were of reagent or higher grade.

RESULTS AND DISCUSSION

Synthesis of Hydroxyectoine by *Halomonas ventosae* DL7

An attempt was made to yield hydroxyectoine through the direct synthesis by moderately halophilic *Halomonas ventosae* DL7. When the strain DL7 was grown in the medium A containing 3 M NaCl at 30°C, both ectoine and hydroxyectoine were synthesized in the cells as major and minor compatible solutes, respectively. Their maximum amounts of the former and the latter resulted in 458.9 and 16.1 mg/L at 44 h of incubation, respectively (Table 1).

Table 1. Maximal amounts of ectoine and hydroxyectoine synthesized by *Halomonas ventosae* DL7 (mg/L)[a]

Temperature (°C)	NaCl (M)	Incubation time (h)	Ectoine	Hydroxyectoine
30	3.0	44	458.9 ± 10.5	16.1 ± 1.5
37	3.0	48	329.4 ± 10.3	44.8 ± 1.9
42	3.0	96	219.5 ± 6.8	81.8 ± 4.4
42	0.5	22	21.7 ± 5.3	3.3 ± 1.3
	1.0	20	113.9 ± 8.4	14.6 ± 2.1
	1.5	38	121.7 ± 6.7	28.4 ± 4.1
	2.0	68	158.4 ± 8.9	43.8 ± 3.0
	2.5	72	191.3 ± 4.9	53.2 ± 2.2
	3.5	120	181.3 ± 5.5	87.0 ± 5.5

[a]The values are the means ± SD from three independent experiments.

Increase of growth temperature from 30°C to 37°C or 42°C in the presence of 3 M NaCl brought about remarkable stimulation on the synthesis of hydroxyectoine in *H. ventosae* DL7, while the amount of ectoine in the cells decreased sharply (Table 1). Then, we examined the changes on the synthetic amounts of hydroxyectoine in the presence of 0.5-3.5 M NaCl at 42°C. As shown in Table 1, amount of hydroxyectoine synthesized in the cells increased with the increase of salinity, *i.e.*, maximal amount of hydroxyectoine reached more than 80 mg/L in the presence of 3-3.5 M NaCl. That is to say, synthesis of hydroxyectoine in the strain DL7 was up-regulated by external salinity as incubation temperature. Similar behavior was also observed for *C. salexigens* [11], indicating that the synthesis of hydroxyectoine might be favourable at higher temperatures rather than normal ones. However, in the present work the incubation time to yield such high amount of hydroxyectoine markedly lengthened, 96-120 h of incubations, which should be shortened in future.

Effect of Incubation Temperatures on the Uptake of Ectoine and Conversion to Hydroxyectoine by Recombinant *E. coli* BL21

The ability of ectoine uptake by recombinant *E. coli* BL21 strain was examined under different temperatures, 25-42°C. The incubation was carried out in the M63 medium (pH 7.2) containing 1 M NaCl, 1 mM ectoine, 1 mM IPTG and 20 mM glucose. Total amount of ectoines in the cells might be regarded as their actual ability of ectoine uptake, since ectoine accumulated in the cells might be possibly converted to hydroxyectoine during incubation. From the amounts of total ectoines shown in Table 2, the most suitable condition was 30°C among the incubation temperatures examined, at least during 48 h of incubation under present condition. Furthermore, the highest amount of hydroxyectoine converted from ectoine was also detected at 30°C among 25-42°C, that is, 58 mg/L hydroxyectoine was obtained at 30°C at 48 h of incubation, but yields at 37°C and 42°C resulted in far lower than that of 30°C, 19 and 9 mg/L, respectively. As a result, the conversion from ectoine to hydroxyectoine by the recombinant strain *E. coli* BL21 was appropriately carried out at lower temperature than direct synthesis of hydroxyectoine by *H. ventosae* DL7, 42°C, as described above.

Table 2. Effect of incubation temperatures on the amounts of ectoine and hydroxyectoine in recombinant *E. coli* BL21 (mg/L)[a]

Temperature (°C)	Time (h)	Ectoine	Hydroxyectoine	Total ectoines
25	24	93.8	40.9	134.7
	48	75.4	51.1	126.5
30	5	127.2	14.6	141.8
	24	94.6	47.4	142.0
	48	79.4	57.8	137.2
37	24	114.4	26.7	141.1
	48	100.9	18.9	119.8
42	24	124.2	10.4	134.6
	48	108.0	9.3	117.3

[a]The values were averages from three independent experiments.

Effect of Carbon Sources on the Uptake of Ectoine and Conversion to Hydroxyectoine by Recombinant *E. coli* BL21

Uptake of ectoine by recombinant *E. coli* BL21 strain was not brought about in the absence of glucose, suggesting that the energy source is absolutely required for the expression of accumulation activity. Thus, an attempt was made to examine in detail which carbon sources are well suited to the uptake of ectoine by *E. coli* BL21. Since the presence of glucose led to the accumulation of ectoine, other carbon sources such as xylose, sucrose, raffinose, succinate, and glutamate were examined under the same condition described above. When these carbon sources adjusted to 20 mM, equivalent with glucose mentioned above, were present in the medium, ectoine uptake was not observed except for xylose (data not shown). The presence of glucose afforded the fast uptake of ectoine during short incubation time, 5-12 h, while gradual increase of uptake rates of ectoine was recognized in the presence of xylose (data not shown).

In this connection, it seems to be meaningful to decide how much concentrations of glucose and xylose are most appropriate for the activities on the ectoine uptake as well as its conversion by *E. coli* BL21. When 10-100 mM glucose was supplemented in the incubation medium, amount of total ectoines, ectoine plus hydroxyectoine, in the cells insignificantly changed with the incubation time, in which hydroxyectoine was sequentially produced by ectoine conversion during incubation (Table 3). As a consequence, conversion rate of ectoine to hydroxyectoine was the highest in the presence of 50 mM glucose after 48 h of incubation, 48 %, among every condition examined. Taking the results after 72 h of incubation as the representative example, maximal values for total ectoines and conversion rates were observed to be 123-125 mg/L and 40-43 % in the presence of 20-50 mM glucose, respectively (Table 3). On the other hand, uptake amounts of ectoine in the presence of 20-50 mM xylose were not so high compared with those of glucose during 144 h of incubation, but the conversion rate was remarkably high, > 90 % after 144 h of incubation, which suggests that the significant portion of total ectoines in the cells is composed of hydroxyectoine. Present result strongly indicates that the combination of glucose and xylose possibly leads to the fast uptake of ectoine and its efficient conversion to hydroxyectoine.

Effect of Glucose and Xylose in Mixture on the Uptake of Ectoine and Conversion to Hydroxyectoine by Recombinant *E. coli* BL21

To clarify the optimal condition of hydroxyectoine acquirement, both uptake of ectoine and its conversion were examined in the presence of glucose and xylose in mixture, since both activities of recombinant *E. coli* BL21 strain were stimulated by these sugars in the different fashion shown above. In the presence of xylose fixed to 50 mM, amounts of ectoine accumulated and its conversion product, hydroxyectoine, in the cells were examined as a function of glucose concentrations. As shown in Figure 2, ectoine supplied decreased sharply with the increase of glucose concentration, except for poor accumulation of ectoine in the presence of 50 mM glucose at 48 h of incubation. Accumulated ectoine was converted to hydroxyectoine one by one with the increase of incubation time and then amount of hydroxyectoine attained to more than 120 mg/L at 72 h of incubation, regardless of the glucose concentrations presented.

Table 3. Effect of glucose or xylose concentrations on the total ectoines (mg/L) and conversion rates (%)

Carbon sources	Concentration (mM)		Time (h)			
			24	48	72	144
Glucose	10	Ectoines	125.8	113.7	112.9	
		Conversion rate	19.5	21.5	20.2	
	20	Ectoines	142.0	137.2	125.0	
		Conversion rate	33.4	42.1	40.2	
	50	Ectoines	121.5	120.8	123.9	122.8
		Conversion rate	23.5	48.0	43.3	42.6
	100	Ectoines	129.6	117.9	114.6	
		Conversion rate	23.1	40.8	42.8	
Xylose	20	Ectoines	21.9	105.5		77.3
		Conversion rate	9.6	37.3		72.7
	50	Ectoines	16.4	66.4	119.2	125.3
		Conversion rate	12.8	25.6	38.0	92.3

[a]The data show the average values from two independent experiments.

For comparison, conversion rates from ectoine to hydroxyectoine during 144 h of incubation were demonstrated in Figure 3, from which we can recognize that the conversion rates increased with the increase of incubation time. In addition, they were dependent on the glucose concentration until 48 h of incubation, that is, maximal conversion was observed in the presence of 10-20 mM glucose. When the cells were incubated for more than 72 h, > 90 % of conversion was achieved for every condition examined. In the absence of glucose, conversion rate at 144 h of incubation was > 90 %, but the actual amount of hydroxyectoine yielded resulted in 35-45 mg/L.

Since the concentration of carbon sources might be better as low as possible for downstream treatments and economical point of view [14,15], we adopted glucose concentration as 10 mM, since similar efficiency of hydroxyectoine acquirement was observed as higher concentrations of glucose than 10 mM, as shown above. Thus, the concentration effect of xylose on both activities of recombinant *E. coli* BL21 strain was examined in the presence of 10 mM glucose. As shown in Figure 4, conversion rate when fixed to 10 mM glucose was optimal in the presence of 50 mM xylose during 24-72 h of incubation, although they were almost same in the presence of 50-75 mM xylose at 144 h of incubation, *i.e.,* about 100 % conversion rates were observed.

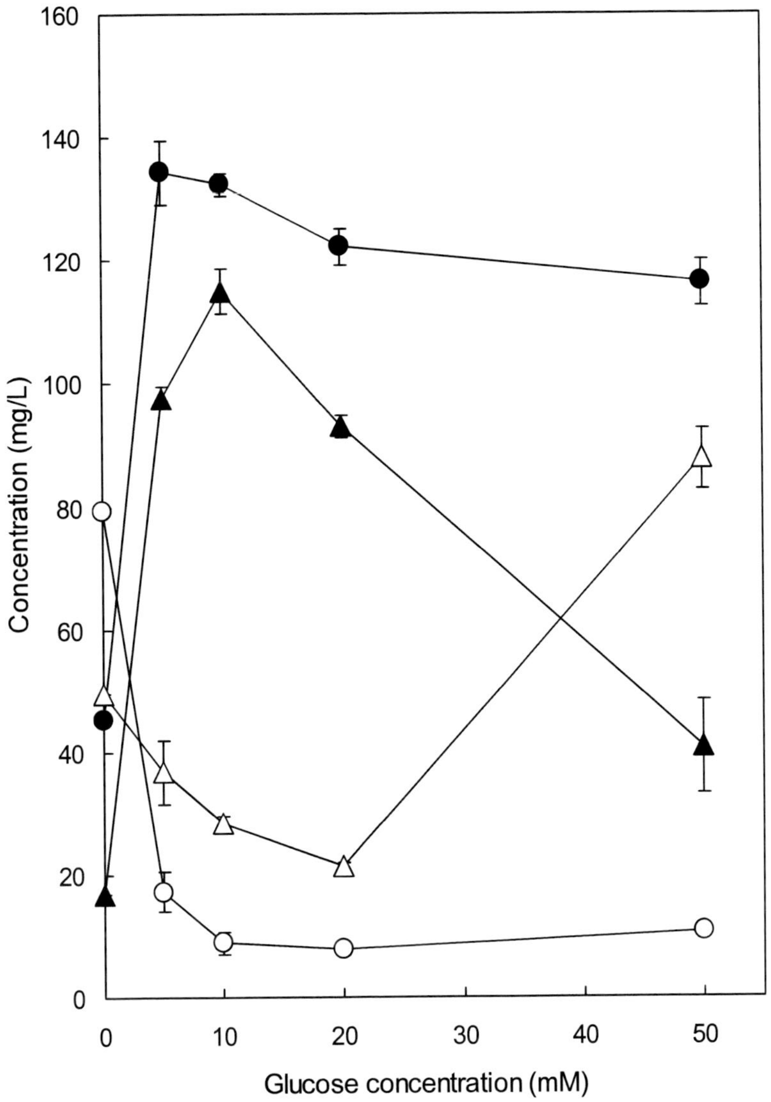

Figure 2. Changes of ectoine and hydroxyectoine in the recombinant *E. coli* BL21 as a function of glucose concentrations. The cells were grown in the medium B for 24 h at 30 °C and then centrifuged (10,000×*g*, 10 min, 4°C). The pellets washed with 50 mM KP_i buffer (pH 7.2) with same concentration of NaCl as the medium were resuspended in M63 medium containing 1 M NaCl, 1 mM ectoine, and 1 mM IPTG (pH 7.2). After cell density was adjusted to $OD_{600} \approx 5.2\text{-}5.3$, cell suspension was incubated in the above medium containing 50 mM xylose and various concentrations of glucose at 30°C. Amount of ectoine (open symbols) and hydroxyectoine (closed symbols) in the cells were followed at 48h (triangles) and 72 h (circles) of incubations. The values are the means ± SD from three independent experiments.

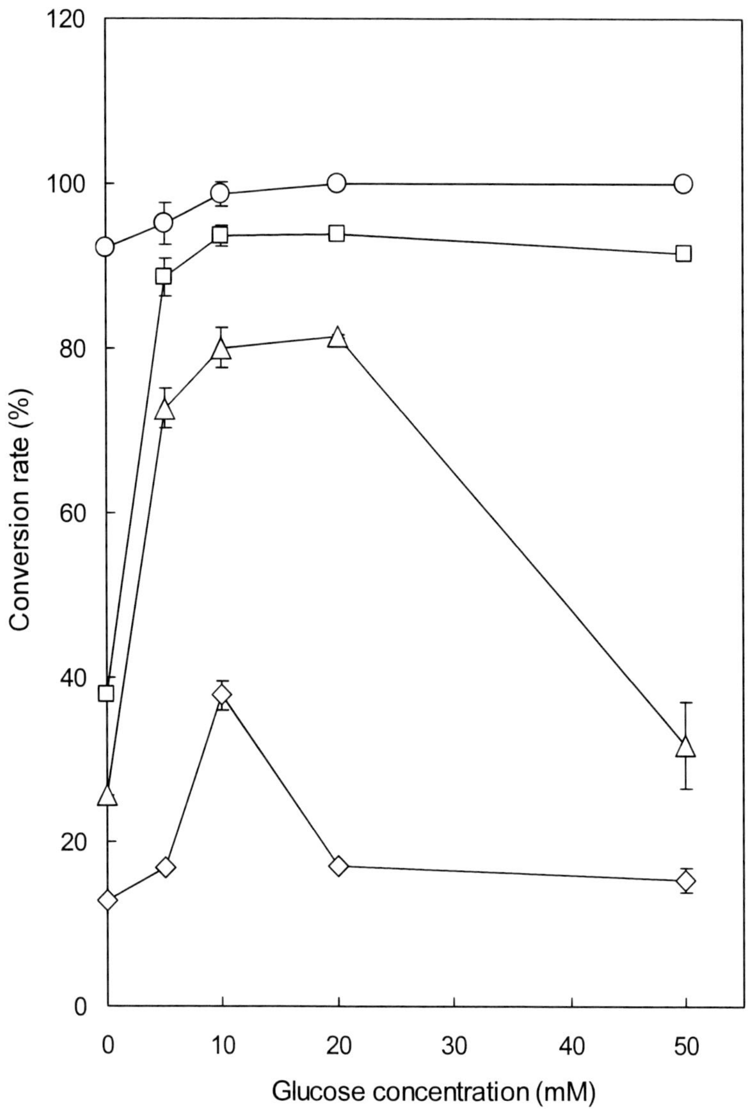

Figure 3. Conversion rates of ectoine to hydroxyectoine by recombinant *E. coli* BL21 as a function of glucose concentrations. Cell preparation and incubation condition were the same as the legend in Figure 2. Changes of ectoine and hydroxyectoine were followed during 144 h. Conversion rates calculated by the equation described in Materials and Methods were depicted at 24 h (diamonds), 48 h (triangles), 72 h (squares) and 144 h (circles) of incubations. The values are the means ± SD from three independent experiments.

Based on the present experimental data, we conclude that the cell incubation should be carried out in M63 medium containing 10 mM glucose and 50 mM xylose at 30°C. Thus, we used this incubation condition to examine other factors to affect the uptake of ectoine and its conversion in the following section.

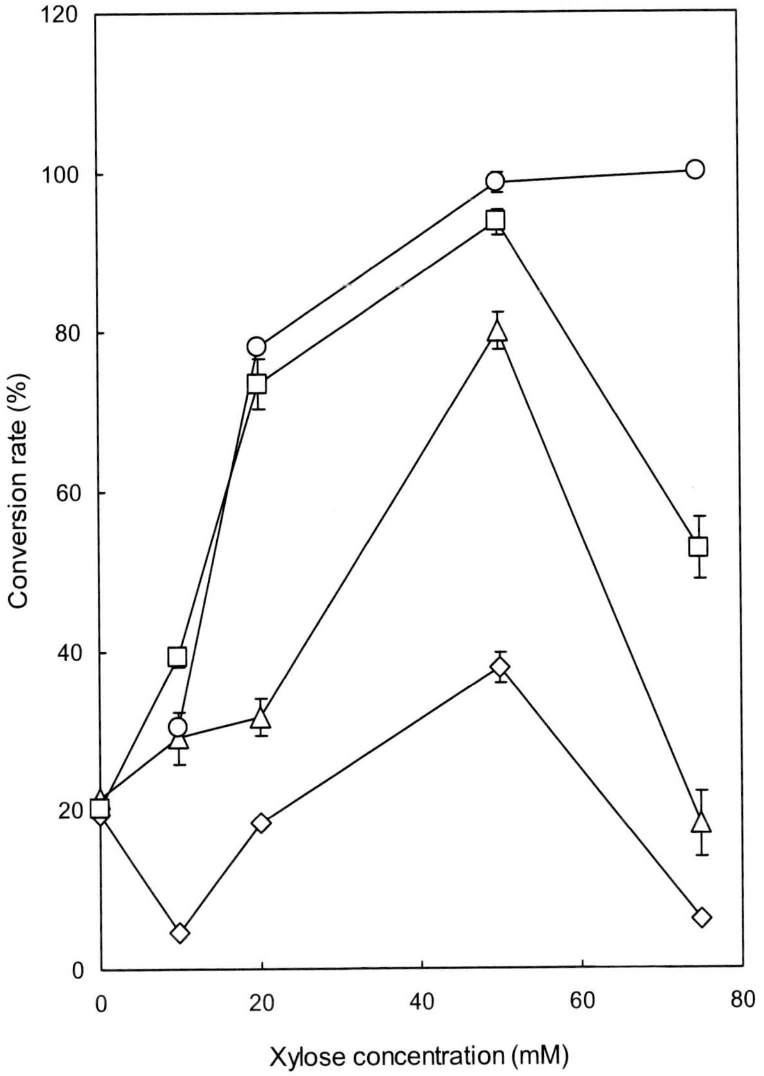

Figure 4. Conversion rates of ectoine to hydroxyectoine by recombinant *E. coli* BL21 as a function of xylose concentrations. Cell preparation and incubation condition were the same as the legend in Figure 2, except that the concentrations of glucose and xylose were fixed to 10 mM and varied from 0 to 75 mM, respectively. Conversion rates calculated were depicted at 24 h (diamonds), 48 h (triangles), 72 h (squares), and 144 h (circles) of incubations. The values are the means ± SD from three independent experiments.

Effect of Salinity, pH, and Ectoine Concentrations on the Uptake of Ectoine and Conversion to Hydroxyectoine by Recombinant *E. coli* BL21

Effect of salinity for the uptake as well as conversion of ectoine by recombinant *E. coli* BL21 was examined in the presence of 0, 0.5, 1 and 1.5 M NaCl. In the absence of NaCl uptake of ectoine was not observed (data not shown), suggesting that the transporters of solutes in non-halophilic bacteria are dependent on the external salinity [25]. In the presence of 0.5-1.5 M NaCl ectoine was certainly accumulated in the cells. After 48 h of incubation we obtained 107, 118, and 7 mg/L of hydroxyectoine in the presence of 0.5, 1, and 1.5 M NaCl, respectively (data not shown), suggesting that ectoine was surely accumulated and converted to hydroxyectoine in the cell cytoplasm. The difference in the synthetic amounts of hydroxyectoine between 0.5 and 1 M NaCl further extended over 50 and 70 mg/L at 96 and 120 h of incubations, respectively. As a consequence, both uptake of ectoine and its conversion were the highest in the presence of 1 M NaCl. It might be plausible to note that the presence of 1.5 M NaCl led to the non-halophilic bacterium functionally inactivated [25]. As a reference, we examined both activities in the presence of sucrose or KCl in place of NaCl as osmolyte, but no activity was observed in the former. Both activities in the latter were reduced to less than half compared with those of NaCl (data not shown).

Next, we tried to refer the pH effect on both activities in the same medium as above, where 1 M NaCl was present. When the cells were incubated under different pH values, 6.0-9.0, higher levels of hydroxyectoine acquirement were recognized at pH 7.2-8.0 in comparison with those of pH 6.0 and 9.0 during 144 h of incubations (Figure 5). Similar pattern on the amounts of ectoine and hydroxyectoine in the cells was observed at pH 7.2 and 8.0, but in detail higher and sustainable acquirement of hydroxyectoine was recognized in the former rather than the latter. In the acidic region, pH 6.0, the amounts of both ectoine and hydroxyectoine were lower than those of neutral pHs. In the alkaline pH, 9.0, poor yield of hydroxyectoine was demonstrated at least until 72 h of incubation, but rapid increase of hydroxyectoine synthesized was observed at 144 h of incubation. This tendency is quite attractive for our purpose, but unfortunately incubation time to reach this level at last was too long at the moment.

For the effect of ectoine concentrations supplied, we attempted to examine both activities by adding 1-5 mM ectoine. As shown in Table 4, increase of ectoine supplied externally led to high accumulation of ectoine in the cells. *e.g.*, at 48 h of incubation accumulated amounts of ectoine resulted in 29, 131, 210, and 241 mg/L when supplied by 1, 2, 3 and 5 mM ectoine, respectively. The amounts of hydroxyectoine converted from ectoine, however, decreased with an increase of ectoine concentrations supplied. Remarkable difference in the amounts of hydroxyectoine was observed at 168 h of incubation, in which they resulted in 123, 81, 34 and 17 mg/L when 1, 2, 3 and 5 mM of ectoine were supplied, respectively. From this result, we conclude that the supply of 1 mM ectoine was the most appropriate for acquirement of hydroxyectoine by the present strain recombinant *E. coli* BL21.

From the present results described above, the most suitable condition to acquire hydroxyectoine efficiently resulted in the following incubation condition; M63 medium containing 1 M NaCl, 1 mM ectoine, 1 mM IPTG, 10 mM glucose and 50 mM xylose (pH 7.2) at 30°C, where ectoine was rapidly accumulated and then 80 and 96 % conversions of ectoine accumulated to hydroxyectoine were observed at 48 and 72 h of incubations, respectively.

Further detailed studies on the effect of carbon sources, behaviors at alkaline pH regions, possibility to use other osmolytes than NaCl, and checking and changing the components in the basic medium M63 should be performed. It is now under progress to yield hydroxyectoine more efficiently and how to set up the downstream treatments.

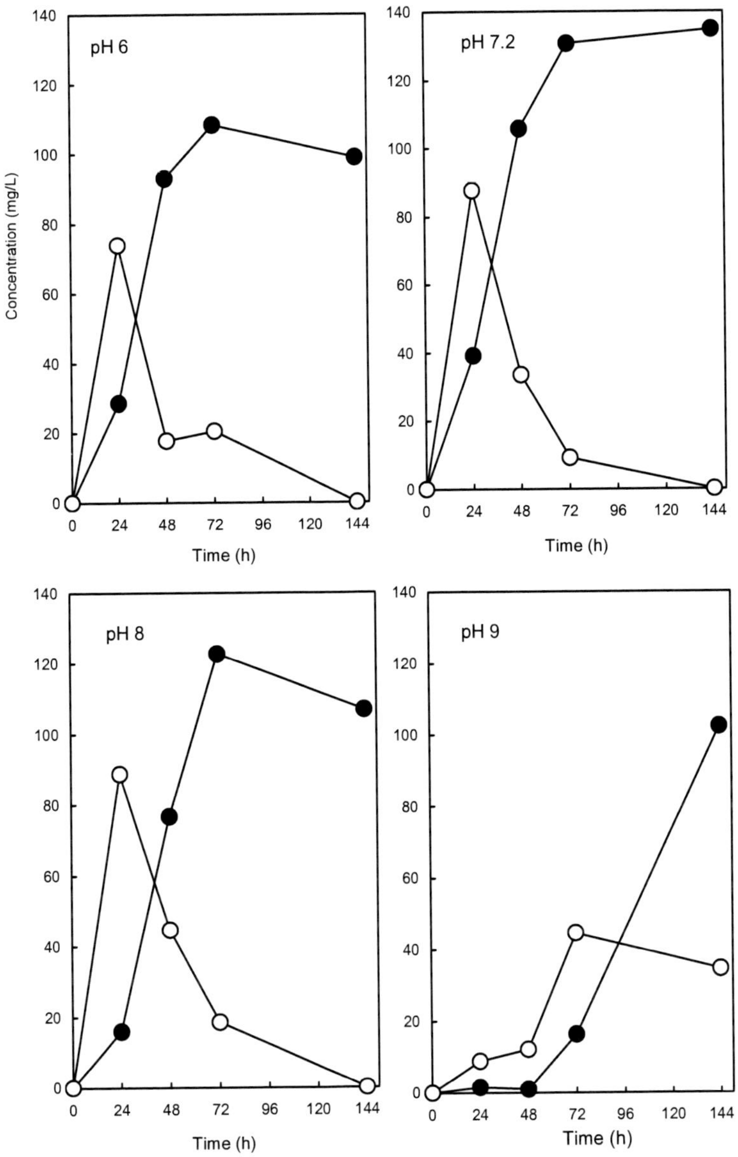

Figure 5. Effect of pH on the uptake of ectoine and its conversion by recombinant *E. coli* BL21. Cell preparation and incubation condition were the same as the legend in Figure 2, except for the external pH values and co-presence of 10 mM glucose and 50 mM xylose as carbon source. Changes of ectoine (open circles) and hydroxyectoine (closed circles) accumulated in the cells were examined with different pH values, 6.0, 7.2, 8.0 and 9.0, during 144 h of incubation. The data show the average values from two independent experiments.

Table 4. Effect of ectoine concentrations on the amounts of ectoine and hydroxyectoine in recombinant *E. coli* BL21 (mg/L)[a]

Ectoine concentration (mM)		Time (h) 24	48	72	168
1	Ectoine	73.7 ± 4.6	28.5 ± 2.9	8.9 ± 2.7	5.0 ± 1.3
	Hydroxyectoine	44.8 ± 3.0	115.0 ± 8.9	132.4 ± 7.7	122.7 ± 9.3
2	Ectoine	157.9 ± 9.0	130.8 ± 6.3	57.9 ± 3.7	30.9 ± 3.4
	Hydroxyectoine	22.4 ± 0.7	112.0 ± 7.9	179.3 ± 2.0	80.9 ± 0.7
3	Ectoine	100.3 ± 2.2	209.5 ± 4.1	109.2 ± 3.5	38.7 ± 0.4
	Hydroxyectoine	14.3 ± 4.8	61.3 ± 4.2	163.7 ± 19.0	34.0 ± 4.4
5	Ectoine	97.6 ± 1.1	241.4 ± 10.7	200.6 ± 9.6	55.8 ± 4.7
	Hydroxyectoine	9.4 ± 1.2	39.4 ± 1.9	67.4 ± 4.9	17.0 ± 6.0

[a]The values are the means ± SD from three independent experiments.

REFERENCES

[1] Galinski, E. A. (1995). Osmoadaptation in bacteria. *Adv. Microb. Physiol.*, *37*, 273-328.

[2] Nagata, S., Adachi, K. & Sano, H. (1996). NMR analyses of compatible solutes in a halotolerant *Brevibacterium* sp. *Microbiology*, *142*, 3355-3362.

[3] Roberts, M. F. (2005). Organic compatible solutes of halotolerant and halophilic microorganisms. *Saline Systems*, *1*, 5.

[4] Zhang, L., Wang, Y., Zhang, C., Wang, Y., Zhu, D., Wang, C. X. & Nagata, S. (2006). Supplementation effect of ectoine on the thermostability of phytase. *J. Biosci. Bioeng.*, *102*, 560-563.

[5] Bayles, D. O. & Wilkinson, B. J. (2000). Osmoprotectants and cryoprotectants for *Listeria monocytogenes*. *Lett. Appl. Microbiol.*, *30*, 23-27.

[6] Manzanera, M., Vilchez, S. & Tunnacliffe, A. (2004). High survival and stability of *Escherichia coli* dried in hydroxyectoine. *FEMS Microbiol. Lett.*, *233*, 347-352.

[7] Zhang, L., Lang, Y., Wang, C. X. & Nagata, S. (2008). Promoting effect of compatible solute ectoine on the ethanol fermentation by *Zymomonas mobilis* CICC10232. *Process Biochem.*, *43*, 642-646.

[8] Kanapathipillai, M., Lentzen, G., Sierks, M. & Park, C. B. (2005). Ectoine and hydroxyectoine inhibit aggregation and neurotoxicity of Alzheimer's β-amyloid. *FEBS Lett.*, *579*, 4775-4780.

[9] Kolp, S., Pietsch, M., Galinski, E. A. & Gütschow, M. (2006). Compatible solutes as protectants for zymogens against proteolysis. *Biochim. Biophys. Acta*, *1764*, 1234-1242.

[10]Manzanera, M., Castr, A. G., Tøndervik, A., R.-Brandes, M., Strøm, A. R. & Tunnacliffe, A. (2002). Hydroxyectoine is superior to trehalose for anhydrobiotic engineering of *Pseudomonas putida* KT2440. *Appl. Environ. Microbiol.*, *68*, 4328-4333.

[11]G.-Estepa, R., Argandona, M., R.-Bueno, M., Capote, N., I.-Guerra, F., Nieto, J. J. & Vargas, C. (2006). The *ectD* gene, which is involved in the synthesis of the compatible solute hydroxyectoine, is essential for thermoprotection of the halophilic bacterium *Chromohalobacter salexigens*. *J. Bacteriol.*, *188*, 3774-3784.

[12]Frings, E., Sauer, T. & Galinski, E. A. (1995). Production of hydroxyectoine: high cell-density cultivation and osmotic downshock of *Marinococcus* strain M52. *J. Biotechnol.*, *43*, 53-61.

[13]Sauer, T. & Galinski, E. A. (1998). Bacterial milking: a novel bioprocess for production of compatible solutes. *Biotechnol. Bioeng.*, *57*, 306–313.

[14]Onraedt, A. E., Walcarius, B. A., Soetaert, W. K. & Vandamme, E. J. (2005). Optimization of ectoine synthesis through fed-batch fermentation of *Brevibacterium epidermis*. *Biotechnol. Prog.*, *21*, 1206-1212.

[15]Nagata, S., Wang, Y., Oshima, A., Zhang, L., Miyake, H., Sasaki, H. & Ishida, A. (2008). Efficient cyclic system to yield ectoine using *Brevibacterium* sp. JCM 6894 subjected to osmotic downshock. *Biotechnol. Bioeng.*, *99*, 941-948.

[16]Canovas, D., Borges, N., Vargas, C., Ventosa, A., Nieto, J. & Santos, H. (1999). Role of Nγ-acetyldiaminobutyrate as an enzyme stabilizer and an intermediate in the biosynthesis of hydroxyectoine. *Appl. Environ. Microbiol.*, *65*, 3774-3779.

[17]Bursy, J., Pierik, A. J., Pica, N. & Bremer, E. (2007). Osmotically induced synthesis of the compatible solute hydroxyectoine is mediated by an evolutionarily conserved ectoine hydroxylase. *J. Biol. Chem.*, *282*, 31147-31155.

[18]Reshetnikov, A. S., Khmelenina, V. N. & Trotsenko, Y. A. (2006). Characterization of the ectoine biosynthesis genes of haloalkalotolerant obligate methanotroph "*Methylmicrobium alcaliphilum* 20Z". *Arch. Microbiol.*, *184*, 286-297.

[19]Schubert, T., Maskow, T., Benndorf, D., Harms, H. & Breuer, U. (2007). Continuous synthesis and excretion of the compatible solute ectoine by a transgenic, nonhalophilic bacterium. *Appl. Environ. Microbiol.*, *73*, 3343-3347.

[20]Prabhu, J., Schauwecker, F., Grammel, N., Keller, U. & Bernard, M. (2004). Functional expression of the ectoine hydroxylase gene (*thpD*) from *Streptomyces chrysomallus* in *Halomonas elongata*. *Appl. Environ. Microbiol.*, *70*, 3130-3132.

[21]Zhu, D., Niu, L., Wang, C. X. & Nagata, S. (2007). Isolation and characterization of moderately halophilic bacterium *Halomonas ventosae* DL7 synthesizing ectoine as compatible solute. *Ann. Microbiol.*, *57*, 401-406.

[22]Nagata, S., Adachi, K. & Sano, H. (1998). Intracellular changes in ions and organic solutes in halotolerant *Brevibacterium* sp. strain JCM 6894 after exposure to hyperosmotic shock. *Appl. Environ. Microbiol.*, *64*, 3641-3647.

[23]Nagata, S. & Wang, C. X. (2006). Effect of duration of osmotic downshock and coexisting glutamate on survival and uptake of ectoine in halotolerant *Brevibacterium* sp. JCM 6894. *J. Biosci. Bioeng.*, *101*, 57-62.

[24]Wang, C. X., Zhu, D. & Nagata, S. (2006). Supplementation effects of hydroxyectoine on proline uptake of downshocked *Brevibacterium* sp. JCM 6894. *J. Biosci. Bioeng.*, *101*, 178-184.

[25]Nagata, S. (2001). Growth of *Escherichia coli* ATCC 9637 through the uptake of compatible solutes at high osmolarity. *J. Biosci. Bioeng.*, *92*, 324-329.

In: Diamino Amino Acids
Editors: H. A. Schäfer, L. M. Wohlbier, pp. 155-170
ISBN: 978-1-60456-892-9

Chapter VI

SIGNIFICANT YIELD OF ECTOINE BY HALOMONAS SP. EG6 DURING OSMOTIC DOWNSHOCK TREATMENT

Omneya Osman, Yaoqiang Wang, Daochen Zhu, Shoko Tanabe-Hosoi and Shinichi Nagata*

Environmental Biochemistry Division, Research Center for Inland Seas, Organization of Advanced Science and Technology, Kobe University, Kobe 658-0022, Japan.

ABSTRACT

Halophilic bacterium strain EG6 was isolated from Burg El Arab solar saltern lake in Alexandria, Egypt. Strain EG6 can grow in a wide range of NaCl concentrations up to 4 M. Phylogenetic position was established by 16S rDNA gene sequencing as a member of the genus *Halomonas*. This strain showed significant yield of ectoine, about 3.70 g/L, when subjected to the osmotic downshock treatment from 2 M to 0.3 M NaCl during 7 days. 0.3 M NaCl was the optimum concentration used for obtaining maximum yield of ectoine in comparison with other NaCl concentrations (0-0.7 M) that were used for osmotic downshock. In addition, same treatment of the type strain *Halomonas elongata* IFO 15536 had a lower yield of ectoine, about 2.89 g/L, than that of strain EG6. Release of ectoine from the cells was carried out within 10 min and its rates were increased by the dilution through downshock treatment. Significant differences were found in the amounts of ectoine synthesized and released by strain EG6 and *H. elongata* IFO 15536. Furthermore, growth of strain EG6 was highly increased in comparison with *H. elongata* IFO 15536 during downshock treatment from 2 M to 0.3 M NaCl.

Keywords: Ectoine; Osmotic downshock; Strain EG6; Halomonas elongata

* Correspondence concerning this article should be addressed to: Shinichi Nagata, nagata@maritime.kobe-u.ac.jp; Telephone/ fax: 81-78-431-6342

Introduction

Halophiles are salt-loving organisms that inhabit hypersaline environments. They include mainly prokaryotic and eukaryotic microorganisms with the capacity to balance the osmotic pressure of the environment and resist the denaturing effects of salts [1]. To cope with hyperosmotic conditions, microorganisms produce large quantities of a particular group of organic osmolytes, so-called compatible solutes [2,3] and they expel these compounds when they are exposed to hypoosmotic circumstances [4,5]. Their accumulation, either through *de novo* synthesis or by direct uptake from the environment, is an evolutionarily well-conserved adaptation strategy in microorganisms for adjusting to high osmolarity surroundings [6]. Compatible solutes are operationally defined as organic osmolytes that can be amassed by the cell in exceedingly high concentrations (up to several moles per liter) without disturbing vital cellular functions, which leads to the correct folding of proteins [7].

Ectoine (1,4,5,6-tetrahydro-2-methyl-4-pyrimidine carboxylic acid), which is one of the best-investigated compatible solutes, was first discovered as a minor component in the phototrophic sulfur bacterium, *Ectothiorhodospira halochloris* [8] and at present, biotechnologically produced with the halophilic *Halomonas elongata* [9]. The physiology and genetics of ectoine biosynthesis in the latter bacterium have been studied in detail [10-15]. Also, Nagata *et al.* [16,17] investigated the stimulation of growth and respiration activity of *Escherichia coli* by the addition of ectoine. Ectoine, common compatible solute in halophilic bacteria [18], acts as an effective stabilizer of protein function [19]. Zhang *et al.* [20] found that ectoine was the most efficient thermoprotectant for PHYA II among other compatible solutes. Exogenously provided ectoine can reverse growth inhibition caused by osmotic stress in *Escherichia coli* [21], *Corynebacterium glutamicum* [22], and the soil bacterium *Rhizobium meliloti* [23]. Recent studies have found that ectoine stabilizes proteins as well as other cellular structures and protects the skin from stresses like UV radiation and dryness. Technical bioprocess for the ectoine synthesis exploits the salt adaptation strategy of *H. elongata*, which is called as "bacterial milking" [9]. When the medium has a high salt concentration, bacteria generally synthesize and accumulate ectoine. After an osmotic downshock, the cells counteract bursting by the sudden release of ectoine. Technical process involves the cyclic increase and decrease of the salt concentration for production and release of ectoine, respectively. On the other hand, the halotolerant *Brevibacterium* sp. JCM 6894 which is a good producer of ectoine [24] can survive efficiently during osmotic downshock treatment [25].

In the course of screening useful microorganisms from hypersaline environments, we isolated some bacterial strains from Burg El Arab solar saltern lake in Egypt. Of these isolates, one Gram-negative, halophilic and rod-shaped bacterial strain attracted our attention because of its high ability of ectoine synthesis and therefore this strain was subjected to taxomomic and biotechnological studies. Strain EG6 might belong to the genus *Halomonas* from the result of 16S rDNA sequence analysis. Our aim in this chapter is to investigate the best condition for high yield of ectoine production using strain EG6 and compare it with type strains *Halomonas elongata* IFO 15536 and *Brevibacterium* sp. JCM 6894 using the same condition applied previously [37].

Materials and Methods

Microorganisms and Growth Conditions

Water samples were collected from Burg el Arab solar saltern lake in Alexandria, Egypt. Isolation was performed aerobically at 30°C by inoculating them to a nutrient medium (g/L) as follows; Polypepton (Nihon Pharmaceutical Co., Ltd., Tokyo, Japan), 5.0, Dried yeast extract (Nihon Pharmaceutical), 3.0, NaCl concentration (1-4 M). Medium pH was adjusted to 7.5 with NaOH before autoclaving (pH 7.4 after autoclaving). Growth was monitored by measuring turbidity at 650 nm (OD_{650}) using a Beckman DU 640 spectrophotometer (Beckman Instruments, Fullerton, CA, USA). The turbidity of cells was measured after appropriate dilution of samples with the medium when OD_{650} was more than 1. Growth on single carbon sources was tested on M63 medium which contains: 100 mM KH_2PO_4, 75 mM KOH, 15 mM $(NH_4)_2SO_4$, 1 mM $MgSO_4$, and 3.9 μM $FeSO_4$, pH 7.2. The tested carbon sources were sterilized by filtration using 0.45 μm PVDF Millex-HV (Millipore corporation, Bedford, USA) and added to the medium with final concentration of 1 % (w/v).

Morphological and Physiological Characterization

Gram staining was determined for cells grown for 24 h in the liquid nutrient medium. Temperature range for growth was determined by incubating the isolate at 4-45°C. Activities of catalase and oxidase were determined by bubble production in a 3 % (v/v) hydrogen peroxide solution and 1 % of tetramethyl-*p*-phenylenediamine hydrochloride, respectively. Acid production from carbohydrates was determined as described by Leifson [26]. H_2S production, hydrolysis of gelatin and Tween 80 were determined according to Lanyi [27]. Sensitivity to antibiotics was tested by the diffusion agar method on nutrient agar medium with ampicillin (10 μg/ml), streptomycin (10 μg/ml), kanamycin (30 μg/ml), or tetracycline (10 μg/ml).

16S rDNA Sequencing and Phylogenetic Analysis

Chromosomal DNA was isolated and purified using DNA extraction kit (Promega, Madison, USA). 16S rDNA was amplified by PCR using universal primers 27F (5'-GAGTTTGATCMTGGCTCAG-3') and 1407R (5'-TACGGYTACCTTGTTACGACTT-3') [28]. PCR reactions were carried out under conditions previously described by Spriner *et al.* [29]. The PCR product was purified with PCR purification kit (Qiagen GmbH, Hilden, Germany). The purified 16S rDNA was sequenced using ABI PRISMA model 3730 sequencer. The alignment of sequences was done with Clustal W software [30] and was used to align the 16S rDNA of our isolate with the sequences of representative rDNA gene of organisms belonging to the genus *Halomonas*.

Sequences are retrieved from the NCBI database, and thus only type strains of species were taken into account. The sequence analyses were performed with software MEGA4 [31].

Phylogenetic trees were constructed by the neighbor-joining method [32] and maximum-likelihood method [33]. Nucleotide sequence was deposited in the NCBI nucleotide sequence database. The accession number of the 16S rDNA nucleotide sequence of strain EG6 is EU273349.

Procedure of the Novel Cyclic System

Strain EG6 and *Halomonas elongata* IFO 15536 used in this chapter were grown aerobically in the nutrient medium. The cells were precultured in the medium containing 1 M NaCl for 48 h at 30°C. 1 % of preculture was added to 6 ml of the medium containing 2 M NaCl. After 24 h of incubation at 30°C, cultures were centrifuged (10,000×*g*, 10 min, 4°C) and the pellets were resuspended in deionized water or different NaCl concentrations such as 0.1, 0.2, 0.3, 0.4, 0.5, and 0.7 M NaCl to induce the release of intracellular ectoine by hypoosmotic shock. The cells were allowed to equilibrate for 10 min at 30°C in water bath, unless otherwise stated, and then harvested by centrifugation (10,000×*g*, 10 min, 4°C). Each pellet was resuspended in the fresh medium with the respective NaCl concentrations and then incubated for further 24 h. The procedure was repeatedly carried out at least 7 times.

HPLC and NMR Analyses of Ectoine

After osmotic downshock for the cells, appropriate amount of cell suspension was sampled and centrifuged (10,000×*g*, 10 min, 4°C). The supernatant was measured to quantify the extracellular concentration of ectoine, while the pellets were extracted with 80 % ethanol. The latter was provided for the measurement of intracellular ectoine concentration.

Ectoine was detected by high performance liquid chromatography (HPLC) system (L-5020; Hitachi, Tokyo, Japan) with a Bio-Sil C8 90-5S reversed phase column (Bio-Rad Laboratories, Hercules, CA, USA) as described previously [16] and NMR spectroscopy JEOL JNM-AL300 (JOEL Ltd., Tokyo, Japan). For the former, 5 μl of samples were eluted with 50 mM potassium phosphate (KP_i) buffer at pH 6.0 with a flow rate of 1 ml/min at 35°C and detected by UV absorbance at 210 nm. For the latter, intracellular samples were evaporated after extraction with 80 % ethanol, while extracelluar samples were evaporated immediately and then suspended in D_2O, both of which were provided for NMR measurements. The identification and quantification of ectoine were carried out using authentic ectoine as a standard. Release rates (RR) of ectoine from the cells subjected to osmotic downshock were calculated by the following equation [37]:

$$RR = [C\ (E)/\ \{C\ (E) + C\ (I)\}] \times 100$$

where C (E) and C (I) indicate the extracellular and intracellular concentrations of ectoine (mg/L), respectively.

Chemicals

Ectoine, which was of > 97 % purity, was provided by Dainippon Sumitomo Pharma Co., Ltd, Tokyo, Japan. Other chemicals used were of reagent or higher grade.

RESULTS AND DISCUSSION

Morphological and Physiological Characterization

Strain EG6 is a Gram-negative and rod-shaped cells and its colony is creamy and circular on nutrient agar medium after 2 days of incubation at 30°C. This strain can grow in the medium containing wide range of NaCl concentrations, 0-25 % NaCl (w/v). Growth was recognized at pH 5-9 and temperatures 10-45°C, in which respective optimum was 7.5 and 30-35°C, respectively. In addition, strain EG6 was positive for catalase and oxidase activities, and could grow on glucose, fructose and mannose as a single carbon source but did not produce acid from sugars (glucose, fructose, galactose and mannose). Gelatin and Tween 80 were not hydrolyzed and neither nitrate reduction nor H_2S production were observed. Cells were not susceptible to each of 10 μg/ml ampicillin, streptomycin and tetracycline as well as 30 μg/ml of kanamycin. Some phenotypic characters of strain EG6 and related *Halomonas* species are summarized in Table 1.

Table 1. Characteristics of stain EG6 and phylogenetically related type strains of genus *Halomonas* [a]

Characterisitic	Strain EG6	*H. elongata* ATCC 33173	*H. eurihalina* ATCC 49336	*H. marina* ATCC 25374	*H. ventosae* DSM 15911	*H. pacifica* DSM 4742
Cell morphology	Rods	Rods	Rods	Rods	Short rods	Rods
Colony pigmentation	Cream	None	Cream	Cream	Cream	ND
Salt range (%, w/v)	0-25	3.5-20	5-20	2-12	3-15	ND
pH range	5-9	5-9	5-10	ND	6-10	ND
Temperature range (°C)	10-40	15-45	15-45	4-35	15-50	5-45
Oxidase	+	+	-	ND	+	ND
Growth on						
Glucose	+	+	-	+	+	+
Fructose	+	ND	-	+	-	+
Sucrose	-	+	+	-	-	-
Mannose	+	+	+	-	-	-
Nitrate reduction	-	+	+	-	+	-
H_2S production	-	-	+	-	-	-
Hydrolysis of						
Gelatin	-	+	+	-	-	-
Tween 80	-	-	+	-	-	-
Reference	This study	[34]	[35]	[34]	[36]	[34]

[a] Characters are scored as: +, positive; -, negative; ND, not determined.

Phylogenetic Analysis

16S rDNA sequence of strain EG6 was determined directly following PCR amplification, which comprised 1006 nucleotides. Phylogenetic tree based on 16S rDNA sequences showed that strain EG6 was placed within genus *Halomonas* and formed a cluster compromising *Halomonas* species (Figure 1). Strain EG6 exhibited levels of 16S rDNA similarity of 95 %, 94 %, 97 %, 93 %, 90 % and 94 % to the type strains of *Halomonas elongata* ATCC 33173 (M93355), *H. ventosae* DSM 15911(AY268080), *H. eurihalina* ATCC 49336 (X87218), *H. pacifica* DSM 4742 (L42616), *H. marina* ATCC 25372 (M93354) and *H. koreensis* JCM 12237 (AY382579), respectively.

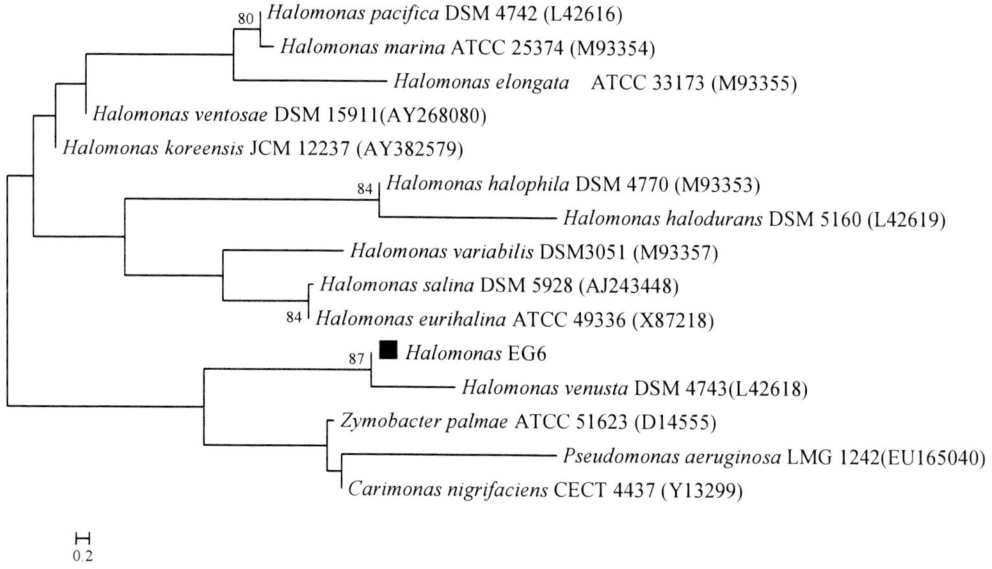

Figure 1. Neighbour- joining tree showing the phylogenetic positions of strain EG6 and *Halomonas* species of related taxa based on 16S rDNA sequences. Scale represents 0.2 substitution per nucleotide position. Bootstrap values (expressed as percentage of 1000 replications) greater than 50 % are shown at branch-points.

Ectoine Yield Synthesized by Strain EG6 During Cyclic System

Strain EG6 was grown in the presence of 2 M NaCl for 24 h at 30 °C and then subjected to osmotic downshock at different NaCl concentrations (0-0.7 M) for 7 days of cyclic system. The amounts of synthesized ectoine and its release from the cells were quantitatively determined every 24 h and tabulated in Table 2. The amounts of ectoine synthesized in the cells decreased after cells were subjected to osmotic downshock in the absence and the presence of 0.1 M NaCl until 72 h of incubation. Thereafter, amounts of ectoine increased with the increase of incubation time. On the other hand, the amounts of ectoine synthesized continued to increase gradually each day, when the cells were subjected to osmotic downshock in the presence of 0.2-0.7 M NaCl.

Table 2. Changes of intracellular ectoine concentrations (mg/L) of strain EG6 and *Halomonas elongata* IFO 15536 [a]

	NaCl used for downshock (M)	Osmotic shock	Incubation time (h) 24	48	72	96	120	144	168	Total yield of ectoine released (mg)[b]
Strain EG6	0.0	BS[c]	315.2 ±1.4	185.0 ± 2.4	279.9 ± 17.1	320.9 ± 16.6	457.1 ± 36.8	538.9 ± 42.2	534.6 ± 40.5	2508.4 ± 64.1
		As[d]	4.0 ± 0.4	19.6 ± 1.8	9.0 ± 0.1	3.6 ± 0.04	16.6 ± 1.1	20.2 ± 1.7	49.9 ± 4.5	
	0.1	BS	315.2 ±1.4	253.8 ± 2.4	267.8 ± 1.9	338.9 ± 1.2	425.8 ± 1.5	550.7	705.4	2539.3
		AS	29.2 ± 0.1	27.6 ± 0.2	21.8 ± 0	36.8 ± 0.3	18.5 ± 0.4	18.9	164.5	
	0.2	BS	315.2 ±1.4	364.4 ± 1.1	442.0 ± 2.2	460.4 ± 7.2	556.1 ± 5.8	654.9 ± 3.3	695.9 ± 2.9	2862.4 ± 28.6
		AS	13.4 ± 0	56.6 ± 0.4	73.6 ± 0.1	90.9 ± 1.0	100.5 ± 1.2	140.7 ± 0.7	149.6 ± 0.6	
	0.3	BS	315.2 ±1.4	463.8 ± 18.8	657.7 ± 42.7	692.9 ± 50.1	803.1 ± 23.9	923.4 ± 36.3	956.4 ± 41.9	3691.5 ± 78.2
		AS	22.1 ± 5.8	99.9 ± 8.5	153.5 ± 4.6	156.7 ± 30.4	183.6 ± 14.1	233.0 ±11.4	304.2 ± 22.1	
	0.4	BS	315.2 ±1.4	484.6± 43.1	621.6± 25.7	730.8 ± 13.2	866.6 ± 35.3	969.4 ± 46.3	1012.0 ±14.3	3378.3 ± 30.5
		AS	60.9 ± 11.7	149.2 ± 61.7	161.1± 34.6	197.3 ± 45.7	277.3 ± 35.3	339.8 ± 39.0	437.1 ± 43.2	
	0.5	BS	315.2 ±1.4	581.4 ± 9.5	769.7 ± 18.3	789.9 ± 38.9	958.4 ± 15.2	1016.3 ± 12.9	1062.8 ± 18.6	3049.9 ± 28.2
		AS	95.2 ± 4.3	241.7 ± 17.2	346.7 ± 40.6	309.4 ± 31.6	384.3 ± 69.2	510.5 ± 60.1	543.0 ± 9.3	
	0.7	BS	315.2 ±1.4	576.1 ± 14.4	823.3 ± 18.0	897.09 ± 57.9	886.4 ± 8.3	1011.9 ± 50.2	1087.3 ± 15.2	2163.0 ± 56.8
		AS	170.4 ± 18.6	313.7 ± 12.5	517.3 ± 31.0	557.3 ± 63.6	497.8 ± 13.7	601.9 ± 55.3	761.2 ± 11.9	
H. elongata IFO15536	0.0	BS	295.8 ± 2.2	230.0 ± 29.1	362.4 ± 46.1	398.9 ± 38.2	436.6 ± 23.4	510.6 ± 25.8	519.0 ± 23.0	2579.7 ± 16.5
		AS	21.2 ± 10.6	9.4 ± 3.9	30.2 ± 9.8	95.3 ± 7.1	28.6 ± 0.2	15.3 ± 5.3	29.8 ± 3.4	
	0.3	BS	295.8 ± 2.2	445.1 ± 7.4	500.8 ± 37.8	556.1 ± 60.8	592.9 ± 66.6	724.3 ± 10.4	711.1 ± 69.4	2896.7 ± 11.7
		AS	43.0 ± 11.3	100.0 ± 17.2	117.4 ± 2.2	144.1 ± 18.1	129.5 ± 22.0	237.3 ± 61.5	205.4 ± 11.8	

[a] The values are the means ± SD from two independent experiments.
[b] Total amounts of ectoine released by seven times of osmotic downshock are given.
[c] Before osmotic downshock.
[d] After osmotic downshock.

Lower yields of total ectoine released were observed when cells were subjected to osmotic downshock from 2 M to 0-0.2 M and 0.7 M NaCl rather than those of 0.3-0.5 M NaCl. While the highest amount of ectoine released was accomplished when cells were subjected to downshock of 0.3 M NaCl, where the total yield of ectoine was 3692 mg/L. It should be noted that osmotic downshock in the presence of 0.4 and 0.5 M NaCl had higher amounts of total yield of released ectoine than those of others, 0, 0.1, 0.2 and 0.7 M NaCl. Nevertheless, total yield of ectoine in the presence of 0.4 and 0.5 M NaCl were still lower than that of 0.3 M NaCl for osmotic downshock. Consequently, 0.3 M NaCl concentration was the optimal to get the highest yield of ectoine compared to the other NaCl concentrations examined for the present osmotic downshock. On the other hand, *Brevibacterium* sp. JCM 6894 released 2469.8 mg/L as the highest yield of ectoine during 7 days of osmotic downshock treatment from 2 M to deionized water [37]. This yield was lower than the total yield of ectoine released by strain EG6 after osmotic downshock treatment from 2 M to 0.3 M NaCl.

Ectoine Yield Synthesized by *Halomonas Elongata* IFO 15536 During Cyclic System

To verify the similarity between strain EG6 and *H. elongata* IFO 15536, we subjected the cells of *H. elongata* IFO 15536 to the same condition on the osmotic downshock, from 2 M to deionized water and 0.3 M NaCl, as strain EG6 described above. As a result, we found out that the total yield of ectoine was almost the same as strain EG6 when cells were subjected to osmotic downshock treatment from 2 M to deionized water (Table 2). However, it was interesting to find out that *H. elongata* IFO 15536 had a total amount of released ectoine, 2897 mg/L, when cells were subjected to downshock from 2 M to 0.3 M NaCl. This amount was far lower than that of strain EG6, 3692 mg/L, in the same condition. This difference might be due to lower amounts of ectoine synthesized by strain IFO 15536 than EG6 after every 24 h of downshock treatment. Present experiment of osmotic downshock strongly indicates that strain EG6 is different from *H. elongata* IFO 15536.

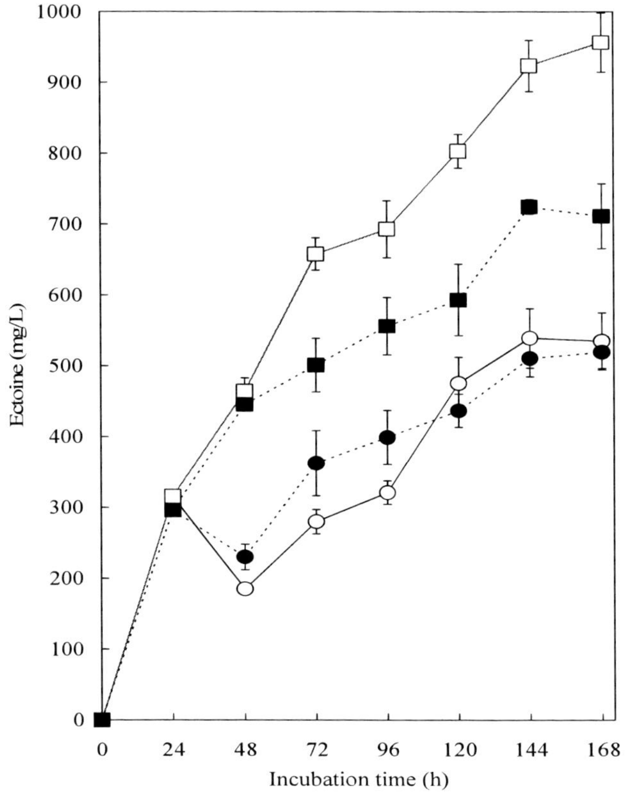

Figure 2. Changes of ectoine concentrations synthesized by strain EG6 and *Halomonas elongata* IFO 15536. Both were grown in the nutrient medium containing 2 M NaCl. Cells of strain EG6 (open symbols) and *H. elongata* IFO 15536 (closed symbols) were subjected to osmotic downshock from 2 M to deionized water (circles) and 0.3 M NaCl (squares), respectively. The values are the averages ± SD from two independent experiments.

In order to clarify the reason beyond the different total yield of ectoine released by strain EG6 and *H. elongata* IFO 15536, changes of ectoine concentrations synthesized by strain EG6 and *H. elongata* IFO 15536 during 7 days of osmotic downshock in the absence and the presence of 0.3 M NaCl were represented in Figure 2. This figure showed that the amounts of ectoine synthesized by strain EG6 and *H. elongata* IFO 15536 were almost the same during osmotic downshock from 2 M to deionized water, but they had big difference in case of downshock treatment from 2 M to 0.3 M NaCl. Concerning the latter condition, the same amounts of ectoine were synthesized at 24 and 48 h of incubations by both of them and then the amounts of ectoine synthesized by strain EG6 sharply increased from 72-168 h of incubations in comparison with *H. elongata* IFO 15536. Accordingly, those data clearly demonstrates the significant increase of ectoine synthesized by strain EG6 and consequently high total yield of ectoine during osmotic downshock treatment in the presence of 0.3 M NaCl, which indicates that there is variation between strain EG6 and *H. elongata* IFO 15536 in both genctic and biotechnological basis.

Release Rate of Ectoine

To determine how much concentration of NaCl was suitable to obtain the optimum release rate of ectoine, we tried to calculate RR values (percentage of release rates of ectoine) during osmotic downshock from 2 M to 0-0.7 M NaCl. As shown in Table 3, the highest release rate in average was observed when cells were subjected to osmotic downshock in the absence of NaCl, which resulted in 95.2 % for strain EG6, but unfortunately this release rate was not sufficient enough to get the maximum total yield of ectoine released (see Table 2).

Table 3. Percentage of release rates of ectoine by strain EG6 and *Halomonas elangota* IFO 15536 [a]

	NaCl used for downshock (M)	Incubation time (h)							Average release rate
		24	48	72	96	120	144	168	
Strain EG6	0.0	98.7 ± 0.4	89.3 ± 0.5	96.5 ± 1.0	98.8 ± 0.1	96.3 ± 1.6	96.2 ± 1.3	90.6 ± 1.1	95.2 ± 0.4
	0.1	90.7 ± 0.2	89.1 ± 0.6	91.8 ± 0.1	89.1 ± 0.9	95.6 ± 0.3	96.5	76.6	89.9
	0.2	95.7 ± 1.9	84.4 ± 0.9	83.3 ± 0.2	80.2 ± 0.1	81.9 ± 1.3	78.5 ± 1.6	78.5 ± 0.3	83.2 ± 1.0
	0.3	92.9 ± 0.0	78.4 ± 0.9	76.6 ± 3.2	77.3 ± 0.4	73.0 ± 0.7	74.7 ± 1.0	68.1 ± 0.6	77.2 ± 1.4
	0.4	80.6 ± 0.2	69.4 ± 0.4	74.0 ± 0.7	73.0 ± 1.4	68.0 ± 0.7	64.9 ± 0.4	56.8 ± 0.6	69.5 ± 0.7
	0.5	69.8 ± 1.4	55.6 ± 6.7	54.9 ± 4.4	60.8 ± 2.0	59.9 ± 0.6	57.9 ± 0.5	48.9 ± 0.1	51.0 ± 4.8
	0.7	45.8 ± 6.0	46.5 ± 8.3	37.1 ± 2.8	37.9 ± 3.1	43.8 ± 2.0	40.5 ± 0.0	30.0 ± 1.2	40.2 ± 0.1
H. elongata IFO15536	0.0	92.7 ± 3.5	95.9 ± 1.1	91.7 ± 1.6	88.0 ± 0.6	96.6 ± 0.2	97.6 ± 0.8	90.2 ± 0.1	93.3 ± 0.0
	0.3	85.4 ± 3.7	81.5 ± 9.1	76.5 ± 1.3	74.1 ± 0.4	78.2 ± 1.2	67.4 ± 3.8	69.6 ± 0.3	76.1 ± 1.2

[a] The values are the means ± SD from two independent experiments.

It was interesting to find out that about 77 % in average of release rate was observed for the cells subjected to osmotic downshock in the presence of 0.3 M NaCl, which showed the highest total yield of ectoine released (Table 2). Higher release rates of ectoine were observed when cells were subjected to osmotic downshock from 2 M to 0.1 and 0.2 M NaCl, 89 % and 83 % in average, respectively, than that of 0.3 M NaCl. Although high release rates were obtained, ectoine yield was not enough to get considerable total yield of ectoine, probably due to the insufficient amounts of ectoine synthesized (Table 2). In the meantime, the ranges of release rates from cells subjected to 0.4-0.7 M NaCl resulted in 40-70 % which were far lower in comparison with those of other cases examined for osmotic downshock. This phenomenon can be explained by the fact that the cells were not affected by a sharp decrease in NaCl concentration to induce high release rate of ectoine. Such results confirmed that the maximal yield of ectoine released not only depends on high ability of ectoine synthesis but also sufficient release rate in the same time. In our case, ectoine synthesis from cells subjected to 0-0.2 M NaCl was low although released rate was high, while the opposite action happened in the presence of 0.4-0.7 M NaCl. Thus, we concluded that 0.3 M was the most suitable NaCl concentration among 0-0.7 M NaCl used for osmotic downshock.

On the other hand, *H. elongata* IFO 15536 showed lower released rate than strain EG6, 93 % and 76 %, in the absence and presence of 0.3 M NaCl, respectively. However, *H. elongata* IFO 15536 did not express the same total yield of ectoine released by strain EG6 during osmotic downshock in the presence of 0.3 M NaCl, owing to the insufficient synthesis of ectoine during osmotic downshock as realized from Figure 2.

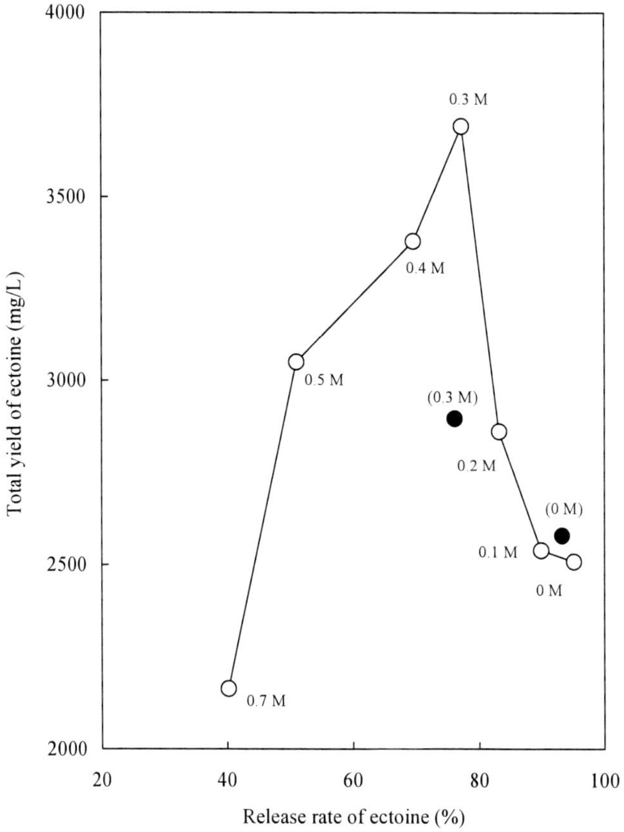

Figure 3. Interrelation between release rates and total yields of ectoine by strain EG6 and *H. elongata* IFO 15536. Cells of strain EG6 were grown in the nutrient medium containing 2 M NaCl and then subjected to osmotic downshock from 2 M to 0-0.7 M NaCl, while *H. elongata* IFO 15536 was subjected to osmotic downshock in the absence and presence of 0.3 M NaCl. Each letter indicates ~~that~~ NaCl concentrations used for osmotic downshock for either strain EG6 (open circles) or *H. elongata* IFO 15536 (closed circles with brackets). The values are the averages from two independent experiments.

Thus, we analyzed the interrelation between release rates and total amounts of ectoine after osmotic downshock from 2 M to 0-0.7 M NaCl for strain EG6 and in the presence and absence of 0.3 M NaCl for *H. elongata* IFO 15536. As shown in Figure 3, this interrelation can be used as a confirmation for our final conclusion that 0.3 M NaCl is the best concentration for the highest yield of ectoine from strain EG6. Figure 3 also clarifies the diverse actions of strain EG6 and *H. elongata* IFO 15536 during osmotic downshock with different NaCl concentrations.

Recent work of Nagata *et al.* [37] revealed that the release rate of ectoine in *Brevibacterium* sp. JCM 6894 was 62 % with 2469.8 mg/L total yield of ectoine after 7-8 days of cyclic system when cells were subjected to osmotic downshock from 2 M to deionized water. Furthermore, Sauer and Galinski [9] were able to harvest 64 % of ectoine from *H. elongata* DSM 142 when cells were subjected to osmotic downshock from 15 % to 3 % NaCl. Those two studies gave us a new approach concerning the strain EG6, because total yield of ectoine released from the three strains, *Brevibacterium* sp. JCM 6894, *H. elongata* DSM 142 and *H. elongata* IFO 15536 have a distinctive difference from strain EG6 and tend to give our strain a new bright vision as a new *Halomonas* species with significant ability of ectoine synthesis.

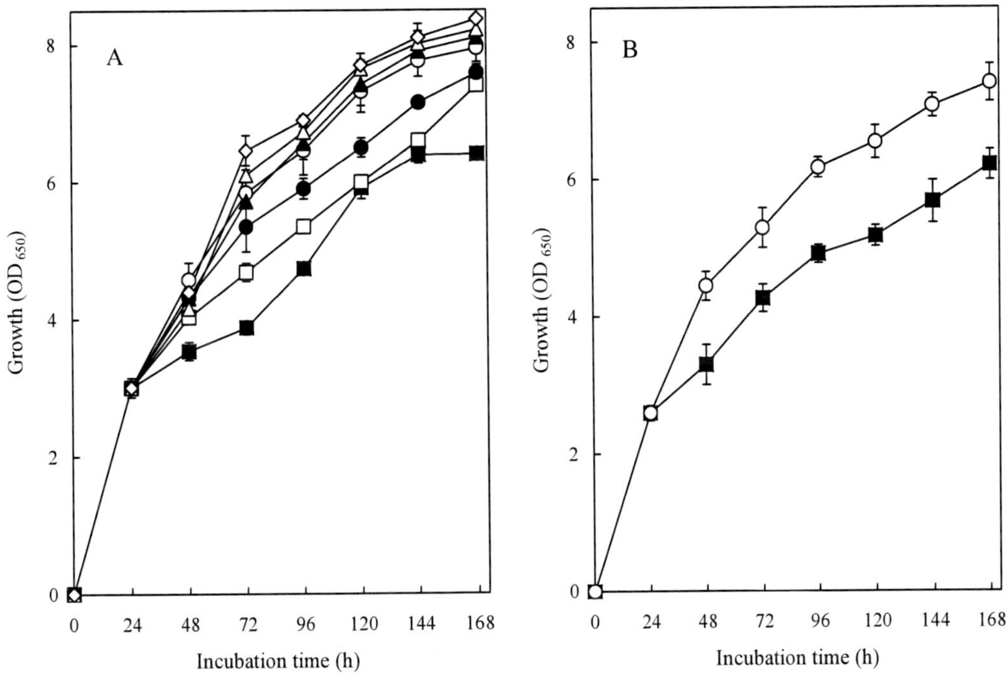

Figure 4. Growth of strain EG6 (A) and *Halomonas elongata* IFO 15536 (B). Both cells were grown in the nutrient medium and their turbidity changes at OD_{650} were measured every 24 h before the cells were subjected to osmotic downshock treatment. Growth measurement was done according to Materials and Methods. The cells were subjected to osmotic downshock from 2 M to deionized water (closed squares), 0.1 M NaCl (opened squares), 0.2 M NaCl (closed circles), 0.3 M NaCl (opened circles), 0.4 M NaCl (closed triangles), 0.5 M NaCl (opened triangles) and 0.7 M NaCl (opened diamonds). The values are the averages ± SD from two independent experiments.

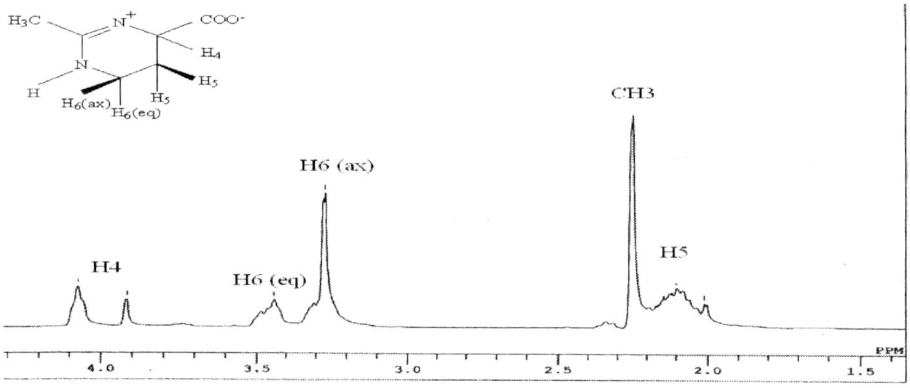

Figure 5. NMR spectrum of solutes released by osmotic downshock from strain EG6. Cell suspension was centrifuged and separated after osmotic downshock from 2 M to 0.3 M NaCl at 168 h of incubation. The supernatant fraction was evaporated at 40°C and suspended in D_2O, for which ^1H NMR spectrum was measured.

NMR Analysis

To investigate the purity of ectoine synthesized by strain EG6, released ectoine was analyzed by NMR spectroscopy for the samples before and after downshock treatment from 2 M NaCl to 0.3 M NaCl. The data revealed that ectoine was the main compatible solute released with high purity after 168 h of incubation (Figure 5). Both intracellular and extracellular solutes after 72 h of incubation were analyzed and almost same spectra were obtained for them (data not shown). This analysis indicates that strain EG6 can be used as a source for ectoine in large scale production with no possibility for the presence of other compatible solutes that could interfere with product purity. From the basis of the present data, we suggests that strain EG6 should be placed in the genus *Halomonas* for a novel species, *Halomonas* sp. EG6, which possesses significant ability of ectoine synthesis and release during osmotic downshock which might have a high impact in the biotechnological applications of ectoine. Further investigation for optimization of ectoine productivity in strain EG6 under large scale will be done in the future.

References

[1] DasSarma, S. & Arora, P. (2002). Halophiles. *In Encyclopedia of Life Sciences*, 8, 458-466.

[2] Csonka, L. (1989). Physiological and genetic responses of bacteria to osmotic stress. *Microbiol Rev*, *53*, 121-147.

[3] Galinski, E. A. & Trüper, H. (1994). Microbial behaviour in salt-stressed ecosystems. *FEMS Microbiol Rev*, *15*, 95-108.

[4] Ajouz, B., Berrier, C., Garrigues, A., Besnard, M., & Ghazi, A. (1998). Release of thioredoxin via the mechanosensitive channel MscL during osmotic downshock of *Escherichia coli* cells. *J Biol Chem, 273*, 26670-26674.

[5] Levina, N., Totemeyer, S., Stokes, N., Louis, P., Jones, M., & Booth, I. (1999). Protection of *Escherichia coli* cells against extreme turgor by activation of MscS and MscL mechanosensitive channels: identification of genes required for MscS activity. *EMBO J, 18*, 1730-1737.

[6] Bremer, E. & Krämer, R. (2000). *Coping with osmotic challenges: osmoregulation through accumulation and release of compatible solutes in bacteria*, (*In* G. Storz and R. Hengge-Aronis). Washington, DC: ASM Press.

[7] Brown, A. (1976). Microbial water stress. *Bacteriol Rev, 40*, 803-846.

[8] Galinski, E. A., Pfeiffer, H., & Trüper, H. (1985). 1,4,5,6-tetrahydro-2-methyl-4-pyrimidine carboxylic acid a novel cyclic amino acid from halophilic phototrophic bacteria of the genus *Ectothiorhodospira*. *Eur J Biochem, 149*, 135-139.

[9] Sauer, T. & Galinski, E. A. (1998). Bacterial milking: a novel bioprocess for production of compatible solutes. *Biotechnol Bioeng, 57*, 306-313.

[10]Canovas, D., Vargas, D., Calderon, M., Ventosa, A., & Nieto, J. (1998). Characterization of the genes for the biosynthesis of the compatible solute ectoine in the moderately halophilic bacterium *Halomonas elongata* DSM 3043. *Syst Appl Microbiol, 21*, 487-497.

[11]Maskow, T. & Babel, W. (2001). Calorimetrically obtained information about the efficiency of ectoine synthesis from glucose in *Halomonas elongata*. *Biochim Biophys Acta, 1527*, 4-10.

[12]Maskow, T. & Harms, H. (2006). Real time insights into bioprocesses using calorimetry: state of the art and potential. *Eng Life Sci, 6*, 266-277.

[13]Maskow, T. & Kleinsteuber, S. (2004). Carbon and energy fluxes during haloadaptation of *Halomonas* sp. EF11 growing on phenol. *Extremophiles, 8*, 133-141.

[14]Ono, H., Sawada, K., Khunajakr, N., Tao, T., Yamamoto, M., Hiramoto, M., Shinmyo, A., Takano, M., & Murooka, Y. (1999). Characterization of biosynthetic enzymes for ectoine as a compatible solute in a moderately halophilic eubacterium, *Halomonas elongata*. *J Bacteriol, 181*, 91-99.

[15]Peters, P., Galinski, E. A., & Trüper, H. (1990). The biosynthesis of ectoine. *FEMS Microbiol Lett, 71*, 157-162.

[16]Nagata, S. & Wang, Y. B. (2001). Accumulation of ectoine in the halotolerant *Brevibacterium* sp. JCM 6894. *J Biosci Bioeng, 91*, 288-293.

[17]Nagata, S., Maekawa, Y., Ikeuchi, T., Wang, Y. B., & Ishida, A. (2002). Effect of compatible solutes on the respiratory activity and growth of *Escherichia coli* K-12 under NaCl stress. *J Biosci Bioeng, 94*, 384-389.

[18]Csonka, L. & Epstein, W. (1996). Osmoregulation. In *Escherichia coli and Salmonella: Cellular and Molecular Biology* (F. C. Neidhardt 2nd ed), Washington, DC: ASM Press.

[19]Yu, I., Jindo, Y., & Nagaoka, M. (2007). Microscopic understanding of preferential exclusion of compatible solute ectoine: direct interaction and hydration alteration. *J Phys Chem B, 111*, 10231-10238.

[20]Zhang, L., Wang, Y., Zhang, C., Wang, Y., Zhu, D., Wang, C., & Nagata, S. (2006). Supplementation effect of ectoine on thermostability of phytase. *J Biosci Bioeng*, *102*, 560-563.

[21]Jebbar, M., Talibart, R., Gloux, K., Bernard, T., & Blanco, C. (1992). Osmoprotection of *Escherichia coli* by ectoine: uptake and accumulation characteristics. *J Bacteriol*, *174*, 5027-5035.

[22]Farwick, M., Siewe, R., & Krämer, R. (1995). Glycine betaine uptake after hyperosmotic shift in *Corynebacterium glutamicum*. *J Bacteriol*, *177*, 4690-4695.

[23]Talibart, R., Jebbar, M., Gouesbet, G., Himdi-Kabbab, S., Wroblewski, H., Blanco, C., & Bernard, T. (1994). Osmoadaptation in *Rhizobia*: ectoine-induced salt tolerance. *J Bacteriol*, *176*, 5210-5217.

[24]Nagata, S., Adachi, K., & Sano, H. (1996). NMR analyses of compatible solutes in a halotolerant *Brevibacterium* sp. *Microbiology*, *142*, 3355-3362.

[25]Nagata, S. & Wang, C. X. (2006). Effect of duration of osmotic downshock and coexisting glutamate on survival and uptake of ectoine in halotolerant *Brevibacterium* sp. JCM 6894. *J Biosci Bioeng*, *101*, 57-62.

[26]Leifson, E. (1963). Determination of carbohydrate metabolism of marine bacteria. *J Bacteriol*, *85*, 1183-1184.

[27]Lanyi, B. (1987). Classical and rapid identification methods for medically important bacteria. *Methods Microbiol*, *19*, 1-67.

[28]Woese, C., Gutell, R., & Noller, H. (1993). Detailed analysis of the higher order structure of 16S-like ribosomal ribonucleic acids. *Microbiol Rev*, *47*, 621-669.

[29]Springer, N., Ludwing, W., Amann, R., Scmidt, H., Gortz, H., & Schieifler, K. (1993). Occurrence of fragmented 16S rRNA in an obligate bacterial endosymbiont of *Paramecium caudatum*. *Proc Natl Acad Sci*, *90*, 9892-9895.

[30]Thompson, J., Higgins, D., & Gibson, T. (1994). CLUSTAL W: improving the sensitivity of progressive multiple sequence alignment through sequence weighting, position- specific gap and weight matrix choice. *Nucl Acids Res*, *22*, 4673-4680.

[31]Tamura, K., Dudley, J., Nei, M., & Kumar, S. (2007). MEGA4: Molecular Evolutionary Genetic Analysis (MEGA) software version 4.0. *Mol Biol Evol*, *24*, 1596-1599.

[32]Saitou, N. & Nei, M. (1987). The neighbor-joining method: a new method for reconstructing phylogenetic trees. *Mol Bio Evol*, *4*, 406-425.

[33]Felsentein, J. (1981). Evolutionary trees from DNA sequences: maximum likelihood approach. *J. Mol Evol*, *17*, 368-376.

[34]Arahal, D. R., Ludwig, W., Schleifer, K. H., & Ventosa, A. (2002). Phylogeny of the family Halomonadaceae based on 23S and 165 rDNA sequence analyses. *Int J Syst Evol Microbiol*, *52*, 241-249.

[35]Mata, J., Martinez-Canovas, J., Quesada, E., & Be′ jar, V. (2002). A detailed phenotypic characterization of the type strains of *Halomonas* species. *Syst Appl Microbiol*, *25*, 360-375.

[36]Martinez-Canovas, M. J., Quesada, E., Llamas, I., & Be′ jar, V. (2004). *Halomonas ventosae* sp. nov., a moderately halophilic, denitrifying, exopolysaccharide-producing bacterium. *Int J Syst Evol Microbiol*, *54*, 733-737.

[37]Nagata, S., Wang, Y., Oshima, A., Zhang, L., Miyake, H., Sasaki, H., & Ishida, A. (2008). Efficient cyclic system to yield ectoine using *Brevibacterium* sp. JCM 6894 subjected to osmotic downshock. *Biotechnol Bioeng*, *99*, 941-948.

In: Diamino Amino Acids
Editors: H. A. Schäfer, L. M. Wohlbier, pp. 171-181
ISBN: 978-1-60456-892-9

Chapter VII

INCREASE IN AMINO ACIDS AND AMMONIA DURING LIQUEFACTION OF SEMEN

Thomas W. Stief [*]
Department of Clinical Chemistry, University Hospital,
Giessen and Marburg, Germany

ABSTRACT

Fresh normal ejaculate clots by action of about 0.05 IU/ml thrombin on a fibrinogen-like compound, and subsequently lyses by action of about 0.1 IU/ml plasmin and about 300 % (of blood plasma norm) carboxypeptidase. Carboxypeptidases sequentially liberate amino acids, that were quantified in the present work: the most abundant amino acids in fresh seminal plasma are serine (6.09 mM), threonine (4.61 mM), lysine (3.45 mM), tyrosine (3.18 mM), glycine (2.79 mM), and arginine (2.29 mM). The concentration of the amino acids approximately doubles within 45 min (37°C), with an exceptional 12fold increase of glutamic acid. When compared to blood plasma, fresh seminal plasma contains 226fold gamma-glutamyl-transferase (γ-GT), 40fold ammonia, 21fold creatinine, and 18fold glutamate dehydrogenase (GLDH). Ammonia (NH_3), generated by glutamate dehydrogenase (GLDH), together with arginine or creatinine might protect sperm cells from female defence cells. Ammonia and natural guanidine derivatives might be physiologic immunsuppressors.

Key words: Semen, liquefaction, GLDH, γ-GT, amino acids, glutamic acid, ammonia.

[*] Correspondence to: PD Dr. med. T. Stief, Department of Clinical Chemistry, University Hospital, D-35043 Marburg, Germany; email: thstief@med.uni-marburg.de; Tel. : +49-6421-28 64471; FAX: +49-6421-28 65594.

INTRODUCTION

Knowledge of the exact composition of human seminal plasma, which is rather unstable due to the presence of proteases, is of importance to understand fertility [1-8]. Recently, high activities of carboxypeptidase in seminal plasma have been described [9-11]. Carboxypeptidases are proteolytic enzymes that cleave peptide bonds at the carboxy terminus of proteins, thereby sequentially liberating amino acids and thus degrading proteins. Physiologically, the fresh ejaculate clots by about 20 ng/ml tissue factor - mediated generation of about 0.05 IU/ml thrombin that acts on about 20-30 g/l of a fibrinogen-like compound (assayed antigenically [12]), generating a coagulum similar to fibrin [13,14]. The seminal clot lyses by action of about 0.1 IU/ml plasmin [3,15,16,80]: during liquefaction of the coagulum, its fibers are transformed into globular bodies that form the homogeneous liquefied semen [13], wherein the spermatozoa can freely move. Seminal clotting and sperm motility correlate negatively [17]. The aim of the present work was the clinical chemical analysis of seminal plasma immediately after ejaculation (S_0) and after a liquefaction time of 45 min (37°C) (S_{45}).

MATERIAL AND METHODS

The experiment was performed 3 times with 3 ml freshly ejaculated normal semen after 3 days of abstinence, 1.5 ml was centrifuged (3000 g for 5 min at 23°C) immediately, and 1.5 ml was centrifuged after 45 min (37°C). The samples were diluted 1+4 with H_2O immediately before duplicate clinical-chemical analysis, which was performed within 1 h (23°C). Amino acids were determined by a cation exchange amino acid analyzer (LC 3000; Eppendorf-Biotronic, Hamburg, Germany). The concentrations found were compared with the respective mean values of normal plasma or with the upper normal value of plasma. The clinical chemical parameters were determined by a Hitachi 917 analyzer (Roche, Basel, Switzerland). The 45 min-results were compared with the basal controls, and were tested for significance ($p< 0.05$) using the $X^2\bar{x}$ test [12]. The mean values were calculated, the standard deviations of all analyte-determinations were < 10 %.

Sodium dodecylsulfate gel electrophoresis (SDS-PAGE): samples of fresh seminal plasma without liquefaction time (sample 1) and with 45 min (37°C) liquefaction time (sample 2) were pre-diluted 1:20 with 0.9 % NaCl, pooled frozen/thawed normal citrated blood plasma was pre-diluted 1:20 and 1:50 with 0.9 % NaCl. These samples were further diluted 1:2 with sample buffer consisting of 10 mM Tris, 2 % SDS, 0.01 % bromphenol blue, 1 mM EDTA, pH 8, and run for 1 h in a PhastSystem with pre-defined molecular weight markers (Amersham Pharmacia Biotech, Freiburg, Germany), using PhastGel Gradient 8-25 (6-300 kilo Dalton (kDa)) and PhastGel buffer strips (0.2 M tricine, 0.2 M Tris, 0.55 % SDS, pH 8.1). Then the SDS-PAGE gels were silver stained for 6.5 min with 0.4 % silver nitrate.

RESULTS AND DISCUSSION

Table 1 demonstrates that all amino acids occur in seminal plasma in an up to about 50fold higher concentration when compared to blood plasma. The exception is methionine that was not found in seminal plasma; methionine in seminal plasma might be metabolized or consumed immediately [18]. The most abundant amino acids in fresh seminal plasma are serine (6.09 mM), threonine (4.61 mM), lysine (3.45 mM), tyrosine (3.18 mM), glycine (2.79 mM), and arginine (2.29 mM). Serine, threonine, and tyrosine have already been described as being the principal amino acids found in seminal plasma [19]. The most abundant amino acids in 45 min (37°C) incubated seminal plasma are serine (9.01 mM), glutamine (6.86 mM), glutamic acid (6.38 mM), tyrosine (5.69 mM), threonine (5.37 mM), glycine (5.35 mM), leucine (4.28 mM), arginine (4.21 mM), histidine (4.19 mM), and lysine (4.1 mM). Overall the concentration of the amino acids approximately doubles within 45 min (37°C), with an exceptional 12fold increase of glutamic acid.

Table 1. Amino acid concentrations in liquefying semen

Amino Acid [mM]	Plasma	$Semen_{0min}$	Ratio S_0/P	$Semen_{45min}$	RatioS_{45}/P	RatioS_{45}/S_0
Alanine	0.15-0.49	0.58	1.8	1.83	5.7	3.2
Arginine	0.03-0.11	2.29	33	4.21	60	1.8
Asparagine	0.03-0.07	1.07	21	3.07	61	2.9
Asp. acid	< 0.01	0.52	52	1.46	146	2.8
Cystin	0.03-0.05	0.09	2.3	0.12	3	1.3
Glutamine	0.34-0.74	1.45	2.7	6.86	13	4.7
Glut. acid	0.01-0.05	0.52	17	6.38	213	12.3
Glycine	0.10-0.38	2.79	12	5.35	22	1.9
Histidine	0.07-0.10	2.08	24	4.19	49	2.0
Isoleucine	0.03-0.08	1.41	26	2.64	48	1.9
Leucine	0.08-0.16	2.08	17	4.28	36	2.1
Lysine	0.11-0.21	3.45	22	4.10	26	1.2
Methionine	0.01-0.03	< 0.01	0	< 0.01	0	-
Ornithine	0.04-0.10	0.23	3.3	0.27	3.9	1.2
PhenAla.	0.04-0.06	0.80	16	1.62	32	2.0
Proline	0.10-0.24	0.60	4.7	1.12	6.6	1.9
Serine	0.07-0.18	6.09	49	9.01	72	1.5
Taurine	0.02-0.08	1.25	25	1.07	21	0.9
Threonine	0.08-0.20	4.61	33	5.37	38	1.2
Tryptophane	0.02-0.05	0.14	4.0	0.32	9.1	2.3
Tyrosine	0.03-0.08	3.18	58	5.69	103	1.8
Valine	0.15-0.25	1.35	6.8	3.52	18	2.6

Table 2 shows the concentrations of important clinical chemical parameters in seminal plasma as compared to blood plasma. Some analytes are severalfold more concentrated (*) in fresh seminal plasma when compared to blood plasma: γ-GT (226*), ammonia (40*), phosphate (29*), creatinine (21*), GLDH (18*), LDH_{1+2} (13*), GOT (10*), Mg (9*), LDH (8*), K^+ (7*), and Ca^{++} (3*), the latter being of importance in sperm motility [20]. Albumin, IgG, IgA, IgM, CRP, and transferrin could not be detected in seminal plasma.

Pseudocholinesterase (PCHE) is found in seminal plasma at only at about 2 % of the blood plasma level. Within 45 min liquefaction time (37°C) ammonia increased from 2.1 mM to 6.8 mM and lactate from 2.2 mM to 3.3 mM; the activity of nearly all clinical chemical enzymes was then approximately 50 % of that before liquefaction, i.e. enzymes are proteolytically degraded and/or inactivated.

Table 2. Clinical chemical parameters in liquefying semen

Parameter	Dimension	Plasma	$Semen_{0min}$	Ratio S_0/P	$Semen_{45min}$	RatioS_{45}/S_0
Na^+	mM	135-145	130	0.93	115	0.88
K^+	mM	3.6-4.8	28.3	6.7	24.8	5.9
Ca^{++}	mM	2.1-2.6	8.0	3.4	7.0	3.0
Mg^{++}	mM	0.7-1.05	7.5	8.6	7.0	8.0
Phosphate	mM	0.84-1.5	33.8	28.9	33.0	28.2
Cl^-	mM	101-111	63	0.6	65	0.6
GPT=ALT	U/l	< 45	58	1.3	38	0.8
GOT=AST	U/l	< 35	335	9.6	195	5.6
γ-GT	U/l	< 55	12428	226	6398	116
GLDH	U/l	< 6.4	118	18	76	12
AP	U/l	< 129	463	3.6	215	1.7
LDH	U/l	< 248	1933	7.8	1145	4.6
LDH_{1+2}	U/l	72-182	1699	13.4	977	7.7
CK	U/l	< 171	555	3.2	268	1.6
P-Amylase	U/l	8-53	25	0.8	18	0.6
Lipase	U/l	< 60	50	0.8	40	0.7
PCHE	U/l	5320-12920	143	0.016	125	0.014
Bilirubin	mg/dl	0.2-1.0	< 0.2	0	< 0.2	0
Glukose	mg/dl	70-110	35	0.4	18	0.2
Creatinine	mg/dl	0.67-1.17	19.7	21.4	19.0	20.7
Urea	mg/dl	20-50	55	1.6	58	1.7
Uric acid	mg/dl	3.4-7.0	13.3	2.6	14.0	2.7
Cholesterol	mg/dl	< 200	30	0.2	10	0.1
Triglyc.	mg/dl	< 150	< 10	0	< 10	0
Iron	µmol/l	11-30	3	0.2	4	0.2
Protein	g/l	66-83	35	0.5	35	0.5
Albumin	g/l	35-52	< 10	0	< 10	0
IgG	g/l	7-16	< 0.3	0	< 0.3	0
IgA	g/l	0.7-4.0	< 0.1	0	< 0.1	0
IgM	g/l	0.4-2.3	< 0.1	0	< 0.1	0
CRP	mg/l	< 5	< 5	0	< 5	0
Myoglobin	µg/l	9.6-67	30	0.8	0	0
Ferritin	µg/l	30-400	183	0.9	85	0.4
Transferrin	g/l	2.0-3.6	< 0.1	0	< 0.1	0
Lactate	mM	0.5-2.2	2.2	1.6	4.4	3.3
Ammonia	µM	< 53	2112	40	6765	128

Figure 1 demonstrates that the protein band of a protein of about 50 kDa in fresh seminal plasma diminishes to some extent whereas the concentrations of high molecular weight proteins of about 200 kDa and low molecular weight proteins of about 10-30 kDa increase.

This might be due to the action of seminal transglutaminase and carboxypeptidase, respectively.

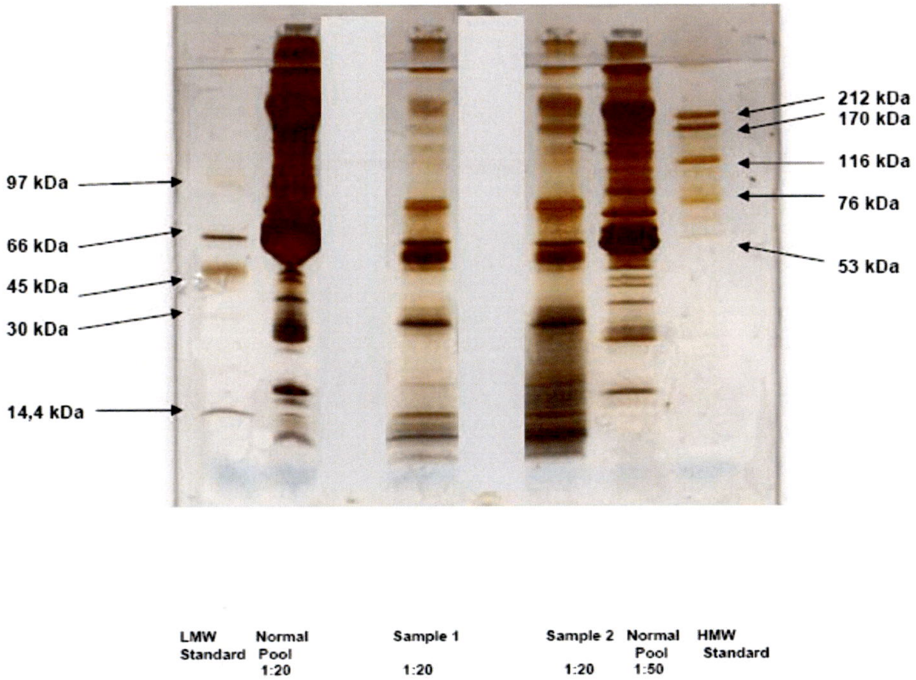

Figure 1. SDS-PAGE of seminal plasma. Seminal plasma without liquefaction time (sample 1) and with 45 min (37°C) liquefaction time (sample 2), both pre-diluted 1:20, and normal blood plasma, pre-diluted 1:20 (lane 2) and 1:50 (lane 5), were run in SDS-PAGE, followed by silver staining of the separated seminal plasma proteins and the low and high molecular weight standards (LMW and HMW; lanes 1 and 6).

Of particular interest is the very high activity of gamma-glutamyl-transferase (γ-GT) [21] and of glutamine/glutamic acid in seminal plasma. γ-GT is an important membrane protein, that shuttles amino acids into cells, whereby an extracellular amino acid is coupled to glutamic acid (out of glutathione) [22-28]. The endothelial membrane ectoenzyme γ-GT is a marker for the blood brain barrier [27]. Seminal carboxypeptidases generate high concentrations of amino acids that might be uptaken by the sperm cells. These high concentrations of amino acids are favourable for the vitality and motility of spermatozoa [7,29-34] and the vitality of the oocyte [35-37]. Taurine seems to be important for sperm survival and sperm/oocyte fusion [37]. 5 mM proline, 10 mM glutamine, and 10-20 mM glycine stabilize spermatozoa [38]. Low liquefaction correlates to low amino-acids content [33].

The actions of γ-GT (induction of redox cycling [28]) or that of glutamate dehydrogenase (GLDH) [39] or the glutamine to glutamic acid decay [40] could generate the cell signals singlet oxygen ($^1O_2$) or ammonia (NH_3) [41]; ammonia at mM concentrations evokes oxidative/nitrosative stress, the mitochondrial permeability transition, and cell swelling [42-45]: it might be toxic for sperm attacking cells of the female immune response. The spermatozoa are also susceptible to high concentrations of reactive oxygen species (ROS) in their ambience [46-51], healthy spermatozoa tolerate some ROS concentrations [47]. Interestingly, spermatozoa via own NADPH-oxidase locally generate ROS, that has physiologic functions in spermatozoa motility and sperma/oocyte fusion [52-61]. NADPH oxidase is mainly located on the head domains of the spermatozoa [60], indicating the importance of oxidants in capacitation.

The ammonia- or arginine- mediated immunsuppression might be of importance for sperm cells to escape from the action of tissue monocytes, granulocytes, or lymphocytes [62-72]. Even high concentrations of the guanidine derivative creatinine do not impair sperm motility [73]. The biochemical composition of seminal plasma might be of interest to develop new physiologic immunsuppressors [74-79,81].

Acknowledgment

Britta Kosche is thanked for performing the miniaturized SDS-PAGE and the subsequent silver staining of the proteins bands on the gels.

References

[1] van der Merwe FH, Kruger TF, Oehninger SC, Lombard CJ. The use of semen parameters to identify the subfertile male in the general population. *Gynecol. Obstet. Invest.* 2005; 59: 86-91.

[2] Balerna M, Medici G, Mazzucchelli L, Bianda T, Marossi L, Colpi GM. Analytical biochemistry of seminal vesicle secretion: a challenge to andrological laboratories. *Andrologia.* 1990; 22 Suppl 1: 166-77.

[3] Lwaleed BA, Goyal A, Delves GH, Cooper AJ. Seminal hemostatic factors: then and now. *Semin. Thromb. Hemost.* 2007; 33: 3-12.

[4] Hirsch I H, Jeyendran R S, Sedor J, Rosecrans R, Staas WE. Biochemical analysis of electroejaculates in spinal cord injured men: comparison to normal ejaculates. *Journal of Urology.* 1991; 145: 73-77.

[5] Kaemmerer H, Mitzkat HJ. Ion-exchange chromatography of amino acids in ejaculates of diabetics. *Andrologia.* 1985; 17: 485-7.

[6] Nissen HP, Schirren C, Kreysel HW, Heinze I. Amino acid composition of human semen from different clinical diagnoses. *Andrologia.* 1979; 11: 109-12.

[7] Silvestroni L, Morisi G, Malandrino F, Frajese G. Free amino acids in semen: measurement and significance in normal and oligozoospermic men. *Arch. Androl.* 1979; 2: 257-61.

[8] Srivastava A, Chopra SK, Dasgupta PR. Biochemical analysis of human seminal plasma. II. Protein, non-protein nitrogen, urea, uric acid and creatine. *Andrologia.* 1984; 16: 265-8.

[9] Lwaleed BA, Goyal A, Greenfield RS, Cooper AJ. Seminal thrombin-activatable fibrinolysis inhibitor: a regulator of liquefaction. *Blood Coagul. Fibrinolysis.* 2007; 18: 449-54.

[10] Willemse JL, Hendriks DF. A role for procarboxypepidase U (TAFI) in thrombosis. *Front. Biosci.* 2007; 12: 1973-87.

[11] Willemse JL, Hendriks DF. Measurement of procarboxypeptidase U (TAFI) in human plasma: a laboratory challenge. *Clin. Chem.* 2006; 52: 30-6.

[12] Stief TW. The fibrinogen antigenic turbidimetric assay (FIATA): the $X^2\bar{x}$ test--the corrected chi-square comparison against the control-mean. *Clin. Appl. Thromb. Hemost.* 2007; 13: 73-100.

[13] Mandal A, Bhattacharyya AK. Some preliminary observations on the liquefaction of human semen. *Andrologia.* 1985; 17: 228-33.

[14] de Lamirande E. Semenogelin, the main protein of the human semen coagulum, regulates sperm function. *Semin. Thromb. Hemost.* 2007; 33: 60-8.

[15] Van Dreden P, Audrey C, Aurélie R. Human seminal plasma fibrinolytic activity. *Semin. Thromb. Hemost.* 2007; 33: 21-8.

[16] Stief TW. Thrombin and plasmin activity in semen. *Blood Coagul. Fibrinolysis.* 2007; 18: 386-7.

[17] Ohta S, Wada H, Gabazza EC, Nobori T, Fuse H. Evaluation of tissue factor antigen level in human seminal plasma. *Urol. Res.* 2002; 30: 317-20.

[18] Laudat A, Lecourbe K, Guéchot J, Palluel AM. Values of sperm thiobarbituric acid-reactive substance in fertile men. *Clin. Chim. Acta.* 2002; 325: 113-5.

[19] Fröhlich JU, Nissen HP, Heinze I, Schirren C, Kreysel HW. Free amino-acid composition of human seminal plasma in different andrological diagnoses. *Andrologia.* 1980; 12: 162-6.

[20] Fakih H, MacLusky N, DeCherney A, Wallimann T, Huszar G. Enhancement of human sperm motility and velocity in vitro: effects of calcium and creatine phosphate. *Fertil. Steril.* 1986; 46: 938-44.

[21] Chen F, Lu JC, Xu HR, Huang YF, Lu NQ. Preliminary investigations on the standardisation and quality control for the determination of γ-glutamyltranspeptidase activity in seminal plasma. *Andrologia.* 2007; 39: 1-43.

[22] Okada T, Suzuki H, Wada K, Kumagai H, Fukuyama K. Crystal structure of the gamma-glutamyltranspeptidase precursor protein from Escherichia coli. Structural changes upon autocatalytic processing and implications for the maturation mechanism. *J. Biol. Chem.* 2007; 282: 2433-9.

[23] Kinlough CL, Poland PA, Bruns JB, Hughey RP. Gamma-glutamyltranspeptidase: disulfide bridges, propeptide cleavage, and activation in the endoplasmic reticulum. *Methods Enzymol.* 2005; 401: 426-49.

[24] Hawkins RA, O'Kane RL, Simpson IA, Viña JR. Structure of the blood-brain barrier and its role in the transport of amino acids. *J. Nutr.* 2006; 136: 218S-26S

[25] Watanabe N, Dickinson DA, Liu RM, Forman HJ. Quinones and glutathione metabolism. *Methods Enzymol.* 2004; 378: 319-40.

[26] Pastore A, Federici G, Bertini E, Piemonte F. Analysis of glutathione: implication in redox and detoxification. *Clin. Chim. Acta.* 2003; 333: 19-39.

[27] Zinke H, Möckel B, Frey A, Weiler-Güttler H, Meckelein B, Gassen HG. Blood-brain barrier: a molecular approach to its structural and functional characterization. *Prog. Brain Res.* 1992; 91: 103-16.

[28] Paolicchi A, Dominici S, Pieri L, Maellaro E, Pompella A. Glutathione catabolism as a signaling mechanism. *Biochem. Pharmacol.* 2002; 64: 1027-35.

[29] Brown-Woodman PD, White IG. Amino acid composition of semen and the secretions of the male reproductive tract. *Aust. J. Biol. Sci.* 1974; 27: 415-22.

[30] Van Thuan N, Wakayama S, Kishigami S, Wakayama T. New preservation method for mouse spermatozoa without freezing. *Biol. Reprod.* 2005; 72: 444-50.

[31] He S, Woods LC. Effects of glycine and alanine on short-term storage and cryopreservation of striped bass (Morone saxatilis) spermatozoa. *Cryobiology.* 2003; 46: 17-25.

[32] Trimeche A, Yvon JM, Vidament M, Palmer E, Magistrini M. Effects of glutamine, proline, histidine and betaine on post-thaw motility of stallion spermatozoa. *Theriogenology.* 1999; 52: 181-91.

[33] Hernvann A, Gonzales J, Troupel S, Galli A. Amino acid content of human semen in normal and infertility cases. *Andrologia.* 1986; 18: 461-9.

[34] Keil M, Wetterauer U, Heite HJ. Glutamic acid concentration in human semen--its origin and significance. *Andrologia.* 1979; 11: 385-91.

[35] Takahashi Y, Kanagawa H. Effects of glutamine, glycine and taurine on the development of in vitro fertilized bovine zygotes in a chemically defined medium. *J. Vet. Med. Sci.* 1998; 60: 433-7.

[36] Ka HH, Sawai K, Wang WH, Im KS, Niwa K. Amino acids in maturation medium and presence of cumulus cells at fertilization promote male pronuclear formation in porcine oocytes matured and penetrated in vitro. *Biol. Reprod.* 1997; 57: 1478-83.

[37] Guérin P, Ménézo Y. Hypotaurine and taurine in gamete and embryo environments: de novo synthesis via the cysteine sulfinic acid pathway in oviduct cells. *Zygote.* 1995; 3: 333-43.

[38] Li Y, Si W, Zhang X, Dinnyes A, Ji W. Effect of amino acids on cryopreservation of cynomolgus monkey (Macaca fascicularis) sperm. *Am. J. Primatol.* 2003; 59: 159-65.

[39] Huijmans JG, Duran M, de Klerk JB, Rovers MJ, Scholte HR. Functional hyperactivity of hepatic glutamate dehydrogenase as a cause of the hyperinsulinism/ hyperammonemia syndrome: effect of treatment. *Pediatrics.* 2000; 106: 596-600.

[40] Murphy S, Munoz S, Parry-Billings M, Newsholme E. Amino acid metabolism during platelet storage for transfusion. *Br. J. Haematol.* 1992; 81: 585-90.

[41] Häussinger D, Schliess F. Glutamine metabolism and signaling in the liver. *Front. Biosci.* 2007; 12: 371-91.

[42] Albrecht J, Norenberg MD. Glutamine: a Trojan horse in ammonia neurotoxicity. *Hepatology.* 2006; 44: 788-94.

[43] Rose C. Effect of ammonia on astrocytic glutamate uptake/release mechanisms. *J. Neurochem.* 2006; 97 Suppl 1: 11-5.

[44] Norenberg MD, Rao KV, Jayakumar AR. Mechanisms of ammonia-induced astrocyte swelling. *Metab. Brain Dis.* 2005; 20: 303-18.

[45] Ziegler HK, Unanue ER. Decrease in macrophage antigen catabolism caused by ammonia and chloroquine is associated with inhibition of antigen presentation to T cells. *Proc. Natl. Acad. Sci. U. S. A.* 1982; 79: 175-8.

[46] Klebanoff S J, Smith DC. The source of H_2O_2 for the uterine fluid-mediated sperm-inhibitory system . *Biology of Reproduction.* 1970; 3: 236-242.

[47] Fraczek M, Sanocka D, Kurpisz M. Interaction between leucocytes and human spermatozoa influencing reactive oxygen intermediates release. *Int. J. Androl.* 2004; 27: 69-75.

[48] Szczesniak-Fabianczyk B, Bochenek M, Smorag Z, Ryszka F. Effect of antioxidants added to boar semen extender on the semen survival time and sperm chromatin structure. *Reprod. Biol.* 2003; 3: 81-7.

[49] Baumber J, Vo A, Sabeur K, Ball BA. Generation of reactive oxygen species by equine neutrophils and their effect on motility of equine spermatozoa. *Theriogenology.* 2002; 57: 1025-33.

[50] Erkkila K, Hirvonen V, Wuokko E, Parvinen M, Dunkel L. N-acetyl-L-cysteine inhibits apoptosis in human male germ cells in vitro. *J. Clin. Endocrinol. Metab.* 1998; 83: 2523-31.

[51] Zalata AA, Christophe AB, Depuydt CE, Schoonjans F, Comhaire FH. White blood cells cause oxidative damage to the fatty acid composition of phospholipids of human spermatozoa. *Int. J. Androl.* 1998; 21: 154-62.

[52] Alabi NS, Whanger PD, Wu AS. Interactive effects of organic and inorganic selenium with cadmium and mercury on spermatozoal oxygen consumption and motility in vitro. *Biol. Reprod.* 1985; 33: 911-9.

[53] de Lamirande E, Yoshida K, Yoshiike TM, Iwamoto T, Gagnon C. Semenogelin, the main protein of semen coagulum, inhibits human sperm capacitation by interfering with the superoxide anion generated during this process. *J. Androl.* 2001; 22: 672-9.

[54] Ball BA, Vo AT, Baumber J. Generation of reactive oxygen species by equine spermatozoa. *Am. J. Vet. Res.* 2001; 62: 508-15.

[55] Griveau JF, Le Lannou D. Reactive oxygen species and human spermatozoa: physiology and pathology. *Int. J. Androl.* 1997; 20: 61-9.

[56] de Lamirande E, Gagnon C. Human sperm hyperactivation and capacitation as parts of an oxidative process. *Free Radic. Biol. Med.* 1993; 14: 157-66.

[57] Babior BM. NADPH oxidase. *Curr. Opin. Immunol.* 2004; 16: 42-7.

[58] Roy SC, Atreja SK. Production of superoxide anion and hydrogen peroxide by capacitating buffalo (Bubalus bubalis) spermatozoa. *Anim. Reprod. Sci.* 2007; [Epub ahead of print]

[59] Aitken RJ, Wingate JK, De Iuliis GN, Koppers AJ, McLaughlin EA. Cis-unsaturated fatty acids stimulate reactive oxygen species generation and lipid peroxidation in human spermatozoa. *J. Clin. Endocrinol. Metab.* 2006; 91: 4154-63.

[60] Shukla S, Jha RK, Laloraya M, Kumar PG. Identification of non-mitochondrial NADPH oxidase and the spatio-temporal organization of its components in mouse spermatozoa. *Biochem. Biophys. Res. Commun.* 2005; 331: 476-83.

[61] Armstrong JS, Bivalacqua TJ, Chamulitrat W, Sikka S, Hellstrom WJ. A comparison of the NADPH oxidase in human sperm and white blood cells. *Int. J. Androl.* 2002; 25: 223-9.
[62] Albina JE, Caldwell MD, Henry WL, Mills CD. Regulation of macrophage functions by L-arginine. *J. Exp. Med.* 1989 169: 1021-9.
[63] Schuwerack PM, Lewis JW, Hoole D, Morley NJ. Ammonia-induced cellular and immunological changes in juvenile Cyprinus carpio infected with the blood fluke Sanguinicola inermis. *Parasitology.* 2001; 122: 339-45.
[64] Minana MD, Corbalan R, Montoliu C, Teng CM, Felipo V. Chronic hyperammonemia in rats impairs activation of soluble guanylate cyclase in neurons and in lymphocytes: a putative peripheral marker for neurological alterations. *Biochem. Biophys. Res. Commun.* 1999; 257: 405-9.
[65] Guerrini VH. Excretion of ammonia by Lucilia cuprina larvae suppresses immunity in sheep. *Vet. Immunol. Immunopathol.* 1997; 56: 311-7.
[66] Hardin-Pouzet H, Krakowski M, Bourbonniere L, Didier-Bazes M, Tran E, Owens T. Glutamate metabolism is down-regulated in astrocytes during experimental allergic encephalomyelitis. *Glia.* 1997; 20: 79-85.
[67] Shapiro MS, DeCoursey TE. Selectivity and gating of the type L potassium channel in mouse lymphocytes. *J. Gen. Physiol.* 1991; 97: 1227-50.
[68] Cossack ZT, Prasad AS. Activities of purine catabolism related enzymes in zinc deficiency: relationship to T-lymphocyte dysfunction and hyperammonemia. *Int. J. Vitam. Nutr. Res.* 1991; 61: 51-6.
[69] Klucinski W, Targowski SP. Ammonia toxicity for mammalian and avian lymphocytes from blood. *Immunopharmacology.* 1984; 8: 47-52.
[70] Targowski SP, Klucinski W, Babiker S, Nonnecke BJ. Effect of ammonia on in vivo and in vitro immune responses. *Infect. Immun.* 1984; 43: 289-93.
[71] Targowski SP, Klucinski W, Jaworek D. Effect of ammonia on viability and blastogenesis of bovine lymphocytes. *Vet. Immunol. Immunopathol.* 1984; 5: 297-310.
[72] Mukhopadhyay NK, Saha AK, Smith W, Dowling JN, Hiserodt J, Glew RH. Inhibition of neutrophil and natural killer cell function by human seminal fluid acid phosphatase. *Clin. Chim. Acta.* 1989; 182: 31-40.
[73] Kim SC, Kim HW. Effects of nitrogenous components of urine on sperm motility: an in vitro study. *Int. J. Androl.* 1998; 21: 29-33.
[74] Miller AL. Therapeutic considerations of L-glutamine: a review of the literature. *Altern. Med. Rev.* 1999; 4: 239-48.
[75] Alevy YG, Mueller KR, Hutcheson P, Slavin RG. Immune response in experimentally induced uremia. V. Description of a model for chronic uremia in the rat that produces a long-term suppressive effect on T cell responses to mitogens. *J. Lab. Clin. Med.* 1983; 101: 717-25.
[76] Alevy YG, Mueller KR, Slavin RG. Immune response in experimentally induced uremia. IV. Characterization of suppressor peritoneal macrophages in the uremic rat. *J. Lab. Clin. Med.* 1982; 100: 735-44.
[77] Alevy YG, Hutcheson P, Mueller KR, Slavin RG. Suppressor alveolar macrophages in experimentally induced uremia. *J. Reticuloendothel. Soc.* 1983; 33: 11-20.

[78] Hoosein NM, Martin KJ, Abdul M, Logothetis CJ, Kaddurah-Daouk R. Antiproliferative effects of cyclocreatine on human prostatic carcinoma cells. *Anticancer Res.* 1995; 15: 1339-42.

[79] Martin KJ, Chen SF, Clark GM, Degen D, Wajima M, Von Hoff DD, Kaddurah-Daouk R. Evaluation of creatine analogues as a new class of anticancer agents using freshly explanted human tumor cells. *J. Natl. Cancer Inst.* 1994; 86: 608-13.

[80] Liu YX. Involvement of plasminogen activator and plasminogen activator inhibitor type 1 in spermatogenesis, sperm capacitation, and fertilization. *Semin. Thromb. Hemost.* 2007; 33: 29-40.

[81] Ogetman Z, Dirlik M, Caglikulekci M, Canbaz H, Karabacak T, Yaylak F, Tamer L, Kanik A, Aydin S. The effect of aminoguanidine on blood and tissue lipid peroxidation in jaundiced rats with endotoxemia induced with LPS. *J. Invest. Surg.* 2006; 19: 19-30.

In: Diamino Amino Acids
Editors: H. A. Schäfer, L. M. Wohlbier, pp. 183-219 ISBN: 978-1-60456-892-9

Chapter VIII

SELECTED DIAMINO ACIDS IN SOILS OF DIFFERENTLY MANAGED MOUNTAIN MEADOW AND FOREST ECOSYSTEMS: ASSESSMENT OF THEIR ROLE IN ECOSYSTEM NUTRITION

Pavel Formanek[1], Klement Rejsek[1], Valerie Vranova[1] and Borivoj Klejdus[2]

[1]Department of Geology and Soil Science and [2]Department of Chemistry and Biochemistry, Mendel University of Agriculture and Forestry, 613 00 Brno, Czech Republic.

ABSTRACT

This chapter sets out to determine the effects of a) long-term abandonment of mown mountain meadows on Oxyaquic Hapludalf soil, and b) differing methods of thinning (2,044 versus 1,652 trees/ha) Norway spruce ecosystems on Typic Haplohumod and Entic Haplorthod soils. This was done by assessing the availability of particular forms of bioavailable nitrogen (N), including 14 proteinaceous monoamino acids (aspartic acid, glutamic acid, serine, histidine, glycine, threonine, alanine, tyrosine, valine, methionine, phenylalanine, isoleucine, leucine and proline) and three diamino acids (arginine, cystine and lysine). The research took place throughout the 2004 vegetation season and during the first half of the 2005 season. All of the study sites represent N-limited ecosystems with ammonium N dominant in plant nutrition.

Abandonment of meadows decelerated N-cycling and increased immobilization of N during the vegetation season. A significant ($P<0.05$) influence of meadow abandonment on concentrations of nitrates and glutamic acid in the Ap horizon was only found when data from 2004 and 2005 were pooled together. Of the amino acids, glutamic acid, arginine, aspartic acid and cystine were the most abundant. The three diamino acids studied composed <23% of total bioavailable N in the Ap horizon of both types of meadow (moderately mown *versus* abandoned). The amount of bioavailable N released in the top 7 cm of the Ap horizon in 2004 and 2005 was too low to supply the needs of the plant community. During the first half of the 2005 vegetation season, arginine and lysine were important as regards nutrition in the plant community, particularly in the moderately mown meadow. During 2004, amino acids were important for plant nutrition in the abandoned meadow at the end of summer. It appears that most N needed by plants is provided from the organic layer. Our results indicate that the plant communities of moderately mown and abandoned meadows can be dependent on mineral only, or mineral plus mono- and diamino acid-N nutrition, in different vegetation seasons. Thinning of mountainous stands of Norway spruce by 392 trees/ha did not significantly affect concentrations of certain forms of bioavailable N, including diamino acids. Glutamic acid, alanine, aspartic acid and cystine were most abundant in the Oe horizon of these ecosystems.

Keywords: arginine, cystine, lysine, soil, diamino acids, mountains, meadows, forests, management, plant nutrition, nitrogen.

INTRODUCTION

Increasing knowledge on the ability of plants to take up amino acids without previous mineralization; about low mineralization rates in arctic, alpine and boreal ecosystems; and on the ability of plants to outcompete microorganisms has opened up a new area of research: the importance of amino acids in ecosystem nutrition (Read, 1991; Chapin *et al.*, 1993; Rehder and Schafer, 1978; Fisk and Schmidt, 1995; Kaye and Hart, 1997; Falkengren-Grerup *et al.*, 2000).

Mountain meadows and forests represent ecosystems in which vegetation is typically nitrogen (N) limited. Both types of ecosystem can be differently managed and the type of management employed can modify N-cycling. Recent socioeconomic changes have strongly influenced land use in mountainous areas of Europe. This has recently led to setting up the EU project Carbomont, which is examining the "effect of land use changes on sources, sinks and fluxes of carbon in European mountain areas" (EVK2-CT2001- 00125, EU, Framework 5). In the mountain regions of southern Europe, pastures degraded in the past through a combination of overgrazing and summer water shortage are currently recovering following a decrease in grazing pressure. In the subarctic tundra, however, a recent increase in reindeer stocks has increased grazing pressure. In recent decades, a rapidly increasing proportion of formerly managed meadows and pastures in the Alps have been abandoned and are now being recolonised by shrub and tree species. In the mountainous areas of Central Europe, large areas have fairly recently been abandoned following re-privatization of land. Abandonment of meadows results in an accumulation of plant biomass, deceleration of N-cycling, and changes in root systems and soil properties, including total carbon content and overall soil quality (Slavikova, 1986; Fiala, 1997; Santruckova *et al.*, 2005; Formanek *et al.*, 2008; Vranova *et al.*, 2008).

Norway spruce (*Picea abies* [L.] Karst.) is one of the most important commercial tree species in European forests. Norway spruce is often grown in even-aged, single-species stands and managed using a clear-cutting system. Thinning is a regular component of the management program if this species is to reach the dimensions needed for the timber trade (Slodicak *et al.*, 2005). The thinning process is accompanied by a decrease in stand density and leaf area and an increase in light and nutrient availability. This can lead to changes in soil temperature, soil water, belowground root density, respiration of roots, leaching of dissolved organic matter from the canopy, and changes in soil microorganisms (Tang *et al.*, 2005; Bird *et al.*, 2002). Thinning may reduce possible competition between plant roots and saprophytic microorganisms for water and nutrients and, in turn, enhance decomposition and nutrient cycling in these ecosystems (Slodicak *et al.*, 2005).

Bioavailable amino acids are present both in solution, as dissolved matter, or as particulate matter loosely bound to organic or inorganic residues. They represent only a very small fraction of the total soil amino acid pool (Kielland, 1994; Lipson *et al.*, 2001). Glutamic acid, glycine and aspartic acid dominate in the soils of many ecosystems (Ivarson and Sowden, 1969; Abuarghub and Read, 1988; Kielland, 1995; Raab *et al.*, 1996, 1999), while basic amino acids (lysine, arginine and histidine), as well as neutral amino acids (alanine, serine, asparagine, glutamine and leucine), are also sometimes present in high concentrations (Formanek *et al.*, 2008; Kielland, 1995; Schmidt and Stewart, 1999; Abuarghub and Read, 1988). As regards meadow ecosystems, both low-productivity or fertilized meadows show a dominance of such amino acids as glycine, lysine and serine (Streeter *et al.*, 2000). Low-productivity meadows tend to have higher amounts of glutamate, histidine, alanine, threonine, leucine, aspartate, valine, arginine, phenylalanine and tyrosine.

Forest trees, grasses and herbs differ in how they use mono- and diamino acids for nutrition. Whereas forest trees take up particular forms of N at ratios related to their availability in the soil (Finzi and Berthrong, 2005), forbs (herbaceous flowering plants), as compared to grasses, strongly prefer N in the mineral form (Falkengren-Grerup *et al.*, 2000). According to Weigelt *et al.* (2005), *Nardus stricta* prefers serine to mineral N, whereas *Lolium perenne* utilizes mineral N. The authors have previously shown that temperate grassland species with high biomass production utilize mineral N forms more effectively for nutrition.

Glycine (or serine) is commonly used in experiments aimed at measuring rates of amino acid uptake by plant roots. In such experiments, amino acids that have previously been labeled with stable or radioactive isotopes (or amino acids without labeling) are taken up by plants and their exudation from roots is not monitored (Falkengren-Grerup *et al.*, 2000; Weigelt *et al.*, 2005; Persson and Näsholm, 2002; Streeter *et al.*, 2000; Kielland, 1994; Näsholm *et al.*, 1998; Raab *et al.*, 1999). Uptake of asparagine, glutamine and glutamic acid by roots of some plants exceeds their exudation (Paynel *et al.*, 2001; Lesuffleur *et al.*, 2007). Lesuffleur *et al.* (2007), on the other hand, found high exudation of glycine, serine and alanine by roots of *Trifolium repens, Lolium perenne, Zea mays, Brassica napus, Lycopersicon esculentum* and *Medicago sativa*, indicating that exudation of these amino acids greatly exceeded their uptake.

Amino acids are important as regards nutrition in low-productivity meadow communities of *Agrostis capillaris* and *Festuca ovina* (Weigelt *et al.*, 2005). In addition to mineral-, mono-, or diamino acid-N, some plants are able to utilize N from peptides and proteins, and mycorrhizal species play an important role in this uptake of organically bound N (Kaye and Hart, 1997). As stated previously, changes in management practice, such as the abandonment of mountain meadows and differing thinning strategies for spruce stands, are likely to affect the availability of particular forms of N, as well as the role of soil mono- and diamino acids in nutrition of plant communities, and produce changes in the microbial community. In studying these issues, we focused our interest on three diamino acids – arginine, cystine and lysine – in relation to their role in managed and abandoned mountain meadows and Norway spruce stands with differing thinning regimes. Arginine and lysine are positively charged diamino acids and are tightly bound to soil particles (Schmidt *et al.*, 1960; Paul and Schmidt, 1960). This strong binding may decrease their availability to plants. On the other hand, arginine is rapidly ammonified in soils, thus providing NH_4^+ for plant nutrition (Alef and Kleiner, 1986). Cystine is a component in sulfur-cycling forms, and its amount in proteins of plants or microbial biomass is low (Lehninger, 1985). The oxidation of cysteine to cystine occurs very rapidly in soils, where it is believed to be catalyzed by trace amounts of iron, copper and manganese (Freney, 1967). The role of cystine in soils has previously been studied, especially as regards its cycling in soil and its role in hydrogen sulfide production (Monreal and McGill, 1989a, b; Morra and Dick, 1989). Cystine is also rapidly ammonified in soil and relatively high amounts are found in bacterial spores, especially in the coat layers (Morra and Freeborn, 1989). According to Abbt-Braun and Jahel (2001), cystine comprised from 4% to 16% of amino acids detected in humic substances of different origin. The basic model accepted for the cycling of cystine in Luvisolic soils comprises soil solution, the adsorbed phase and two microbial pools – cytoplasmic free amino acids and cell proteins (Monreal and McGill, 1989 a, b). Cystine and cysteine are transported into microbial cells where cystine is reduced into cysteine, both forms accumulating in the cytoplasm of microbial cells in a free state. Alternatively, cystine can be released into soil solution through cell lysis, protease activity, root exudation, metabolic excretion or leakage of soil microorganisms from cytoplasm. Once cystine is released from microorganisms or humus, it may be subjected to the following basic reactions: a) electrostatic adsorption onto colloidal particles and microbial cell wall surfaces, b) active microbial uptake to satisfy cell nutritional and energy needs, or c) concurrent desorption from electrically charged surfaces (Harter, 1977; Gabler, 1978; Anraku, 1980; Paul and Smith, 1960). Arginine and lysine are also present in root exudates of plants (Grayston *et al.*, 1996) and, conversely, can be taken up directly by plant roots (Jones and Darrah, 1994; Lipson and Näsholm, 2001; Weigelt et al., 2005). Evidence for the direct uptake of cystine by plant roots is presently lacking.

In this chapter, we determine the effect of long-term abandonment of previously mown mountain meadows on: a) seasonal availability of three diamino acids (arginine, cystine and lysine) and their N in relation to other forms of bioavailable N in topsoil, b) stand heterogeneity of bioavailable N forms, and c) concentration of particular bioavailable N forms in organic soil. In addition, we compare concentrations of particular forms of bioavailable N, as well as concentrations of arginine, cystine and lysine, in the Oa (sapric, decomposing) and Oe (hemic, intermediate) horizons of mountainous Norway spruce stands under differing thinning regimes.

MATERIAL AND METHODS

Experimental Plots

The “Bílý Kříž” (“White Cross”) experimental research station, where the experiments took place, is located in the Moravian-Silesian Beskidy Mountains in the northeastern part of the Czech Republic. The region has a subcontinental climate with a typical mean annual air temperature of 4.9°C and mean annual precipitation of 1,100 mm. The number of days with snow cover is 160 per year.

The 1-hectare experimental meadow (N 49°30'17", E 18°32'28") is located at an elevation of 825–860 meters a.s.l. on a slope with a southeasterly orientation. Two study plots were located within the meadow, each differing in current management. The meadow was originally mowed regularly, the hay being removed and stored as feed for livestock (Formanek *et al.*, 2008). This traditional management practice ceased in 1993 on one-half of the meadow (unmown or abandoned meadow), while the remaining half has been moderately mown once per season since then. For example, the plot was mown on 28 July in 2004, and over 25 and 26 July in 2005. Before abandonment, the original area covering the two study plots maintained uniform plant populations and soil properties.

The moderately mown meadow plant community is representative of the *Nardo-Callunetea* class, while the abandoned meadow, characterized by a higher representation of forbs, is more typical of the *Molinio-Arrhenatheretea* class (Formanek *et al.*, 2008). The soil is classified as an Oxyaquic Hapludalf with the following sequence of horizons Oi-Oa-Oe-Ap-BAt-Bt1-Bt2-Cg (USDA-NRCS, 1999).

Holub and Tuma (2005) found that, by 2004, between-meadow differences in total (aboveground plus belowground) biomass production were still insignificant, being 1,889 ± 427 g/m^2 versus 1,758 ± 394 g/m^2 on the unmown and mown meadows, respectively (mean ± SD; n = 5–6). Despite this, belowground biomass production was higher in the moderately mown section, while the abandoned section had greater aboveground production. Plant uptake of N, P and K (nitrogen, phosphorous and potassium) was higher in the unmown section, with N, P and K concentrations in aboveground live biomass as follows (mown/unmown): N = 1.14/1.67%, P = 0.18/0.24% and K = 1.47/2.11%. The N:P ratio in aboveground biomass ranged from 6.8 to 7.9 for both meadows. Plant cover properties found by Holub and Tuma (2005) are listed in Table 1.

The second part of the study was undertaken in an experimental 23-year-old Norway spruce stand located at 908 meters a.s.l. (N 49°30'10", E 18°32'20") and approximately 500 m from the experimental meadow. Though the stand comprises approximately 99% Norway spruce and 1% fir (*Abies alba* Mill.), it is herein termed a Norway spruce stand for clarity. Within the experimental stand, two 0.25 ha study plots were established, each differing in intensity of thinning: the FD-dense plot at 2,044 trees/ha, and the FS-open plot at 1,652 trees/ha. These densities were established in 2005. The plots are located side by side and a fence surrounds each. The experimental Norway spruce stand was originally established on an area created by clear-cutting the previous 120-year-old Norway spruce forest. Before thinning in 1995, the experimental stand was of the same initial density of 2,830 trees/ha throughout (Formanek and Grunda, 2000). Thinning of the FD and FS plots was then performed on a number of separate occasions as shown in Table 2. Both plots were limed on three occasions (in 1983, 1985 and 1987), using dolomitic limestone with a total application rate of 9 t/ha (3 x 3 t/ha).

Table 1. Plant cover on moderately mown and abandoned meadows

Mown meadow		Abandoned meadow	
Ec = 95%		Ec = 95%	
E1 = 89-91%		E1 = 95%	
E0 = 10%		E0 = 0%	
Festuca rubra agg.	5	*Rumex acetosa*	4
Veronica officinalis	3	*Hypericum maculatum*	4
Hieracium laevigatum	2b	*Holcus mollis*	4
Nardus stricta	2a	*Achillea millefolium*	3
Luzula campestris	2a	*Potentilla erecta*	2a
Agrostis capillaris	2a	*Veronica chamaedrys*	2a
Potentilla erecta	2a	*Hieracium lachenalii*	2m
Avenella flexuosa	2a	*Cardaminopsis halleri*	2m
Carex pilulifera	1	*Jacea pseudophrygia*	2m
Holcus mollis	1	*Carex ovalis*	1
Hieracium pilosella	1	*Ranunculus repens*	1
Hieracium lachenalii	1	*Festuca rubra* agg.	1
Vaccinium myrtillus	1	*Agrostis capillaris*	1
Anthoxanthum odoratum	+	*Deschampsia cespitosa*	1
Lychnis flos-cuculi	+	*Galeopsis pubescens*	1
Campanula patula	+	*Lychnis flos-cuculi*	1
Hypericum maculatum	+	*Carex nigra*	1
Rumex acetosa	+	*Campanula patula*	+
Carex nigra	+	*Leontodon hispidus*	+
Carex pallescens	+	*Juncus effusus*	+
Gnaphalium sylvaticum	+	*Agrostis stolonifera*	r
Leucanthemum ircutianum	+	*Galeopsis tetrahit*	r
Achillea millefolium	+	*Carex pallescens*	r
Betula pendula	+	*Leontodon autumnalis*	r
Salix caprea	+	*Hieracium aurantiacum*	(r)
Taraxacum sec.Ruderalia	r		
Fragaria vesca	r		

Plant cover was estimated on 25 m^2 plots according to Barkman *et al.* (1964): EC - total coverage, E1 - herb layer, EO - moss layer; 5 = 75-100% coverage; 4 = 50-75%; 3 = 25-50%; 2b = 15-25%; 2m = 15%; 2a = 5-15%; 1 = < 5%; + negligible; r - rare; (r) - rare, out of plot.

Table 2. Tree density (trees/ha) in experimental Norway spruce stands (FD-dense plot and FS-open plot), before and after experimental thinning (thinning, * thinning after winter)**

year plot	1997	1998	1999	2000	2001	2002	2003	2004	2005
FD-dense plot	2600	2600	2600	2600	2600	2500**	2440**	2048**	2044
FS-open plot	2100*	2100	2100	2100	1880*	1820**	1820	1652**	1652

In comparison with the FD plot, the FS plot has a relatively rich herb layer with grasses, shrubs, rowan seedlings, etc. Both plots maintain Typic Haplohumod (sequence horizons: Oi-Oa-Oe-A-E-Bhs-Bs-BC-C) and Entic Haplorthod (sequence horizons: Oi-Oa-Oe-A-Bsw-BC-C) soil types and subtypes (USDA-NRCS, 1999).

Though meadows can be influenced in various ways by neighboring forests, and especially moderately mown meadow, there was no direct impact from the experimental forest plots on the experimental meadows.

The procedures of Raison *et al.* (1987) were used to measure the net release of different forms of N, as well as their net changes over time, within the moderately mown and abandoned meadows. Soil sampling and incubation were performed throughout the 2004 vegetation season, starting from mid-May and finishing in mid-October, and through a period from mid-May to mid-July 2005. Sampling took place until the middle of the 2005 vegetation season only, in order to allow us to verify a hypothesis that suggested the release of bioavailable N in the Ap horizon was insufficient to supply plant needs during the period of intensive growth. We did not take nitrate release during this period into account, in contrast to previous research, due to its negligible significance (Formanek *et al.*, 2008). Though the results from the 2004 season have previously been published (Formanek *et al.*, 2008), diamino acids were not separated from the sum of proteinaceous amino acids in that publication, and figures for cystine were not included. This chapter, therefore, presents new data and a new interpretation on the availability of diamino acids (arginine, cystine and lysine) in the Ap horizon of the experimental meadows over the course of the 2004 and first half of the 2005 vegetation seasons.

As described in Formanek *et al.* (2008), pairs of PVC cylinders (15 cm x 5.9 cm) were inserted to a depth of 13 cm at each of five randomly selected replicate points (marked by posts) within each of the two treatment areas (10 cylinders per area; 20 cylinders in total). Throughout the experiment, the cylinders were inserted close to the selected points to ensure equivalent conditions. One of each pair (A = beginning of incubation, actual concentration) was immediately removed for analysis. Cylinder B (with soil) was also removed, then wrapped in a plastic bag and re-inserted into the original hole and left for 30 days, after which it was again removed for analysis. This procedure was repeated at 30-day intervals. The soil in the cylinders was undisturbed and plants, and their roots, remained in the incubation tubes. This treatment, whereby plant roots are cut by insertion of the cylinder, restricts plant N uptake, while N leaching was avoided by wrapping the cylinders in plastic bags. Upon removal, the cylinders were transported to the laboratory at low temperature. Only that part of the soil corresponding to the Ap horizon (from 3 to 10 cm) was used further, after preparation by sieving through a 5 mm mesh and storing for 12 h at 4°C. The >5 mm fraction constituted 14.9% (w_t/w_t) of the soil in the moderately mown meadow, and 22.9% of that in the abandoned meadow (Formanek *et al.*, 2008). Mean soil moisture throughout the 2004 and 2005 experimental seasons for the A samples ranged from 17.8% to 39.3% (w_t/w_t). The water content in "B" samples was up to 19.6 percentage points (w_t/w_t) higher in comparison with "A" samples (36.7% versus 17.1%). The water content in "B" samples taken from mid-July 2004, however, was up to 21.1 percentage points (w_t/w_t) lower in comparison to the beginning of the incubation (36.2% versus 15.1%). Selected properties of the Ap horizon are shown in Table 3. Relative stand heterogeneity of total carbon in the Ap horizon was slightly higher in the moderately mown meadow (a coefficient of variance of 16.01% versus 14.85%), whereas heterogeneity of total N in the same horizon was higher (27.70% versus 20.45%) in the abandoned meadow (Vranova *et al.*, 2007). There were no significant differences found between properties of the Ap horizons in either meadow, with the exception of significantly higher (P<0.05) cation exchange capacity (CEC) in the abandoned meadow (Formanek *et al.*, 2006).

Table 3. Selected properties of the Ap horizon on moderately mown and abandoned mountain meadows (Mean ± SD; n = 3-8) (*significant differences, P<0.05)

Soil properties	Moderately mown	Abandoned meadow
Clay (%)	18.60 ± 3.4	19.93 ± 2.3
Silt (%)	26.0 ± 3.7	27.1 ± 7.0
Sand (%)	55.40 ± 6.5	52.93 ± 9.1
pH (H_2O)/pH 0.01 M $CaCl_2$	4.30 ± 0.45/3.80 ± 0.19	4.27 ± 0.42/3.78 ± 0.18
CEC ($mmol_c\,kg^{-1}$)	159.34 ± 2.28*	184.85 ± 10.01*
C_t (%)	4.76 ± 0.76	4.77 ± 0.71
N_t (%)	0.29 ± 0.06	0.33 ± 0.09
C/N	16.91 ± 3.58	15.24 ± 3.14
C-HA and C-FA (%)	0.50 ± 0.05 and 0.72 ± 0.05	0.57 ± 0.11 and 0.77 ± 0.05

Table 4. Total carbon (C_t), nitrogen (N_t) and their ratio (C/N) in the Oi, Oa and Oe horizons of moderately mown and abandoned mountain meadows. Asterisks mark significant differences between meadows (P<0.05)

Soil properties	Moderately mown Oi / Oa / Oe	Abandoned meadow Oi / Oa / Oe
C_t (%)	30.67 / 26.32 / 24.63	35.22 / 35.52 / 22.30
N_t (%)	1.65* / 1.67* / 1.60	2.08* / 2.10* / 1.53
C/N	18.86* / 15.93 / 15.48	16.92* / 15.45 / 14.57

Organic soil was only included in this experiment in June 2005. This allowed comparison of concentrations of particular N forms and the qualitative composition of diamino and monoamino acids. Composite samples were only prepared in the case of organic soil; in all other cases, samples were prepared and tested separately. Organic soil samples were not separated into Oi, Oa and Oe horizons when assessing bioavailable forms of N. Selected properties of the organic soil samples are shown in Table 4.

Net microbial release/immobilization of particular N forms (C) was calculated as the difference between the N form concentration at the end of incubation (B) and the concentration at the beginning of incubation (A), whereby C = B – A. The net change in total bioavailable N forms in uninfluenced soil (D) was calculated as the difference between the N form concentration in fresh samples from the beginning of the following incubation (A2) minus the concentration at the beginning of the previous incubation (A1), whereby D = A2 – A1 (Formanek *et al.*, 2008). This net change (D) is the net result of all processes ongoing in the uninfluenced soil, including microbial release/immobilization, root uptake, leaching, gaseous loss and atmospheric deposition. We did not separate these processes, in contrast to the procedures of Raison *et al.* (1987). The results for bioavailable amino acids obtained using such calculations should be interpreted with care due to their very short half-life (Jones, 1999).

Soil sampling in the forest took place twice, in June and October of 2005. Three mixed samples (one mixed sample = five random subsamples) were taken from the Oa horizon (1-3 cm depth) and the Oe horizon (3-9 cm depth) on each of the forest plots. Samples from the Oa horizon were taken in June only, in order to compare concentrations of particular N forms and qualitative composition of diamino acids. Differences between Typic Haplohumod and Entic Haplorthod were not investigated. As for the meadows, the samples were sieved through a 5 mm mesh in the laboratory and stored for 12 h at 4°C until extraction. Properties of the Oa and Oe horizon of the forest plots are shown in table 5. The pH_{H2O} for the Oe horizon of the FD and FS forest plots was 5.14 and 4.95, respectively (Formanek *et al.*, 2006).

Table 5. Selected properties of Oa and Oe horizons of the FD-dense (2,044 trees/ha) and FS-open (1,652 trees/ha) forest stands. All figures were not significant (P>0.05)

Soil properties	FD-dense stand Oa / Oe horizon	FS-open stand Oa / Oe horizon
C_t (%)	32.43 / 23.07	35.73 / 27.23
N_t (%)	1.29 / 1.01	1.48 / 1.21
C/N	25.36 / 23.15	24.06 / 22.51

Soil Extraction and Analysis

Diamino and monoamino acids were extracted from the soil using ammonium acetate (0.5 M, pH = 6.8). The use of ammonium acetate, as well as the advantages and disadvantages of extraction by this method, are discussed in Formanek *et al.* (2004, 2005, 2007, 2008), Abuarghub and Read (1988), and Lojkova *et al.* (2006). Briefly, individual soil samples (50 g) were extracted with 0.5 M ammonium acetate (200 ml) in 500 ml polyethylene bottles. For organic soil samples (Oa and Oe horizons of the forest stands), 25 g was extracted using 200 ml of ammonium acetate. After shaking for 1 h, samples were filtered through paper and glass fiber GF 30 filters (55 mm, 1 μm, Schleicher & Schuell, Germany) at 6°C, and frozen for storage at –18°C. When needed, the frozen extracts were reduced to 20 times their original concentration through lyophylization. The lyophylized extracts were then dissolved in 0.1 M HCl (500 μl) and filtered through a nylon membrane filter (13 mm, 0.45 μm, Chromatography Research Supplies, Adison, IL, USA). Concentrations of 14 monoamino acids (aspartic acid, glutamic acid, serine, histidine, glycine, threonine, alanine, tyrosine, valine, methionine, phenylalanine, isoleucine, leucine, and proline) and 3 diamino acids (arginine, cystine and lysine) were determined using the HP 1100 liquid chromatograph (Hewlett Packard, Wilmington, DE, USA) with a FLD HP 1100 fluorometric detector (operating at 450 nm (Ex = 340nm)) and a Zorbax Exlipse AAA Rapid Resolution column (4.6 x 150 mm, 3.5 μm particle size, Agilent Technologies, USA). Proline was not assessed in extracts from soil samples obtained in 2005. A linear gradient mobile phase profile, consisting of 40 mM Na_2HPO_4, pH 7.8 (solvent A) and ACN/MeOH/water at 45:45:10 (v/v) (solvent B), was applied at a flow rate of 2.0 ml/min. The procedure and gradients produced were as follow: 0% B (1.9 min), 0–57% B (16.2 min), 57–100% B (0.7 min), 100% B (3.5 min), 100–0% B (0.9 min) and 0% B (2.8 min). The column was equilibrated for 5 min under initial conditions prior to injection of each sample. Column temperature was 40°C. Amino acids from soil extracts were then determined using precolumn derivatization with *o*-phthalaldehyde (OPA) (Formanek *et al.*, 2005). Mineral N forms (NH_4^+ and NO_3^-) were extracted using 1% K_2SO_4 (soil/extractant = 10/50 w/v) and concentrations of NH_4^+ were determined by spectrophotometry, in which the ammonium ions react with sodium salicylate and chlorate ions in the presence of sodium nitroprusside (CSN ISO 7150-1, 1994). NO_3^- was determined by spectrophotometry with sulfosalicylic acid (CSN ISO 7890-3, 1994). Analysis of NO_3^- in soil extracts was only included in samples from June in 2005, in order to compare differences between horizons (Oi plus Oa plus Oe versus Ap for meadows; Oa versus Oe for forest plots). Other than in June, NO_3^- was not included for analysis during 2005 due to its negligible significance in the soil.

Statistical Analysis

Statistical analysis of the data was performed through one-way ANOVA (analysis of variance) and Fisher's LSD test (least significant difference). When preconditions for parametric testing were not fulfilled (i.e. non-normality of data), even after transformation, the nonparametric Kruskal-Wallis ANOVA and Dannet's *t*-test were used. Parametric t-tests were used to compare two independent groups of data and, if preconditions for parametric testing were not fulfilled, the nonparametric Mann-Whitney U test was used instead. All statistical analyses were undertaken using the Statistica 7.0 program.

Results and Discussion

Ammonium N (NH_4^+-N) was the dominant form of bioavailable N in the Ap horizon of both the moderately mown and abandoned mountain meadow during 2004 (Figures 1 and 2), as previously reported by Formanek *et al.* (2008). In this chapter, we present additional results (based on the addition of data for cystine and separation of a pool of three diamino acids – arginine, cystine and lysine – from the sum of 17 amino acids), showing that concentrations of mono- and diamino acid-N and nitrate-N remained broadly similar in the Ap horizon in both meadows during the 2004 vegetation season. Nitrogen of arginine, cystine and lysine comprised from 4.05% to 21.70% of total bioavailable N in the Ap horizon of the moderately mown meadow, and from 2.98% to 15.07% on the abandoned meadow. Statistical analysis of the data comparing the moderately mown and abandoned meadows indicated that N concentrations of the three diamino acids and the 14 other monoamino acids were not significantly different. In 2005, when measurements were restricted to the first half of the vegetation season (May–July) and NO_3^--N was not included, NH_4^+-N dominated bioavailable N on both meadows and the proportions of the three diamino acids decreased, ranging from 0.14% to 22.22% for the mown meadow and from 1.19% to 8.69% for the abandoned meadow (Figures 3 and 4). We found no significant differences between the meadows for concentrations of NH_4^+-N and for monoamino and diamino acid-N in the Ap horizon. When all data from 2004 and 2005 were taken into account, however, a significantly higher amount of NO_3^--N ($P<0.05$) was found in the Ap horizon of the abandoned meadow.

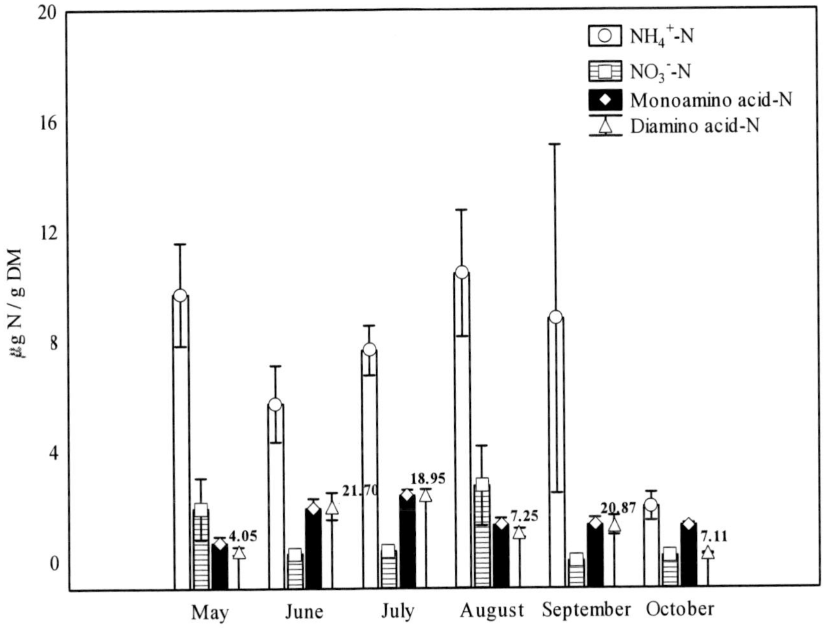

Figure 1. Seasonal dynamics of concentrations of bioavailable forms of nitrogen in the Ap horizon of the moderately mown mountain meadow. Mean ± 1SE (n = 3-5).

In June 2005, the organic soil (Oi plus Oa plus Oe horizons) was characterized by high concentrations of NH_4^+-N, with diamino acid-N representing only the third highest source of bioavailable N and amounting to just 4.23% of the total in the mown meadow and 6.21% in the abandoned meadow (Figure 5).

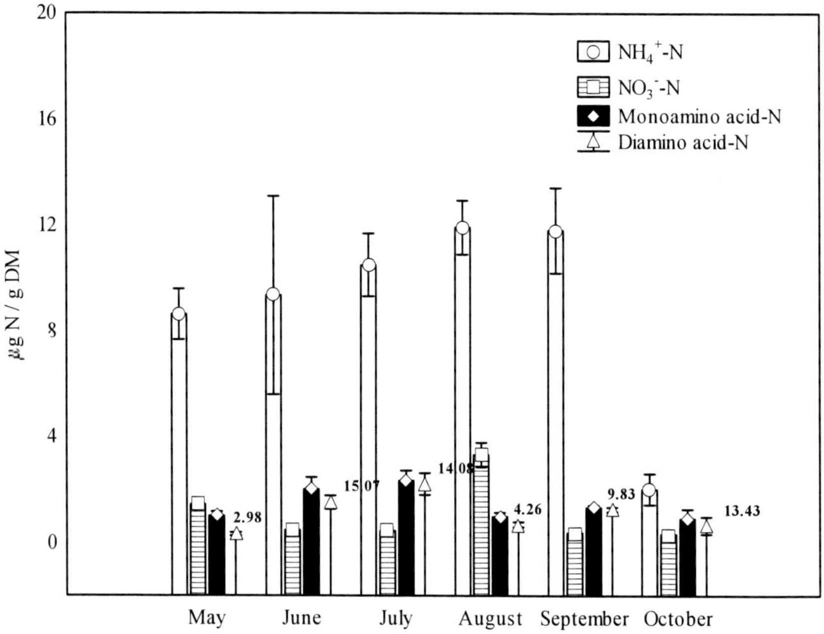

Figure 2. Seasonal dynamics of concentrations of bioavailable forms of nitrogen in the Ap horizon of the abandoned mountain meadow. Mean ± 1SE (n = 3-5).

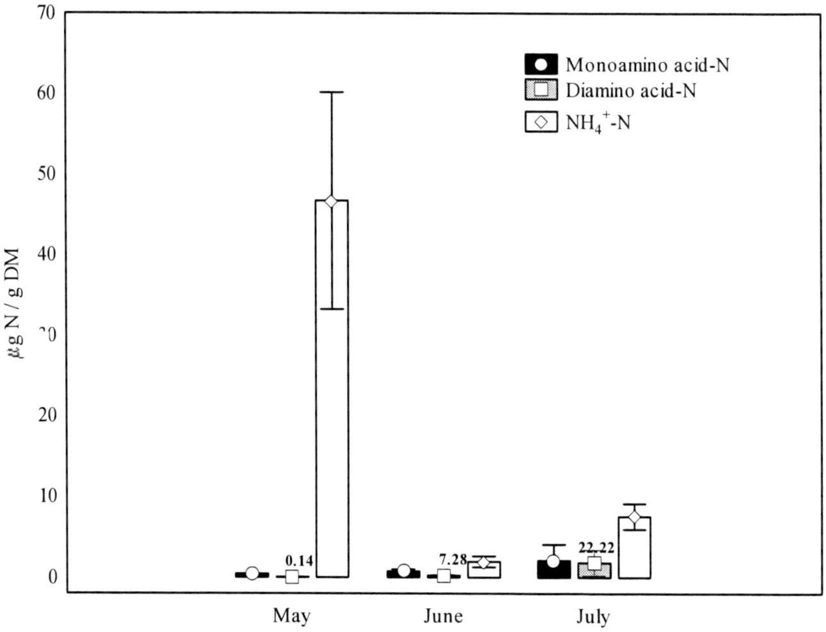

Figure 3. Concentrations of bioavailable forms of nitrogen in the Ap horizon of the moderately mown mountain meadow between May and July 2005. Mean ± 1SE (n = 1-5).

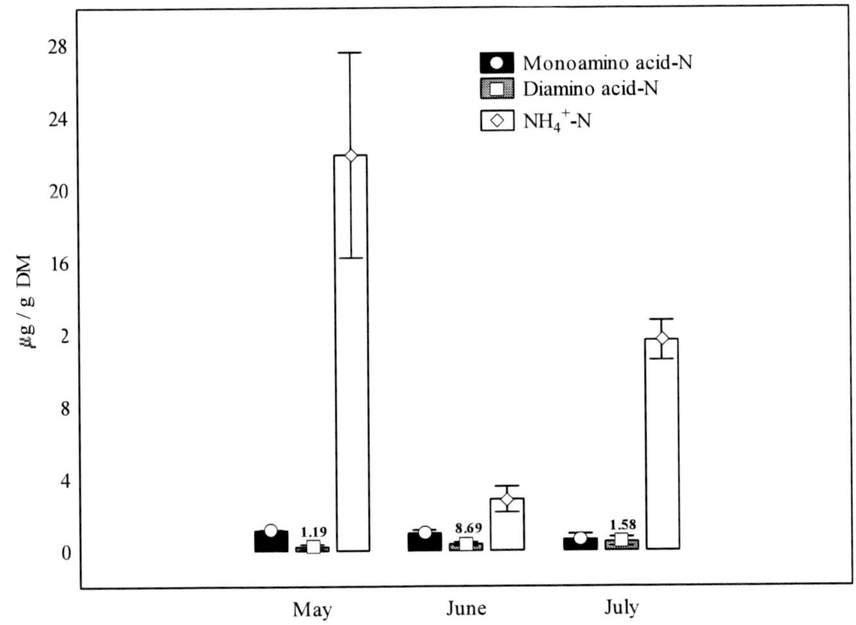

Figure 4. Concentrations of bioavailable forms of nitrogen in the Ap horizon of the abandoned mountain meadow between May and July 2005. Mean ± 1SE (n = 2-5).

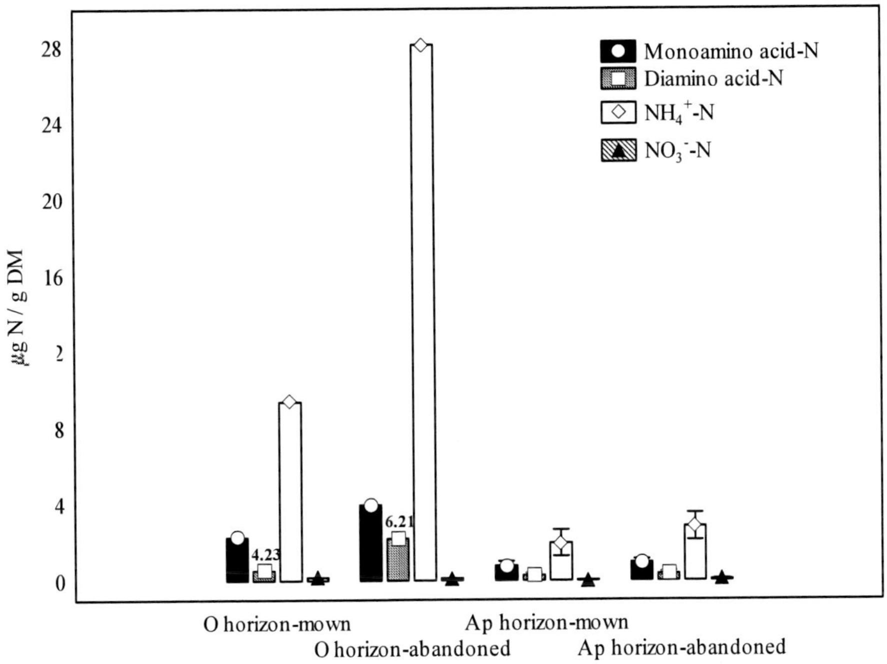

Figure 5. Concentrations of bioavailable forms of nitrogen in the O (Oi plus Oa plus Oe) and Ap horizons of the moderately mown and abandoned mountain meadows in June 2005. Mean ± 1SE (n = 1-5).

Average concentrations of the three diamino acids in the Ap horizons of the moderately mown and abandoned mountain meadows from 2004 and 2005 decreased in the following order: arginine > cystine > lysine. When both mono- and diamino acids were examined, the most abundant in the moderately mown meadow were glutamic acid > arginine > cystine > aspartic acid; and in the abandoned meadow these were glutamic acid > arginine > aspartic acid > cystine (Table 6). Apart from glutamic acid, which was significantly higher ($P<0.05$) in the moderately mown meadow, we found no other significant differences in concentrations of amino acids between the Ap horizons of either meadow[1].

Table 6. Concentrations of bioavailable diamino and monoamino acids (µg/g dry soil) in the Ap horizon of moderately mown and abandoned mountain meadows, from May to October 2004 and from May to July 2005 (n = 36-37)

Amino acid	Moderately mown	Abandoned meadow
Aspartic acid	1.57 ± 1.19	1.81 ± 2.00
Glutamic acid	4.60 ± 3.20	3.45 ± 2.73
Serine	0.58 ± 0.51	0.54 ± 0.42
Histidine	0.10 ± 0.08	0.13 ± 0.20
Glycine	0.66 ± 0.39	0.90 ± 0.93
Threonine	0.45 ± 0.42	0.54 ± 0.54
Alanine	1.25 ± 1.06	0.97 ± 0.65
Tyrosine	0.20 ± 0.29	0.19 ± 0.23
Valine	0.39 ± 0.53	0.33 ± 0.32
Methionine	0.54 ± 0.51	0.49 ± 0.48
Phenylalanine	0.54 ± 0.64	0.49 ± 0.46
Isoleucine	0.67 ± 0.80	0.83 ± 0.96
Leucine	0.41 ± 1.07	0.37 ± 0.44
Arginine	2.53 ± 2.52	2.06 ± 2.33
Cystine	2.04 ± 4.59	1.36 ± 1.35
Lysine	0.27 ± 0.20	0.52 ± 0.97

The concentration of total nitrogen and total carbon / total nitrogen within the organic horizon differed slightly between meadows due to changes in the vegetation cover (see Table 4). This may well have affected the composition of amino acids released to the topsoil. Indeed, this has been reported by a number of authors, whereby "treatments leading to changes in the concentration of total N in plant litter cause changes in the composition of amino acids" (McGuire *et al.*, 1995; Curtis and Wang, 1998; Norby *et al.*, 1999).

Microbial release of mineral N showed a positive balance throughout the 2004 vegetation season on the moderately mown meadow, whilst the abandoned meadow showed decelerated mineralization of N throughout the same period, with immobilization of mineral N halfway through the vegetation season (Figures 6 and 7). As previously reported in Formanek *et al.*

[1] The results from 2004, previously published by Formanek et al. (2008), showed no significant differences in concentrations of glutamic acid in the Ap horizon of the meadows. In this chapter, we present updated calculations including new data from the first half of the 2005 vegetation season.

(2008), ammonium was the dominant form of mineral N in both ecosystems during this vegetation season. Examination of the role of diamino acids as regards net release/immobilization over 30-day periods indicates a net release of diamino acid-N in the first half of the 2004 vegetation season on the moderately mown meadow, as well as during May to June and August to September in the abandoned meadow (Figures 6 and 7). Between May and June, diamino acid-N represented 50.29% of total released bioavailable N in the abandoned meadow, and between August and September this was 51.75%. During these periods, we can assume that diamino acids are playing a significant role in the nutrition of the plant community. There remains, however, an unanswered question over the role of cystine, which is included in the sum of diamino acid-N. No information exists on the ability of plants to use this diamino acid in a direct way. If we disregard cystine, therefore, and produce a total for diamino acid-N based only on figures for arginine and lysine, then net release of diamino acid-N (which can be used directly for nutrition by the plant community) is 48.12% of total plant-available N between May and June, and between August and September it is 48.93%. The net release of diamino acid-N over the first 30-day interval of the 2004 vegetation season is 17.81% in the moderately mown meadow, and over the second interval in the Ap horizon is 0.45%. Diamino acid-N showed immobilization during the second half of the 2004 vegetation season (Figure 6).

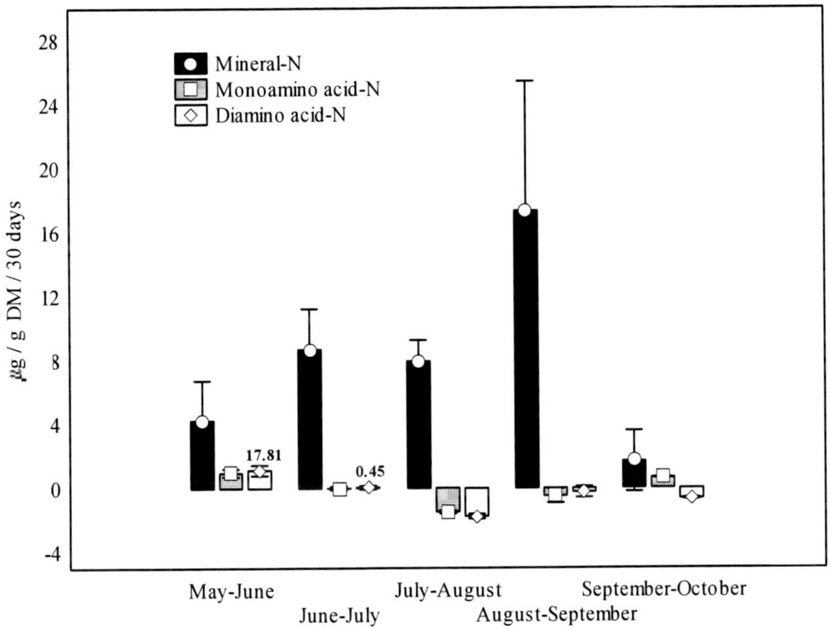

Figure 6. Net microbial release/immobilization of bioavailable forms of nitrogen in the Ap horizon of the moderately mown mountain meadow over the course of the 2004 vegetation season. Mean ± 1SE (n = 2-5).

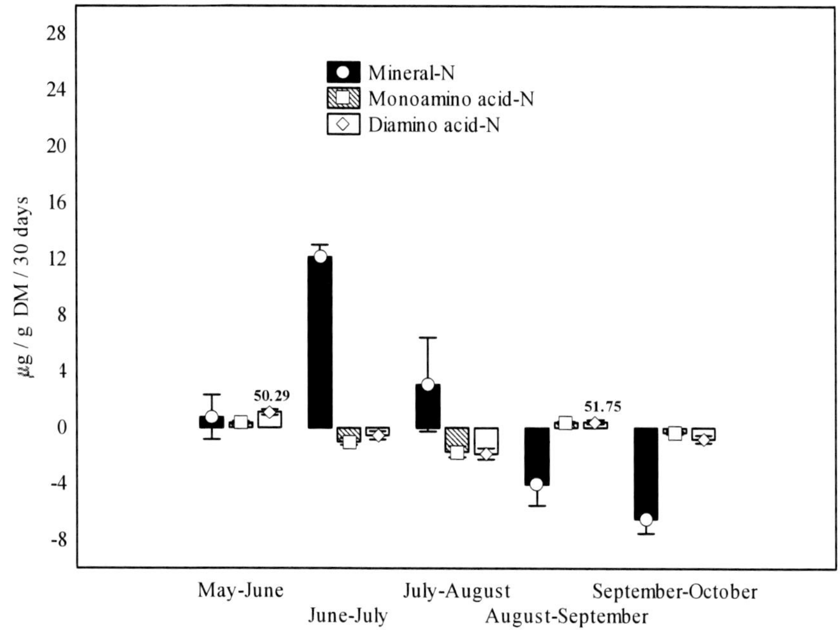

Figure 7. Net microbial release/immobilization of bioavailable forms of nitrogen in the Ap horizon of the abandoned mountain meadow over the course of the 2004 vegetation season. Mean ± 1SE (n = 3-5).

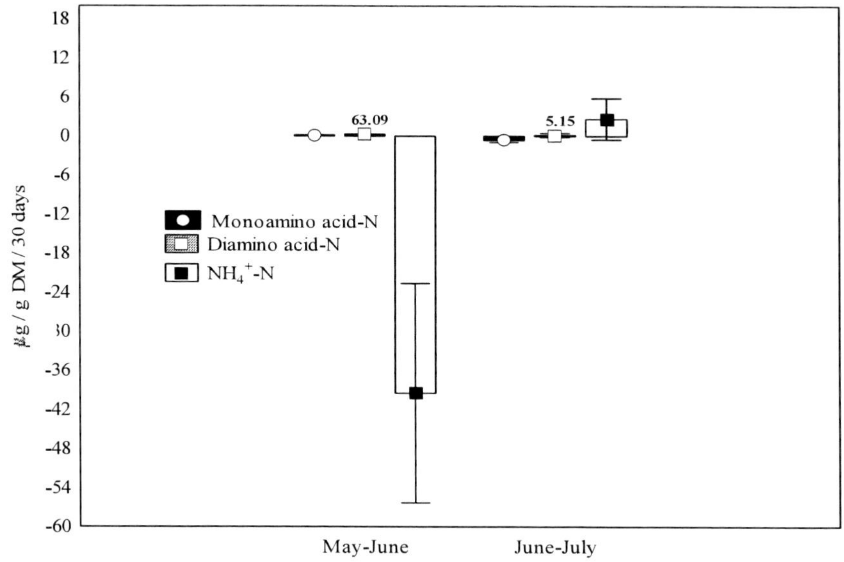

Figure 8. Net microbial release/immobilization of bioavailable forms of nitrogen in the Ap horizon of the moderately mown mountain meadow between May and July 2005. Mean±1SE (n=1-5).

A positive balance for the release of total bioavailable N was only observed from June to July in the moderately mown meadow over the beginning of the 2005 vegetation season. Between May and June, diamino acid-N as a proportion of bioavailable N released was 63.09%. This proportion was reduced to 5.15% between June and July, the dominant proportion comprising NH_4^+-N (Figure 8). Similarly, there was a net immobilization of all the studied forms of bioavailable N over May to June 2005 in the abandoned mountain meadow, whilst over June to July the positive balance was dominated by NH_4^+-N and diamino acid-N made up only 0.60% of total bioavailable N (Figure 9). In total, there was a significantly higher amount ($P<0.05$) of mineral-N released in the Ap horizon of the moderately mown meadow, as compared with the abandoned meadow, over the 2004 vegetation season, but there was no significant difference in the net release of the 14 monoamino and three diamino acid-N forms.

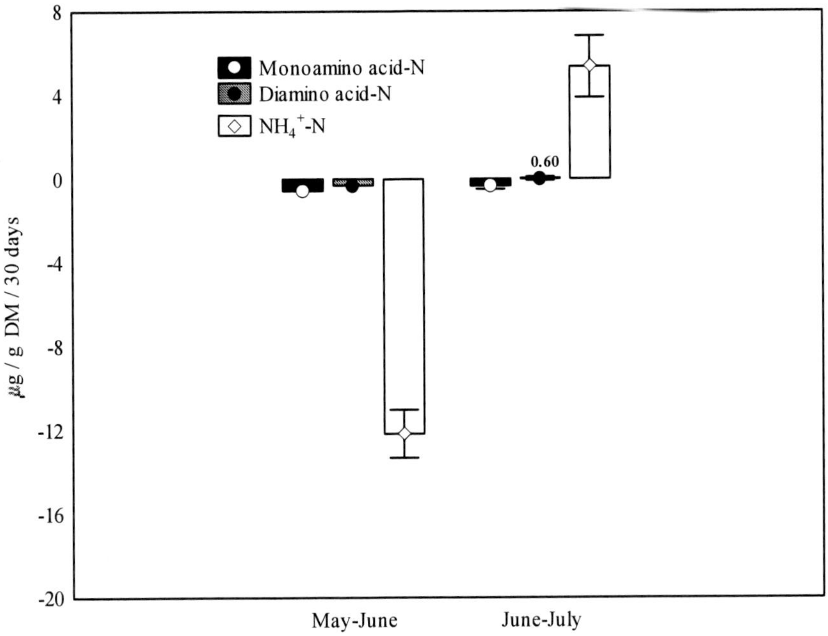

Figure 9. Net microbial release/immobilization of bioavailable forms of nitrogen in the Ap horizon of the abandoned mountain meadow between May and July 2005. Mean ± 1SE (n = 1-5).

There are some disadvantages related to the procedures used in this work. For example, plant roots clipped in PVC cylinders can still take up mineral and mono- or diamino acid-N (Bloom and Candwell, 1988; Falkengren-Grerup *et al.*, 2000). Such a method, therefore, may result in an underestimation of net N mineralization. Alternatively, root clipping can cause the release of mono- and diamino acids to the soil, and thus overestimate their actual concentrations and mineralization. Further, a 30-day incubation period may be very long, especially for mono- and diamino acids for which turnover may be very high (Lipson and Näsholm, 2001). The differences of up to 21.1 percentage points (w_t/w_t) of actual soil moisture content observed between samples at the beginning of this study and at the end of incubation would, due to an increase in partial oxygen pressure, affect the rate of soil denitrification (Tiedje, 1988; Parkin, 1987). The methodology of Raison *et al.* (1987), adopted for this study, does not include analysis of the role of denitrification as it was considered to be of negligible significance. As mentioned in Formanek *et al.* (2008), methods based on the sieving of soil samples before the extraction of mono- and diamino acids result in the release of higher amounts of organic compounds into the soil. Nevertheless, we had no other method available for removing skeletal particles and roots from soil samples.

The amount of bioavailable N released in the top 7 cm of the Ap horizon over the first half of the vegetation season 2004 was previously calculated by Formanek *et al.* (2008). Over this period, plant uptake of N was equivalent to 20.0 g/m^2 in the moderately mown meadow, and 28.8 g/m^2 in the abandoned meadow (Holub and Tuma, 2005). According to Formanek *et al.* (2008), microbial release of bioavailable N in 2004 was very low and insufficient to supply the needs of the plant community over this period. Using data from 2005 for the same period and soil layer, we found that 11.52 mg/m^2 of diamino acid-N was released in the moderately mown meadow. Release of monoamino acid-N and NH_4^+-N over this period was negative, i.e. these forms were immobilized. Thus, diamino acid-N was the only source for plant uptake but covered only 0.06% of the plants' needs (note that this figure includes cystine, which cannot be taken into account due to a lack of knowledge as to its direct use by plants). In the abandoned meadow, 0.31 g of NH_4^+-N/m^2 and 2.41 mg of diamino acid-N/m^2 were released over the same period (May to July 2005). Once again, monoamino acid-N showed a negative balance, i.e. immobilization. The amount of N released in the top 7 cm of the Ap horizon in the abandoned meadow supplied only 10.7% of plant needs. We believe, therefore, that the Oa and Oe organic horizons are likely to play the most important role in nutrition of meadow communities. The organic horizons were not included in the seasonal study of N due to its very low thickness following mowing treatment. According to results from 2004 and 2005, it is evident that plant communities of moderately mown and abandoned meadows can, in different vegetation seasons, be dependent on mineral only or mineral plus mono- and diamino acid-N nutrition.

Very few plants in the study meadows are thought to utilize peptides and proteins. Those that do are the ericoid-mycorrhizal *Vaccinium myrtillus* and the ectomycorrhizal *Betula pendula* in the moderately mown meadow (Bajwa and Read, 1985; Abuzinadah and Read, 1986; Abuzinadah and Read, 1989). Aside from *V. myrtillus*, there should not be any other ericoid-mycorrhizal plants in the meadows. Both *V. myrtillus* and *B. pendula* are present in the moderately mown meadow at very low or negligible levels. Of the other plants, though covering <5% in both meadows and in case lacking mycorrhizal symbiosis, *Carex* sp. are able to utilize amino acids (particularly glycine) to obtain N for nutrition (Kielland, 1994; Schimel and Chapin, 1996). Most other plant species in both test meadows are likely to support arbuscular mycorrhiza (Eriksen *et al.*, 2002; Titus and Leps, 2000). In some of these plants, e.g. *Nardus stricta* in the moderately mown meadow (coverage 5–15%), we can expect a high uptake of some amino acids, e.g. serine (Weigelt *et al.*, 2005). As regards nutrition of plant communities by mineral N, plant communities tend to grow better when NO_3^- is available (Falkengren-Grerup, 1995). Apart from mineralization of N from organic matter, biological N-fixing activity will also contribute to N nutrition. According to Rychnovska *et al.* (1993), N fixation in natural mown meadows can account for up to 5 kg N/ha/year. Nodulating and N fixing microbes of the *Rhizobiacea* and *Frankiaceae* families are also likely to play a role in N nutrition of meadow communities, and actinomycetes of the latter family in particular interact with *Betulaceae* plant species, which are present in the moderately mown meadow (Squartini, 2001).

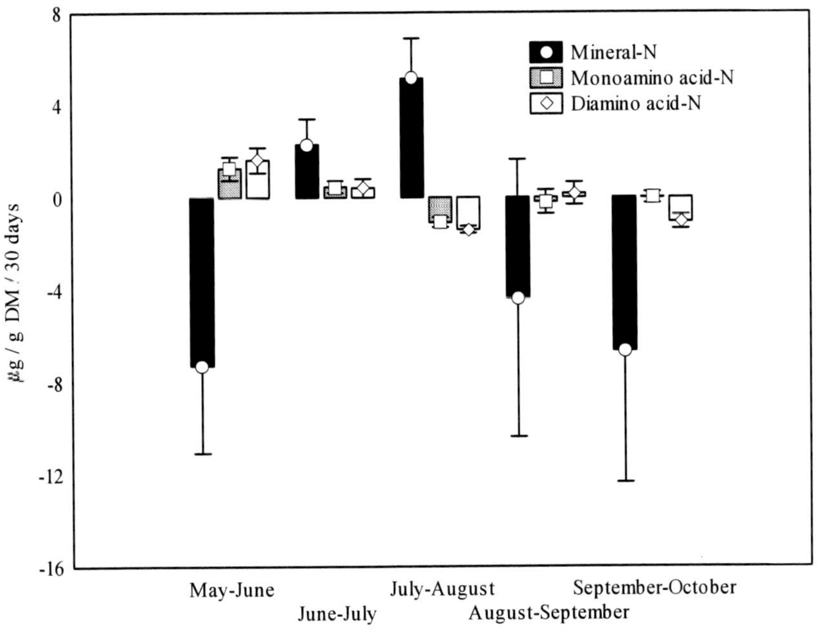

Figure 10. Net changes in bioavailable forms of nitrogen in the Ap horizon of the moderately mown mountain meadow over the course of the 2004 vegetation season. Mean ± 1SE (n = 3-5).

The total net change in forms of N over 30-day intervals represents the net result of all processes running in the uninfluenced soil, including microbial release/immobilization, root uptake, leaching, gaseous losses and atmospheric deposition. Their evaluation over the vegetation season in the moderately mown and abandoned meadows, previously described in Formanek *et al.* (2008), was performed without distinguishing pools of mono- and diamino acids. Total net changes in mineral-N were higher than those for both mono- and diamino acids. Evaluation of net release/immobilization and total sum of net changes in forms of N indicates that microbiologically released mineral-N is most important for plant community nutrition in the moderately mown meadow, as indicated by highest consumption over May to June (before mowing) and August to September (approximately one month after mowing) (Figures 6 and 10). Total net changes in monoamino and diamino acid-N in the Ap horizon of the moderately mown meadow were low in comparison to NH_4^+-N and, therefore, the sum of processes such as root uptake, leaching, gaseous losses and atmospheric deposition, at most, only slightly increased the amount of monoamino and diamino acid-N already being released by soil microorganisms. On the abandoned meadow, the first half of the vegetation period was the most important period as regards plant community nutrition, as indicated by the significance of NH_4^+-N. Diamino acid-N, as a result of root uptake, leaching, gaseous losses and atmospheric deposition, sometimes increased slightly in the 2004 vegetation season (Figures 7 and 11). Free amino acids make up a considerable amount of the N leached from leaf litter (Chapin *et al.*, 1986). Therefore, the increased amount of mono- and diamino acid-N observed in spring, and caused by root uptake, leaching, gaseous losses and atmospheric deposition, as compared to their net microbial release, may be a result of their leaching from the previous autumn's plant litter. Formanek *et al.* (2008), when examining the sum of 16 proteinaceous amino acids (without cystine), has already shown that amino acids were possibly the most important for plant community nutrition for the *Molinio-Arrhenatheretea* class in the abandoned meadow by the end of summer (Figure 7).

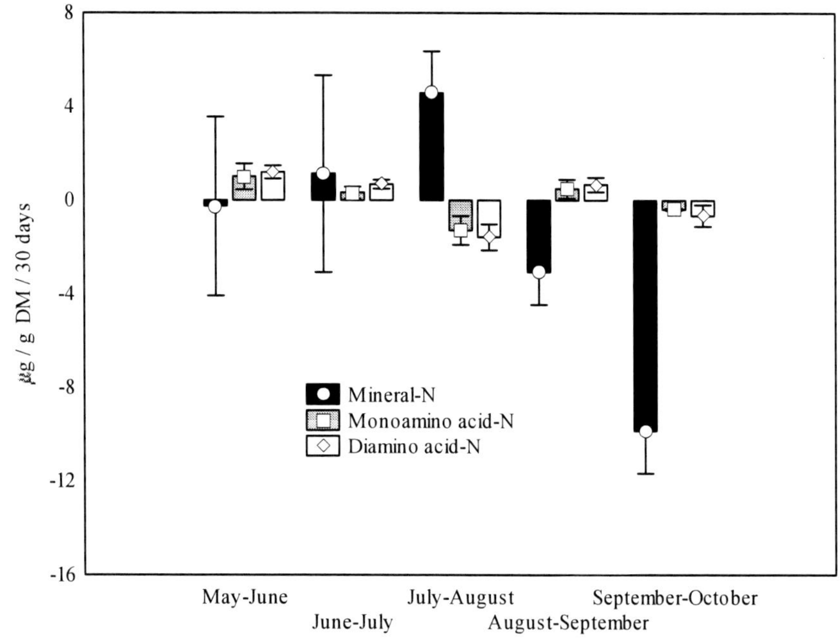

Figure 11. Net changes in bioavailable forms of nitrogen in the Ap horizon of the abandoned mountain meadow over the course of the 2004 vegetation season. Mean ± 1SE (n = 2-5).

The rate of mono- or diamino acid uptake by plants is affected by many factors. According to Persson and Näsholm (2002), amino acid uptake is greater in N-deficient plants. Both study meadows are N-limited ecosystems (Holub and Tuma, 2005) and, therefore, the significance of amino acid uptake will be of some importance. Aside from the availability of mineral nitrogen, mono- and diamino acids, the preferences of plants for particular N forms, and their ability to outcompete soil microorganisms, also has to be taken into account (Falkengren-Grerup *et al.*, 2000; Weigelt *et al.*, 2005; Kielland, 1994; Jackson *et al.*, 1989; Groffman *et al.*, 1993). We did not use labeling of mono- and diamino acids to indicate their participation in particular soil and plant compartments. Although the introduced grass communities of the study meadows are genetically predisposed to nitrate uptake (Larcher, 1988), their availability at this locality is low. Plants also show seasonal fluctuations in amino acids uptake (Gupta and Rorison, 1974; Abuarhub and Read, 1988), with increasing temperatures promoting increased amino acid uptake and assimilation (Jones, 1999). The results presented in this chapter indicate that mono- or diamino acids were important for direct nutrition only in some periods of the vegetation season (i.e. end of summer 2004 in the abandoned meadow or beginning of the 2005 vegetation season in the moderately mown meadow), when only bioavailable organic N remained for the plant community. The concentration dynamics of arginine, cystine and lysine over the 2004 vegetation season in the Ap horizon of the moderately mown and abandoned mountain meadows are displayed in Figure 12. The concentration of arginine was highest in July in both meadows. Cystine concentrations were highest in June on the mown meadow and in June and July in the abandoned meadow. Lysine was highest at the end of the vegetation season (October) on the abandoned meadow, but it remained relatively stable throughout the vegetation season on the moderately mown meadow.

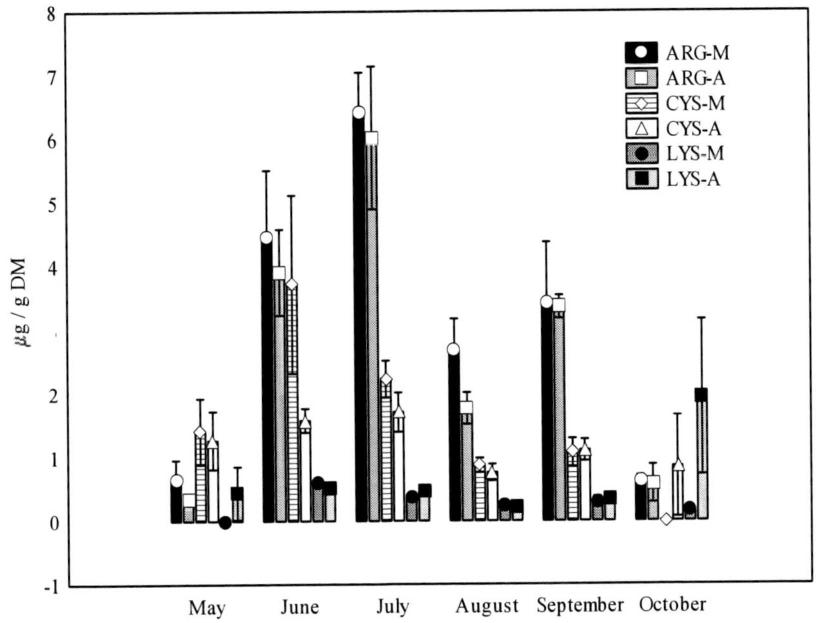

Figure 12. Seasonal changes in concentrations of arginine, cystine and lysine in the Ap horizon of the moderately mown (M) and abandoned (A) mountain meadows over the course of the vegetation season 2004. Mean±1SE (n=3-5).

The highest net release of arginine and cystine was observed at the beginning of the 2004 vegetation season in both meadows, whilst highest net microbial immobilization of these diamino acids was found from mid-July to mid-August (Figure 13). Net microbial release of lysine was highest at the end of the vegetation season on the abandoned meadow. Over the rest of the season, levels changed very little. Slight changes in lysine level observed in the moderately mown meadow were due to microbial release/immobilization.

Comparison of arginine, cystine and lysine in the organic soil and topsoil (Oi plus Oa plus Oe versus Ap horizons) of the two meadow types, undertaken in June 2005, indicated that concentrations of arginine, and especially cystine, were higher in the organic layer (Oi plus Oa plus Oe horizons) of the abandoned meadow (Figure 14). Concentrations of lysine did not differ greatly.

The relative stand heterogeneity of the diamino acids studied, expressed using coefficient of variance (CV), differed with vegetation development (Table 7). Highest heterogeneity was found for lysine in both meadows (21.48%–223.61% (mown); 20.03%–223.61% (abandoned)). When monoamino acids or mineral forms of N were taken into account, relative stand heterogeneity of methionine, proline and mineral forms of N in both meadows, as well as threonine and cystine in the abandoned meadow, were high over parts of the vegetation season.

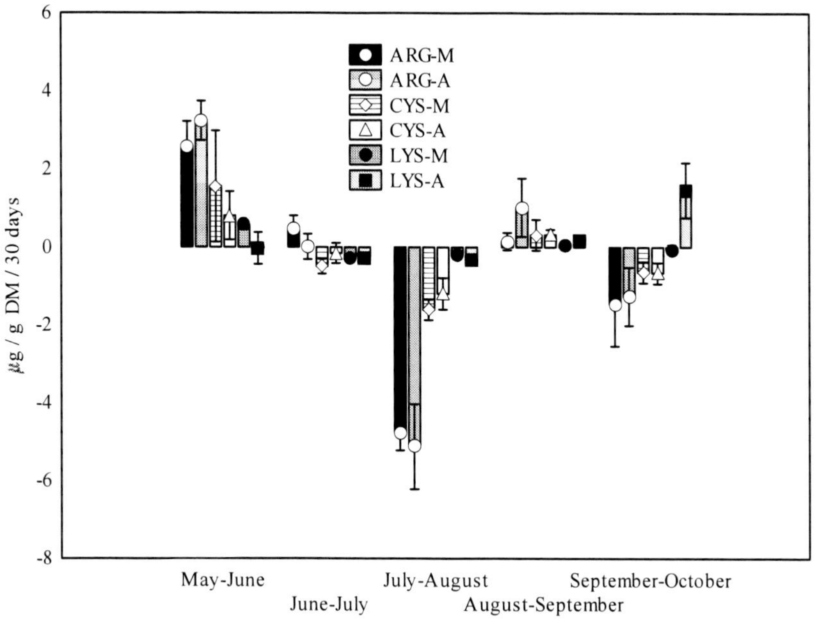

Figure 13. Seasonal net release/immobilization of arginine, cystine and lysine in the Ap horizon of the moderately mown (M) and abandoned (A) mountain meadows over the course of the vegetation season 2004. Mean±1SE (n=3-5).

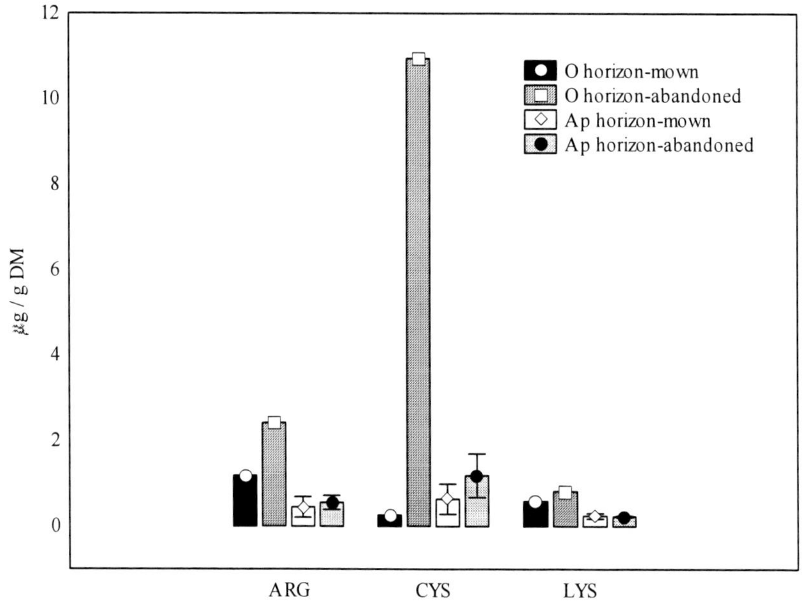

Figure 14. Concentration of arginine, cystine and lysine in the O (Oi plus Oa plus Oe) and Ap horizons of moderately mown and abandoned mountain meadows in June 2005. Mean ± 1SE (n = 1-5).

Table 7. Range of Coefficients of Variance (CV, %) for selected diamino and monoamino acids and ammonium ions or nitrates in the Ap horizon of moderately mown and abandoned mountain meadows in the Moravian-Silesian Beskidy Mts. over the course of the 2004 vegetation season

Amino acid	Moderately mown	Abandoned meadow
Arginine	22.89 - 112.21	9.79 - 107.00
Cystine	30.47 - 88.50	27.66 - 195.56
Lysine	21.48 - 223.61	20.03 - 223.61
Aspartic acid	27.21 - 113.92	16.81 - 108.28
Glutamic acid	15.02 - 108.91	18.71 - 104.53
Serine	15.03 - 79.77	6.66 - 74.17
Histidine	20.92 - 111.27	31.49 - 173.03
Glycine	24.12 - 84.25	30.94 - 132.86
Threonine	22.98 - 49.34	14.43 - 223.61
Alanine	22.28 - 108.21	10.43 - 92.51
Tyrosine	19.05 - 129.51	3.73 - 118.54
Valine	20.25 - 109.60	22.92 - 92.56
Methionine	25.75 - 221.97	22.52 - 223.61
Phenylalanine	26.69 - 78.35	32.29 - 82.03
Isoleucine	20.09 - 119.72	30.13 - 123.30
Leucine	13.09 - 98.98	20.00 - 74.17
Proline	36.84 - 181.28	10.37 - 200.00
NH_4^+	27.61 - 169.74	20.12 - 94.13
NO_3^-	48.94 - 156.03	14.87 - 111.05

Huhta (2001) has shown that the mowing of meadows can influence plant species diversity, microclimate, biomass production, colonization by new species, stature, shoot density, patchiness, litter accumulation and soil properties. Most primary production of meadows is accumulated underground, where changes occur in total underground biomass, stratification of soil edaphon, and composition of microbial biomass, and where microbial dying out occurs (Fiala, 1997; Ulehlova, 1988; Tesarová and Simek, 1989). Changes in plant coverage on mown and abandoned meadows results in differing qualities and quantities of root exudation (Vancura, 1988; Aulakh *et al.*, 2001). For example, the quantity of proteins and carbohydrates in root exudates of herbaceous plants decreases with increasing plant age (Juo and Stotzky, 1970; Hamlen *et al.*, 1972). Both mowing and abandonment of meadows can alter evapotranspiration over the course of a vegetation season (Rychnovska *et al.*, 1993). Plant roots have free amino acids sap concentrations in the range of 1–5 mM, as reported by Jones and Darrah (1994), and greater dying of roots due to mowing could lead to increased release of amino acids into soil, which, as our results seem to indicate, are quickly mineralized by soil microorganisms. There should be no direct effect of mowing on protease activity, as shown by the experiments of Kuzyakov *et al.* (2002) on *Lolium perenne*. Thus, the immobilization of N observed in the abandoned meadow was probably caused by higher inputs of organic matter at this site and a consequent increase in the activity of soil microorganisms. We found no effect of either mowing or abandonment on soil water content in the meadows. We did find, however, that the moderately mown meadow showed deficiencies in some elements, including phosphorus (Formanek *et al.*, unpublished).

Concentrations of bioavailable forms of N in FD-dense (2,044 trees/ha) and FS-open (1,652 trees/ha) Norway spruce stands were determined in June and October of 2005. In June, we took measurements from the Oa and Oe horizons, and the results from these observations are shown in Figure 15. In both forest stands, and in both horizons studied, NH_4^+-N was the dominant form of bioavailable N, whilst monoamino acid-N formed the second biggest pool and third was diamino acid-N. The amount of NO_3^--N in the Oa and Oe horizons of both forest stands was negligible. Concentration of NH_4^+-N and monoamino acid-N was higher in the Oa horizon as compared with the Oe horizon.

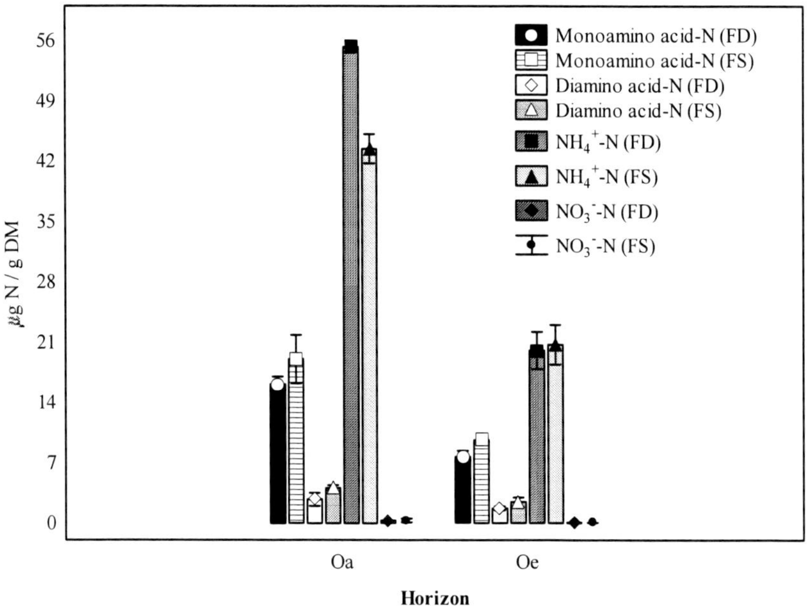

Figure 15. Concentration of forms of bioavailable nitrogen in the Oa and Oe horizons of FD-dense (2,044 trees/ha) and FS-open (1,652 trees/ha) forest stands in June 2005. Mean ± 1SE (n = 3).

Of the three diamino acids studied, cystine was present in the highest concentration, and it was higher in the Oa horizon than the Oe horizon in both the FD and FS plots (Figure 16). Concentrations of lysine were slightly higher than were those of arginine in the Oa and Oe horizons. Of the 13 monoamino and 3 diamino acids, highest concentrations were found in the order glutamic acid > alanine > aspartic acid > cystine in the Oe horizon of both the FD and FS forest stands (Table 8). No significant effect was observed on concentrations of mono- and diamino acids in the Oe horizon after thinning the forest stands by 392 trees/ha. A similar situation was observed by McGuire *et al.* (1995), Curtis and Wang (1998) and Norby *et al.* (1999), who found that, even after thinning, there was no significant change in the concentration of total N in needles and in the Oa and Oe organic horizons.

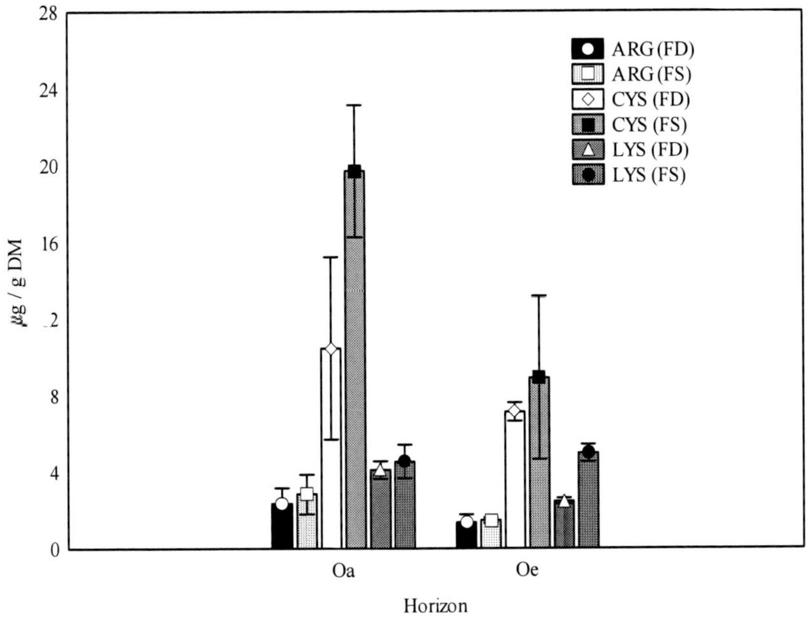

Figure 16. Concentration of arginine, cystine and lysine in the Oa and Oe horizons of FD-dense (2,044 trees/ha) and FS-open (1652 trees/ha) forest stands in June 2005. Mean ± 1SE (n = 3).

The concentrations for the sum of mono- and diamino acid-N in the Oa and Oe horizons of both forest stands were approximately within the range reported by Abuarghub and Read (1988) for the litter layer of pure stands of 6-year-old *Calluna vulgaris* (299 m altitude). Concentrations of mono- and diamino acid-N in the Oi plus Oa plus Oe horizons and the Ap horizon of both meadow types were within the range reported by Abuarghub and Read (1988) for soils of 0–10 cm depth under the litter of pure stands of *Calluna vulgaris*. The sum of mono- and diamino acid-N was also within the range reported by Monreal and McGill (1985) for cultivated and virgin soils (0.32–4.72 µg/g soil). Contrary to our study, these authors found alanine, phenylalanine and threonine to be the most abundant amino acids. Arginine, cystine and lysine were found to be least abundant by Monreal and McGill (1985) and were frequently not detected at all in these soils. The amount of mono- plus diamino acid-N in the Ap and Oi plus Oe plus Oa horizons in the study meadows, as well as the Oa and Oe horizons of the forest plots, were within the range reported by Kielland *et al.* (2006) for soils of Alaskan ecosystems across successional sequences (from 2.7 to over 24 µg N/g). According to Hertenberger *et al.* (2002), the relative abundance of free soil amino acids in plant roots did not correlate with their relative abundance in the soil when plants were cultivated for a short period. Later, there was a strong increase in the abundance of free soil amino acids, depending on soil type and plant species, which reflected differing microbial metabolisation of amino acids in the different soils. In our case, both meadows were on the same soil type and subtype, as were both forest plots.

Table 8. Concentrations of bioavailable diamino and monoamino acids (μg/g dry soil) in the Oe horizon of FD-dense and FS-open forest stands in June and October 2005 (n = 6).

Amino acid	Dense stand (2,044 trees/ha)	Open stand (1,652 trees/ha)
Aspartic acid	8.50 ± 4.81	6.64 ± 1.63
Glutamic acid	23.75 ± 3.16	23.81 ± 2.05
Serine	3.19 ± 1.66	2.08 ± 2.29
Histidine	0.41 ± 0.23	0.43 ± 0.14
Glycine	3.35 ± 0.51	4.52 ± 2.17
Threonine	3.55 ± 1.58	3.78 ± 0.92
Alanine	11.92 ± 3.56	10.44 ± 0.89
Tyrosine	0.57 ± 0.68	1.22 ± 0.76
Valine	3.65 ± 2.84	2.92 ± 3.04
Methionine	0.74 ± 0.85	0.77 ± 0.89
Phenylalanine	3.00 ± 0.29	3.03 ± 0.31
Isoleucine	2.10 ± 1.88	1.58 ± 1.62
Leucine	4.50 ± 2.41	5.03 ± 1.33
Arginine	2.06 ± 0.92	2.14 ± 0.80
Cystine	4.85 ± 3.81	6.27 ± 6.04
Lysine	1.77 ± 0.82	3.10 ± 2.10

An abundance of glutamic and aspartic acids was found in the soils of both the meadows and the differently thinned forest stands in our study, and this has been found in many other ecosystems (Ivarson and Sowden, 1969; Abuarghub and Read, 1988; Kielland, 1995; Raab *et al.*, 1996, 1999; Turnbull *et al.*, 1996). In this study, cystine was the fourth dominant amino acid in the Oe horizon of the forest ecosystems, as well as in the Ap horizon of the abandoned meadow, and the third dominant amino acid in the Ap horizon of the moderately mown meadow. Johnson and Pregitzer (2007) found lower concentrations of mono- and diamino acids in soils of three forest communities with elevated CO_2 and O_3. In their case, cystine was the fourth to sixth most abundant amino acid. On the other hand, Sowden and Ivarson (1966) did not find any free cystine in Brown Forest soil (Ah horizon), Podzol soil (H horizon) or in peat soil.

The microbial communities of the forest stands studied tend to be more fungal-dominated (Högberg *et al.*, 2003). Trees of both the forest plots are N-limited (Formanek, 2000). Ectomycorrhizae, representing a symbiotic relationship between basidiomycete and ascomycete fungi and their host plant, improve N nutrition of trees by conversion of litter and complex soil N into forms more readily utilized by either fungus or the host trees, as well as by absorption, assimilation, and translocation of inorganic N compounds from the soil to plant roots (Martin *et al.*, 2007). Absorption of proteins requires their enzymatic degradation to peptides and then to amino acids. Ectomycorrhizal fungi produce a wide range of proteolytic enzymes when grown on animal proteins (e.g. casein, gelatin and albumin) and protein fractions from beech forest litter as a substrate (Botton and Chalot, 1995). The ectomycorrhizal fungal community is influenced through management practices such as clear-cutting, thinning, fertilization and liming (Dahlberg and Stenström, 1991; Kårén and

Nylund, 1997; Andersson and Söderström, 1995; Korb *et al.*, 2001). Nitrogen fertilization increases fine-root growth but negatively affects mycorrhizal development (Ahlström *et al.*, 1988; Holopainen and Heinonen-Tanski, 1993). Using available data, we assessed the level of N deposition in our Norway spruce study plots as reaching 12 kg/ha/year on the FD-plot and 9 kg/ha/year on the more sparse FS plot (Formanek, 2000). We believe that such a small difference would have a negligible effect on mycorrhizal colonization and mycorrhizal types, as reported by Nilsen *et al.* (1998). Any changes in mycorrhizae occurring following forest thinning are likely to be due to a shift in understorey vegetation (Korb *et al.*, 2001).

Liming of Norway spruce stands can, in the short-term (0–10 years after liming), reduce aboveground growth in low-production stands, while enhancing it in more productive stands (Staaf *et al.*, 1996). In some cases, liming of these stands increases mortality of mycorrhizal root tips (Lehto, 1994). Changes due to liming have also been demonstrated as regards the composition of mycorrhizal populations associated with tree roots (Erland and Andersson, 1996). Wallender (2002) found that liming can, in some cases, reduce uptake of N from ^{15}N labeled fungal mycelium by mycorrhizal fungi. This is likely to be due to microbial and chemical immobilization of N at higher pH levels, as reported by Anderson *et al.* (1997). Uptake of amino acids by mycorrhizal fungi is dependent on pH. For example, Chalot *et al.* (1995) found that optimal uptake of glutamate and glutamine occurred at pH 4.0, and uptake declined steeply between pH 4.5 and 6.4. In the case of alanine and aspartate, uptake was not influenced in this pH range. Mycorrhizal fungi generally absorb amino acids with high affinity and improve N nutrition of host plants by increasing absorptive area and proteolytic activity (Chalot and Brun, 1998).

In our study, liming of the forest plots was performed before thinning, and pH_{H2O} in the Oe horizon was 5.14 (FD) versus 4.95 (FS) (Formanek *et al.*, 2006). Twelve years after liming (in 1999), the pH of the Oa and Oe horizons of both the forest plots had increased by approximately 1 pH point when compared with the same horizons of an unlimed forest area located ca 500 m from these plots (Formanek, 2000). Even though liming and altered pH is known to influence mycorrhizal types on study plots, as reported by Erland and Söderström (1990) and Lehto (1992), we believe there was no difference between our study plots. According to Kisza (2005), a higher thinning intensity on the FS plot decreased both microbial activity, measured by basal soil respiration in the Oe horizon, and the amount of easily utilizable carbonaceous compounds released after drying and re-wetting of the soil. On both forest plots, the carbon from microbial biomass in the Oe horizon was similar throughout most of the 2004 vegetation season one year before our experiment (Kafka, 2005). Increased pH due to previous liming could have had an effect on the activity of proteases, which are involved in the release of N from natural humus. This activity can decline when soil pH falls below 4.5 (Bending and Read, 1995). According to Flaig *et al.* (1975), liming can cause stabilization of proteins and amino acids in organic soils by reacting with chinons to form complexes protected against microbial decomposition. Liming could, therefore, have influenced the availability of N, its uptake, and the presence of ectomycorrhizal fungi on both the forest plots studied. Most important in the thinned FS plot will have been the development of a relatively rich herb layer, with grasses, ericoid-mycorrhizal dwarf shrubs such as *Vaccinius myrtillus* L., and rowan seedlings. These species differ in their efficiency for utilizing available N and, therefore, will contribute differing levels of availability.

CONCLUSIONS

Ammonium-N is the dominant form of bioavailable N in mountain meadows and forests under differing management regimes. Abandonment of a meadow significantly decreased the N-mineralization level. Microbial release of plant-available N, including mono- and diamino acids, in the topsoil during the first half of the 2004 and 2005 vegetation seasons was not sufficient to meet the needs of the *Molinio-Arrhenatheretea* plant community on the abandoned meadow and of the *Nardo-Callunetea* plant community on the moderately mown meadow. Arginine and lysine can be important in plant nutrition during certain periods of some vegetation seasons. Glutamic acid, arginine, cystine and aspartic acid were the most abundant in the Ap horizon of the meadows. Thinning of mountainous Norway spruce stands by 392 trees/ha did not significantly affect the concentration of particular forms of bioavailable N, including of diamino acids, in the Oe horizon. Glutamic acid, alanine, aspartic acid and cystine were the most abundant of 17 mono- and diamino acids in the Oe horizon of the forest stands.

ACKNOWLEDGEMENTS

This work was supported by the Grant Agency of the Czech Republic (Postdoctoral Project No. 526/03/D058), by the Ministry of Education of the Czech Republic (Project No. 4/2/2, Part II "The Management Strategy of Nature Conservation Areas" (Grant VZ MSM 6215648902)), by the European Union project "Carbomont - Effect of Land Use Changes on Sources, Sinks and Fluxes of Carbon in European Mountain Areas" (EVK2-CT2001-00125, Framework 5), and by the project Czech-Carbo No. VaV/640/18/03.

REFERENCES

Abbt-Braun,G.& Jahel, J. (2001). Organically bound sulfur in refractory organic substances. *Fresenius' Journal of Analytical Chemistry, 371*, 682-687.

Abuarghub, S.M. & Read, D.J. (1988). The biology of mycorrhiza in the Ericaceae XI. The distribution of nitrogen in soil of a typical upland *Callunetum* with special reference to the 'free' amino acids. *New Phytology, 108*, 425-431.

Abuzinadah R, A. & Read, D. J. (1986). The role of proteins in the nitrogen nutrition of ectomycorrhizal plants III. Protein utilization by *Betula*, *Picea*, and *Pinus* in mycorrhizal association with *Hebeloma crustuliniforme*. *New Phytologist, 103*, 507-514.

Abuzinadah R, A. & Read, D. J. (1989). The role of proteins in the nitrogen nutrition of ectomycorrhizal plants IV. The utilization of peptides by birch (*Betula pendula* L.) infected with different mycorrhizal fungi. *New Phytologist, 112*, 55-60.

Ahlström, K., Persson, H. & Börjesson, I. (1988). Fertilization in a mature Scots pine (*Pinus sylvestris* L.) stand - effects on fine roots. *Plant and Soil, 106*, 179-190.

Alef, K. & Kleiner, D. (1986). Arginine ammonification, a simple method to estimate microbial activity potentials in soils. *Soil Biology & Biochemistry, 18*, 233-235.

Andersson, S. Sk. H. & Söderström, B. (1995). Effect of lime ($CaCO_3$) on ectomycorrhizal colonization of *Picea abies* (L.) Karst. seedlings planted in a spruce forest. Scandinavian *Journal of Forest Research, 10*, 149-154.

Andersson, S. Sk. H. & Söderström, B. (1997). Effect of liming on the uptake of organic and inorganic nitrogen by mycorrhizal (*Paxillus involutus*) and non-mycorrhizal *Pinus sylvestris* plants. *New Phytologist, 135*, 763-771.

Anraku, Y. (1980). Transport and utilization of amino acids by bacteria. In Payne, J.W. (Ed.), *Microorganisms and Nitrogen Sources*, pp. 9-33. New York, US: John Wiley.

Aulakh, M. S., Wassmann, R., Bueno, C., Kreuzwieser, J. & Rennenberg, H. (2001). Characterization of root exudates at different growth stages of ten rice (*Oryza sativa* L.) cultivars. *Plant biology (Stuttg), 3*, 139-148.

Barkman, J.J., Doing H. & Segal S. (1964). Kritische Bemerkungen und Vorschläge zur quantitativen Vegetationsanalyse. *Acta Botanica Neerlandica, 13*, 394-419.

Bajwa, R. & Read, D. J. (1985). The biology of mycorrhiza in the Ericaceae IX. Peptides as nitrogen sources for ericoid endophyte and for mycorrhizal and non-mycorrhizal plants. *New Phytologist, 101*, 459-467.

Bending, G. D. & Read, D. J. (1995). The structure and function of the vegetative mycelium of ectomycorrhizal plants: V. The foraging behaviour of ectomycorrhizal mycelium and the translocation of nutrients from exploited organic matter. *New Phytologist, 130*, 411-417.

Bird, M. I., Santruckova, H., Arneth, A., Grigoriev, S., Gleixner, G., Kalaschnikov, Y. N., Lloyd, J. & Schulze, E. D. (2002). Soil carbon inventories and carbon - 13 on a latitude transect in Siberia. *Tellus, 54B*, 631-641.

Bloom, A.J. & Caldwell, R.M. (1988). Root excision decreases nutrients absorption and gas fluxes. *Plant Physiology, 87*, 794-796.

Botton, B. & Chalot, M. (1995). Nitrogen assimilation: enzymology in ectomycorrhizae. In Hock, B. & Varma, A. (Eds.), *Mycorrhiza: Structure, Function, Molecular Biology and Biotechnology*, pp. 325-363. Berlin, DE: Springer-Verlag.

Chapin, F.S. III., Vitousek, P.M. & Van Cleve, K. (1986). The nature of nutrient limitation in plant communities. *The American Naturalist, 127*, 48-58.

Chapin, F.S., Moilanen, L. & Kielland, K. (1993). Preferential use of organic nitrogen for growth by a non-mycorrhizal arctic sedge. *Nature, 36*, 150-153.

Chalot, M., Kytoviita, M.M., Brun, A., Finlay, R.D. & Söderström, B. (1995). Factors affecting amino acid uptake by the ectomycorrhizal fungus *Paxillus involutus*. *Mycology Research, 99*, 1131-1138.

Chalot, M. & Brun, A. (1998). Physiology of organic nitrogen acquisition by ectomycorrhizal fungi and ectomycorrhizas. *FEMS Microbiology Reviews, 22*, 21-44.

CSN ISO 7150-1. (1994). Jakost vod. Stanovení amonných iontů. Část 1: Manuální spektrometrická metoda. *Vydal Český normalizační institut*, 3 - 11.

CSN ISO 7890-3. (1994). Jakost vod. Stanovení dusičnanů. Část 3: Spektrometrická metoda s kyselinou sulfosalicylovou. *Vydal Český normalizační institut*, 2-7.

Curtis, P. S. & Wang, X. (1998). A meta-analysis of elevated CO_2 effects on woody plant mass, form,and physiology. *Oecologia, 113*, 299-313.

Dahlberg, A. & Stenström, E. (1991). Dynamic changes in nursery and indigenous mycorrhiza of Pinus sylvestris seedlings planted out in forest and clear cuts. Plant and Soil, 136,73-86.

Eriksen, M., Bjureke, K. E. & Dhillion, S. S. (2002). Mycorrhizal plants of traditionally managed boreal grasslands in Norway. *Mycorrhiza, 12*, 117-123.

Erland, S. & Söderström, B. (1990). Effect of liming on ectomycorrhizal fungi infecting Pinus sylvestris L. II growth rates in pure culture at different pH values compared to growth rates in symbiosis with the host plant. *New Phytologist, 115*, 683-688.

Erland, S. & Andersson, S. (1996). Ektomykorrhiza och skogsmarkskalkning. In Staaf, H., Persson, T. & Bertills, U. (Eds.), *Skogsmarkskalkning: resultat och slutsatser från naturvårdsverkets försöksverksamhet: Rapport 4559*, Naturvårdsverkets förlag. Stockholm, SE: IVL Swedish Environmental Research Institute.

Falkengren-Grerup, U. (1995). Interspecies differences in the preference of ammonium and nitrate in vascular plants. *Oecologia, 102*, 305-311.

Falkengren-Grerup, U., Månson, K.F., & Olsson, M.O. (2000). Uptake capacity of amino acids by ten grasses and forbs in relation to soil acidity and nitrogen availability. *Environmental and Experimental Botany, 44*, 207-219.

Fiala, K. (1997). Underground plant biomass of grassland communities in relation to mowing intensity. Acta Scientiarum Naturalium Academiae Scientiarum Bohemicae. Brno, CZ: Institute of Landscape Ecology.

Finzi, A.C. & Berthrong, S.T. (2005) The Uptake of Amino Acids by Microbes and Trees in Three Cold-Temperate Forests. *Ecology, 86*, 3345-335.

Fisk, M.C. & Schmidt, S.K. (1995). Nitrogen mineralization and microbial biomass nitrogen dynamics in three alpine tundra communities. *Soil Science Society of America Journal, 59,* 1036-1043.

Flaig, W.J.A., Beutelspacher, H. & Rietz, E. (1975). Chemical composition and physical properties of humic substances. In Gieseking, J.E. (Ed.), *Soil Components 1*, pp. 1-211. New York, US: Springer.

Formanek, P. (2000). Nitrogen transformation in anthropogenically influenced Norway spruce ecosystems of the Moravian-Silesian Beskids Mts (PhD thesis, in Czech). Brno, CZ: Mendel University of Agriculture and Forestry.

Formanek, P. & Grunda, B. (2000). Seasonal changes of nitrogen and biological activity in the forest soil of the Moravskoslezské Beskydy Mts. *(In Czech). Journal of Forest Science, 46*, 425-435.

Formanek, P., Vranova, V. & Marek, M.V. (2004). Methods of soil's free amino acids extraction: a review. *Phytopedon (Bratislava), 3*, 40-43.

Formanek, P., Klejdus, B. & Vranova V. (2005). Bioavailable amino acids extraction from soil by demineralized water and 0.5 M ammonium acetate. *Amino Acids, 28*, 427-429.

Formanek, P., Kisza, L. & Vranova, V. (2006a). Soil heterotrophic respiration potential and maximum respiration rate of differently managed meadows. *Soil and Water Research, 1*, 153-157.

Formanek, P., Vranova, V. & Rejsek, K. (2006b). Bioavailable amino acids in soils of differently managed meadow and forest ecosystems (In Czech). In Simoncicova, A., Dlapa, P. & Micuda, R. (Eds), *Život v pôde VII*, pp. 73-85. Bratislava, SK: Comenius University, Science Faculty.

Formanek, P., Klejdus, B. & Vranova V. (2007). The effect of extractants on degradation of L-glutamate and L-arginine in the course of shaking and filtration at low temperature. *Amino Acids, 32*, 489-491.

Formanek, P., Rejsek, K., Vranova, V. & Marek, M. V. (2008). Bioavailable amino acids and mineral nitrogen forms in soil of moderately mown and abandoned mountain meadows. *Amino Acids, 34*, 301-306.

Freney, J. R. (1967). Sulphur containing organics. In McLaren, A. D. & Peterson, G.H. (Eds.). *Soil Biochemistry*, pp. 229-259. New York, US: Marcel Decker.

Gabler, R. (1978). *Electrical Interactions in Molecular Biophysics*. New York, US: Academic Press.

Grayston, S. J., Vaughan, D. & Jones, D. (1996). Rhizosphere carbon flow in trees, in comparison with annual plnts: The importance of root exudation and its impact on microbial activity and nutrient and nutrient availability. *Applied Soil Ecology, 5*, 29-56.

Groffman, P.M., Zak, D.R., Christensen, S., Moiser, A. & Tiedje, J.M. (1993) Early spring nitrogen dynamics in a temperate forest landscape. *Ecology, 74*, 1579-1585.

Gupta, P. L. & Rorison, I. H. (1974). Seasonal difference in the availability of nutrients down a podzolic profile. *Journal of Ecology, 63*, 521-534.

Hamlen, R. A., Lukezic, F. L. & Bloom, J. R. (1972). Influence of age and stage of development on the neutral carbohydrates components in root exudates from alfalfa plants grown in a gnotobiotic environment. *Canadian Journal of Plant Science, 52*, 633.

Harter, R. D. (1977). Reactions of minerals with organic compounds in soils. In Dixon, J.B. (Ed), *Minerals in soil environments*, pp. 709-739. Madison, US: SSSA.

Hertenberger, P., Zampach, P. & Bachmann, G. (2002). Plant species affect the concentration of free sugars and free amino acids in different types of soil. *Journal of Plant Nutrition and Soil Science, 165*, 557-565.

Holub, P. & Tuma, I. (2005). Biomass production and nutrient uptake of mountain meadows with different management. *Beskydy, 18*, 69-72.

Holopainen, T. & Heinonen-Tanski, H. (1993). Effect of different nitrogen sources on the growth of Scots pine seedlings and the ultrastructure and evelopment of their mycorrhizae. *Canadian Journal of Forest Research, 23*, 362-372.

Högberg, M. N., Bååth, E., Nordgren, A., Arnebrant, K. & Högberg, P. (2003). Field evidence of opposing effect of nitrogen availability on carbon supply to ectomycorrhizal fungi and decomposers in boreal forest. *New Phytologist, 154*, 791-795.

Huhta, A. (2001). Restorative mowing on semi-natural grasslands: community-level changes and species-level responses. *Acta Universitatis Ouluensis A 365*. Oulu, FI: University of Oulu.

Ivarson, K.C. & Sowden, F.J. (1969). Free amino acid composition of the plant root environment field conditions. *Canadian Journal of Soil Science, 49*, 121-127.

Jackson, L.E., Schimel, J.P., Firestone, M.K. (1989) Short-term partitioning of ammonium and nitrate between plants and microbes in an annual grassland. *Soil Biology & Biochemistry, 21*, 409-415.

Johnson, R.M. & Pregitzer, K. S. 2007. Concentration of sugars, phenolic acids, and amino acids in forest soils exposed to elevated atmospheric CO_2 and O_3. *Soil Biology & Biochemistry, 39*, 3159-3166.

Jones, D.L. & Darrah, P.R. (1994). Amino-acid influx at the soil-root interface of *Zea mays* L. and its implications in the rhizosphere. *Plant and Soil, 163*, 1-12.

Jones, D.L. (1999). Amino acid biodegradation and its potential effects on organic nitrogen capture by plants. *Soil Biology & Biochemistry, 3*1, 613-622.

Juo, P. & Stotzky, G. (1970). Electrophoretic separation of proteins from roots and root exudates. *Canadian Journal of Botany, 48*, 713.

Kafka, O. (2005). *Assessment of carbon of microbial biomass on selected localities (in Czech)*. B.Sc thesis in Soil Science. Brno, CZ: Mendel University of Agriculture and Forestry.

Kaye J. P. & Hart, S. (1997). Competition for nitrogen between plants and soil microorganisms. *Trees, 12*, 139-143.

Kårén, O. & Nylund, J. E. (1997). Effects of ammonium sulphate on community structure and biomass of ectomycorrhizal fungi in a Norway spruce stand in South West Sweeden. *Canadian Journal of Botany, 75*, 1628-1643.

Kielland, K. (1994). Amino acid absorption by arctic plants: implications for plant nutrition and nitrogen cycling. *Ecology, 75*, 2373-2383.

Kielland, K. (1995). Landscape patterns of free amino acids in arctic tundra soils. *Biogeochemistry, 31*, 85-98.

Kielland, K., McFarland, J. & Olson, K. (2006). Amino acid uptake in deciduous and coniferous taiga ecosystems. *Plant and Soil, 288*, 297-307.

Kisza, L. (2005). *The effect of different managements of forest and meadow stands of selected localities on heterotrophic respiration and amount of easily utilizable carbon (in Czech)*. M.Sc thesis in Soil Science. Brno, CZ: Mendel University of Agriculture and Forestry.

Korb, J. E., Johnson, N. C. & Covington, W. W. (2001). Effect of restoration thinning on mycorrhizal fungal propagules in a Northern Arizona ponderoza pine forest: preliminary results. *USDA Forest Service Proceedings RMRS-P-22*, 74-79.

Kuzyakov, Y., Biryukova O. V., Kaznetzova T.V., Mölter K., Kandeler E. & Stahr K. (2002). Carbon partitioning in plant and soil, carbon dioxide fluxes and enzyme activities as affected by cutting ryegrass. *Biology and Fertility of Soils, 35*, 348-358.

Larcher, W. (1988). *Physiological Plant Ecology (in Czech)*. Academia, CZ: Praha.

Lehninger, A. L. (1985). *Biochemie. Weinheim*, DE: Verlag Chemie.

Lehto, T. (1992). Effect of drought on *Picea sitchensis* seedlings inoculated with mycorrhizal fungi. *Scandinavian Journal of Forest Research, 7*, 171-182.

Lehto, T. (1994). Effect of soil pH and calcium on mycorrhizas of *Picea abies*. *Plant and Soil, 163*, 69-75.

Lesuffleur, F., Paynel, F., Batalié, M.P., Le Deunf, E. & Cliquet, J.B. (2007). Root amino acid exudation: measurement of high efflux rates of glycine and serine from six different plant species. *Plant and Soil, 294*, 235-246.

Lipson , D.A. & Näsholm, T. (2001). The unexpected versatality of plants: organic nitrogen use and availability in terrestrial ecosystems. *Oecologia, 128*, 305-316.

Lipson, D., Raab, T.K., Schmidt, S.K. & Monson, R.K. (2001). An empirical model of amino acid transformations in an alpine soil. *Soil Biology & Biochemistry, 33*, 189-198.

Lojková, L., Klejdus, B., Formánek, P. & Kubáň, V. (2006). Supercritical fluid extraction of bioavailable amino acids from soils and their liquid chromatographic determination with fluorometric detection. *Journal of Agricultural and Food Chemistry, 54*, 6130-6138.

Martin, F., Kohler, A. & Duplessis, S. (2007). Living in harmony in the wood underground: ectomycorrhizal genomics. *Current Opinion in Plant Biology, 10*, 204-210.

McGuire, A. D., Melillo, J.M. & Joyce, L.A. (1995). The role of nitrogen in the response of forest net primary production to elevated atmospheric carbon dioxide. *Annual Review of Ecology & Systematics, 26*, 473-503.

Monreal, M C. & McGill, W. B. (1985). centrifugal extraction and determination of free amino acids in soil solutions by TLC using tritiated 1-fluoro-2,4-dinitrobenzene. *Soil Biology & Biochemistry, 17*, 533-539.

Monreal, M C. & McGill, W. B. (1989a). The dynamic of free cystine cycling at steady-state through the solutions of selected cultivated and uncultivated chernozemic and luvisolic soils. *Soil Biology & Biochemistry, 21*, 689-694.

Monreal, M C. & McGill, W. B. (1989b). The effect of soil amendments on the dynamics of free cystine cycling at steady-state through the solutions of a black chernozemic and an andept soil. *Soil Biology & Biochemistry, 2121*, 695-701.

Morra, M. J. & Freeborn, L. L. (1989). Catalysis of amino acid deamination in soils by pyridoxal-5'-phosphate. *Soil Biology & Biochemistry, 21*, 645-650.

Morra, M.J. & Dick, W. A. (1989). Hydrogen sulfide production from cysteine (cystine) in soil. *Soil Science Society of America Journal,* 53, 440-444.

Näsholm, T. (1998) Qualitative and quantitative changes in plant nitrogen acquisition induced by anthropogenic nitrogen deposition. *New Phytology, 139*, 87-90.

Nilsen, P., Børja, I., Knutsen, H. & Brean, R. (1998). Nitrogen and drought effect on ectomycorrhizae of Norway spruce [*Picea abies* L. (Karst.)]. *Plant and Soil, 198*, 179-184.

Norby, R. J., Wullschleger, S. D., Gunderson, C. A., Johnson, D. W. & Ceulemans, R. (1999). Tree response to rinsing CO_2 in field experiments: implication for future forest. *Plant, Cell and Environment, 22*, 683.

Parkin, T.B, (1987). Soil microsites as a source of denitrification variability. *Soil Science Society Journal of America, 51*, 1194-1197.

Paul, E. A. & Schmidt, E. L. (1960). Extraction of free amino acids from soil. Soil. *Proceedings of Soil Science Society of America, 24*, 195-198.

Paynel, F., Murray, P.J. & Cliquet, J.B. (2001). Root exudates: a pathway for short-term N transfer from clover and ryegrass. *Plant and Soil, 229*, 235-243.

Persson, J. & Näsholm, T. (2002). Regulation of amino acid uptake in conifers by exogenous and endogenous nitrogen. *Planta, 215*, 639-644.

Raab, T.K., Lipson, D.A. & Monson, R.K. (1996). Non-mycorrhizal uptake of amino acids by roots of the alpine sedge *Kobresia myosuroides*: implications for the alpine nitrogen cycle. *Oecologia, 108*, 488-494.

Raab, T.K., Lipson, D.A. & Monson, R.K. (1999). Soil amino acid utilization among the Cyperaceae: plant and soil processes. *Ecology, 80*, 2408-2419.

Raison, R.J., Connell, M.J. & Khanna, P.K. (1987). Methodology for studying of soil mineral-N *in situ*. *Soil Biology & Biochemistry, 19*, 521-530.

Read, D.J. (1991). Mycorrhizae in ecosystems. *Experentia, 47*, 376-391.

Rehder, H. & Shafer, A. (1978). Nutrient turnover studiesin alpine ecosystems. IV. Communities of the Central Alps and comparitive survey. *Oecologia, 34*, 309-327.

Rychnovska, M., Balatova-Tulackova, E., Bar, I., Fiala, K., Gloser, J., Halva, E., Hejny, S., Hrabe, F., Jakrlova, J., Klaska, F., Makusova, Z., Pelikan, J., Pradkova, E., Sibl, V., Smid, P., Tesarova, M., Ulehlova, B. & Zelena, V. (1993). *Structure and functioning of seminatural meadows: A case study on the Man and Biosphere Project No 91: function of meadows in a spring region*. Praha, CZ: Academia.

Santruckova, H., Kurbatova, J. A., Shibistova, O. B., Smejkalova, M. & Uhlirova, E. (2005). Short-term kinetics of soil microbial respiration - a general parameter across scales? Chapter 13. In Binkley, D. & Menyailo, O. (Eds). *Tree Species Effects on Soils: Implications for Global Change*, pp. 229-246. Dodrecht, NL: Kluwer Academic Publishers.

Schimel, J. P. & Chapin, F. S. (1996). Tundra plant uptake of amino acids and NH_4^+ nitrogen *in situ*: plants compete well for amino acid N. *Ecology, 77*, 2142-2147.

Schmidt, E.I., Putnam, H.D. & Paul, E.A. (1960). Behaviour of Free Amino acids in Soil. Division III - Soil Microbiology. *Proceedings of Soil Science Society of America, 24*, 107-109.

Schmidt, S. & Stewart, G.R. (1999). Glycine metabolism by plant roots and its occurrence in Australian plant communities. *Australian Journal Plant Physiology, 26*, 253-264.

Slavíková, J. (1986). *Plant Ecology (In Czech)*. Praha, CZ: SPN.

Slodicak, M., Novak, J. & Skovsgaard, J. P. (2005). Wood production, litter fall and humus accumulation in a Czech thinning experiment in Norway spruce (*Picea abies* (L.) Karst.). *Forest Ecology & Management, 9*, 157-166.

Sowden, F.J. & Ivarson, K. C. (1966). The 'free' amino acids of soil. *Canadian Journal of Soil Science, 46*, 109-114.

Squartini, A. (2001). Functional ecology of the rhizobium-legume symbiosis. In Pinton, R. at al. (Eds), *The rhizosphere: biochemistry and organic substrates at the soil-plant interface*, pp. 297-326. New York, US: Marcel Dekker.

Staaf, H., Persson, T. & Bertills, U. (Eds.). (1996). *Skogsmarkskalkning. Resultat och slutsatser från Naturvårdsverkets försöksverksamhet (In Swedish)*. Rapport 4559, Naturvårdsverkets förlag, Stockholm, SE: IVL Swedish Environmental Research Institute.

Streeter, T.C., Bol, R. & Bardgett, R.D. (2000). Amino acids as a nitrogen source in temperate upland grasslands: the use of dual labelled (^{13}C, ^{15}N) glycine to test for direct uptake by dominant grasses. *Rapid Communication in Mass Spectrometry, 14*, 1351-1355.

Tang, J., Qi, Y., Xu, M., Misson, L. &. Goldstein, A. H. (2005). Forest thinning and soil respiration in a ponderosa pine plantation in the Sierra Nevada. *Tree Physiology, 25*, 57-66.

Tesarova, M. & Simek, M. (1989). Rhizosphere microflora of managed grasslands. In Vancura, V. & Kunc, F. (Eds.), *Interrelationship between microorganisms and plants in soil. Proceedings of an International Symposium Liblice*, Czechoslovakia, June 22-27, 1987, pp. 307-313. Amsterdam, NL: Elsevier.

Tiedje, J.M. (1988). Ecology of denitrification and dissimilatory nitrate reduction to ammonium. In Zehnder, A.J.B. (Ed.), *Biology of Anaerobic Microorganisms*, pp. 179-244. New York, US: Wiley.

Titus, J.H. & Leps, J. (2000). The response of arbuscular mycorrhizae to fertilization mowing and removal of dominant in a diverse oligotrophic wet meadow. *Americal Journal of Botany, 87*, 392-401.

Turnbull, M.H., Schmidt, S., Erskine, P.D., Richards, S. & Stewart, G.R. (1996). Root adaptation and nitrogen source acquisition in natural ecosystems. *Tree Physiology, 16*, 941-948.

Ulehlova, B. (1988). Residues of organisms. In Vancura, V. & Kunc, F. (Eds.), *Soil microbial associations: control of structures and functions*, pp. 29-56. Praha, CZ: Academia.

USDA-NRCS. (1999). *Soil Taxonomy: Agricultural Handbook*. Washington, US: US government Printing Office.

Vancura, V. (1988). Plant metabolites in soil. In Vancura, V. & Kunc, F. (Eds.), *Soil microbial associations: control of structures and functions*, pp. 57-144. Praha, CZ: Academia.

Vranova, V., Formanek, P., Rejsek, K. & Janous, D. (2007). Stand heterogeneity of total carbon and nitrogen, and C/N in soil of mountain meadows. *Soil and Water Research, 2*, 149-155.

Vranova, V., Formanek, P., Rejsek, K. & Kisza, L. (2008). Selected kinetic parameters of soil microbial respiration in the A horizon of differently managed mountain forests and meadows of Moravian-Silesian Beskids Mts. *Euroasian Soil Science, (in press)*.

Wallender, H. (2002). Utilization of organic nitrogen at two different substrate pH by different ectomycorrhizal fungi growing in symbiosis with *Pinus sylvestris* seedlings. *Plant and Soil, 243*, 23-30.

Weigelt, A., Bol R. & Bardgett, R.D. (2005). Preferential uptake of soil nitrogen forms by grassland species. *Oecologia, 142*, 627-635.

Peer-Reviewer:

Dr. Keith R. Skene, Lecturer in Plant Ecology

University of Dundee, Department of Biological Sciences, Dundee DD1 4HN, the United Kingdom, telephone: (0)1382 344257, fax: (0)1382 344275, E-mail: k.r.skene@dundee.ac.uk

In: Diamino Amino Acids
Editors: H. A. Schäfer, L. M. Wohlbier, pp. 221-235
ISBN: 978-1-60456-892-9

Short Communication A

Arginine Conserves the Hemostasis Activation State of Plasma Even Against Freezing/Thawing

Thomas W. Stief *

Department of Clinical Chemistry, University Hospital, Marburg, Germany.

Abstract

Clinical hemostasis activation can be very dangerous, e.g. in pathologic disseminated intravascular coagulation (PDIC). Long transport times of this unstable plasma into the hemostasis laboratory or freezing/thawing of the samples induces artefactual hemostasis changes. 308 unselected EDTA plasmas of patients were analyzed for plasmatic amidolytic thrombin activity. EDTA-plasmas of 94 patients were stabilized with 1.25 M plasma concentration arginine and analyzed for plasmatic amidolytic thrombin activity directly or after freezing at -24°C/thawing at room temperature (RT). EDTA-plasmas were supplemented with 0-1100 mM arginine, stored for 24h at RT, and analyzed for plasmatic thrombin activity.

The normal range of amidolytic thrombin activity in arginine-stabilized EDTA plasma is 100 ± 20 % of normal, the overall patient range was here 131 ± 40 % of normal (mean value ± 1 standard deviation); the thrombin activity in patient plasmas distributed normally according to Gauss. Analysis of arginine-stabilized EDTA-plasma prior to or after freezing/thawing correlated with $r=0.980$, however thrombin activity in unstabilized frozen EDTA-plasma correlated only with $r=0.835$ to that of arginine stabilized EDTA-plasma. Addition of > 50 mM arginine to EDTA plasma increased the stability of thrombin activity after 24h storage at RT in individual plasmas, with an increase of correlation from $r=0.824$ to $r > 0.9$.

* Correspondence concerning this article should be addressed to: Priv.-Doz. Dr. med. T. Stief, Department of Clinical Chemistry, University Hospital, D-35033 Marburg, Germany. FAX: +49-6421-5865594; Email: thstief@med.uni-marburg.de.

Arginine stabilization of EDTA-plasma enables the analysis of the true hemostasis activation state of an individual patient even in frozen plasma. This might be of great importance in both diagnosis and therapy monitoring of hemostasis disorders.

Keywords: Hemostasis Activation, Thrombin, Arginine, Chromogenic Substrate, Laboratory Medicine.

ABBREVIATIONS:

IIa,	thrombin;
ΔA,	increase in absorbance;
Act,	activity;
CS,	chromogenic substrate;
CV,	coefficient of variation;
DIC,	disseminated intravascular coagulation;
K_3-EDTA,	ethylene diamine tetra acetic acid, tri-potassium salt;
HD-CHG-Ala-Arg-pNA,	HD-cyclohexylglycyl-alanyl-arginyl-paranitroanilide;
IU,	international units;
mA,	milli absorbance units;
max,	maximal;
MV,	mean value;
NIBSC,	National Institute for Biological Standards and Control;
r,	correlation coefficient;
pl,	plasmatic;
RT,	room temperature (23°C);
SD,	standard deviation.

1. INTRODUCTION

Diagnosing the activation state of hemostasis is of great clinical importance: hemostasis activation can be life threatening, e.g. in pathologic disseminated intravascular coagulation (PDIC). Here the pre-analytic is problematic, since long sample transport times into the hemostasis laboratory or freezing/thawing of the samples induce artefactual changes of the plasma, resulting in assay results that are not interpretable. Guanidine-compounds such as arginine competitively inhibit serine proteases [1]. Therefore, in the present study it was investigated, if arginine conserves the real blood activity of an important parameter of hemostasis activation: plasmatic amidolytic thrombin activity [2].

2. MATERIAL AND METHODS

2.1. Calibration of Thrombin Activity

The 2nd international thrombin standard (110 IU/vial; 1 mg = 2525 IU) was obtained from the National Institute for Biological Standards and Control (NIBSC), Blanche Lane,

South Mimms, Potters Bar, Hertfordshire EN6 3QG, United Kingdom (http://www.nibsc.ac.uk). The lyophilized 110 IU human thrombin were dissolved in 1 ml 6.7 % human albumin (Kabi, Stockholm, Sweden), 1.7 M arginine, pH 8.7 (Sigma, Deisenhofen, Germany). This reagent was the calibrator for commercially available bovine thrombin (B4233-25, DadeBehring, Marburg, Germany) tested to have 130 IU/vial. For calibration, the lyophilized vial of bovine thrombin was also reconstituted with 1 ml 6.7 % human albumin, 1.7 M arginine, pH 8.7. Calibrator and bovine thrombin were diluted 1:100 with 6.7 % human albumin, 1.7 M arginine, pH 8.7 (final thrombin activity of the international calibrator = 1.1 IU/ml). In 4fold determination (intra-assay coefficients of variation < 2 %), 50 µl of both thrombin solutions were incubated with 50 µl 6.7 % human albumin, 1.7 M arginine, pH 8.7, and with 10 µl 3.85 mM (final: 0.64 mM) chromogenic substrate (CS) for thrombin (IIa) HD-CHG-Ala-Arg-pNA in H_2O (Pentapharm, Basel, Switzerland) in U-well polystyrol microtiterplates (NUNC, Wiesbaden, Germany). After strongly shaking for 10 seconds the absorbance at 405 nm was determined immediately and after 12 min (room temperature (RT) = 23°C) using a microtiterplate reader with a 1 mA resolution (Milenia; DPC, Los Angeles, USA). 1 IU/ml thrombin in the sample resulted in a linear absorbance increase (ΔA) of 214 mA/12 min RT = 17.8 mA/min RT.

2.2. Range of Thrombin Activity in EDTA Plasma of Unselected Patients

Plasma with EDTA (1.6 mg K_3-salt/ml blood; Sarstedt, Nümbrecht, Germany) of n=308 unselected patients was analyzed for basal thrombin. Therefore, 50 µl plasma that remained after routine analysis (stored for 3 h at RT; or NaCl=zero control) were incubated in duplicate with 50 µl 2.3 M arginine, pH 8.6, containing 0.77 mM CS-IIa for 2 h at 37°C in U-wells microtiter polystyrol plates. ΔA at 405 nm was determined; approximate maximal ΔA=1400 mA, linear ΔA range = up to about 560 mA (40 % of maximal ΔA); the assay was calibrated against 100 % normal EDTA-plasma (at -80°C stored 500 µl aliquots of a pooled plasma of 39 healthy donors that gave informed consent). Intra- and inter-assay coefficients of variation were determined by 10fold determinations of a pathological pool – standard (patho-standard; from n=38 unselected patients). This 126 % of normal standard had a specific absorbance increase of 96 mA/2 h 37°C. Lyophilized control plasma (control plasma N®, DadeBehring, Marburg, Germany) supplemented with 3.2 mg/ml EDTA (approximate concentration of EDTA in plasma derived from EDTA-blood) had also a thrombin activity of 126 % of normal.

2.3. Plasmatic Thrombin Activity in Frozen/Thawed Plasma

50 µl unselected EDTA-plasma of n=94 patients (stored for 3 h at RT) was added to 50 µl 2.5 M arginine, pH 8.7, in U-well microtiter plates in duplicate. The plates were incubated for 0 h or 2 h at RT or they were frozen at -24°C and thawed at RT. 10 µl 3.85 mM CS-IIa in H_2O was added, the plates were shaked strongly for 10 s, and ΔA/h (37°C) was determined at 405 nm, using an incubation time of 4 h (37°C). Alternatively, the arginine was added after freezing/thawing.

2.4. Stabilization of Plasmatic Thrombin Activity by Arginine Against Storage

2.6 ml pooled normal citrated plasma (n=25 patients; 500 µl aliquots frozen at -80°C, thawed for 2 min at 37°C) was supplemented with 3.2 mg/ml K_3 - EDTA and with 0 mIU/ml or 150 mIU/ml thrombin (DadeBehring). After 30 min RT 0 IU/ml or 1 IU/ml (final) heparin (Sarstedt) were added (in 26 µl vehicle = physiol. NaCl). Pooled normal citrated plasma was also supplemented with 0 mIU/ml or 300 mIU/ml thrombin. 100-300 mIU/ml thrombin is the approximate thrombin activity, when normal citrated plasma is recalcified and begins to clot [3]. Within 30 min or after 24h at RT, 37°C, or 6°C or after freezing at -24°C and thawing at RT 50 µl samples thereof and of 10 mIU/ml fresh bovine thrombin standard in 7 % human albumin [4] were incubated in duplicate with 50 µl 0-2.3 M arginine, pH 8.6 containing 0.91 mM CS-IIa (CS-IIa in arginine: 10 µmoles CS-IIa dissolved in 1 ml H_2O, then addition of 10 ml 0-2.5 M arginine). The linear ΔA/h (37°C) was determined at 405 nm.

2.5. Stabilization of Individual Plasmas by Different Concentrations of Arginine

50 µl samples of unselected EDTA-plasmas (n=23; stored for 3h at RT) were analyzed in duplicate for plasmatic thrombin activity by addition of 50 µl 0.77 mM CS-IIa in 2.3 M arginine, pH 8.6, and determination of ΔA/h (37°C) (final incubation time = 3h).

These unselected EDTA-plasmas were also supplemented with 10 µl 313 mM arginine, pH 8.7, 625 mM arginine pH 8.7, 1250 mM arginine pH 8.7, 2.5 M arginine, pH 8.6, or with 50 µl 2.2 M arginine, pH 8.6, or was not supplemented with arginine, resulting in final plasmatic arginine concentrations of 52 mM, 104 mM, 208 mM, 417 mM, 1100 mM, or 0 mM, respectively. The samples were stored for 24h at RT and then analyzed for plasmatic thrombin activity [% of normal]. For control, plasma samples without arginine or with 1100 mM arginine were analyzed after 2h at RT.

2.6. Statistics

The intra-assay coefficients of variation (CV), as defined as standard deviation divided by mean value and multiplied by 100 %, of the amidolytic thrombin activities were < 4 %.

The data points were compared to the respective controls and tested for significance ($p< 0.05$), using the non-categorical to categorical transformed two-fold Yates-corrected chi-square comparison against the control-mean ($X^2\bar{x}$ Test) [5].

3. Results and Discussion

The normal range of plasmatic amidolytic thrombin activity is 100 ± 20 % of normal [2,13], the overall patient average measured here was 131 ± 40 % of normal (mean value

(MV=x̄) ± 1 standard deviation (SD)). Figure 1 demonstrates nearly Gaussian distribution around the 100 % of normal level and around the 130 % of normal level. The thrombin activity in unselected patient samples analyzed by addition of 50 µl of the sample to 50 µl 2.5 M arginine, pH 8.6 (138 ± 35 % of normal) correlated with r=0.980 with the thrombin activity in these samples added to arginine, frozen at -20°C, thawed at RT, and analyzed (136 ± 35 % of normal) (Figure 2). The plasmatic amidolytic thrombin activity in freshly collected unselected EDTA-samples with and without arginine supplementation was identical, demonstrating absence of any artefactual effect introduced by arginine supplementation.

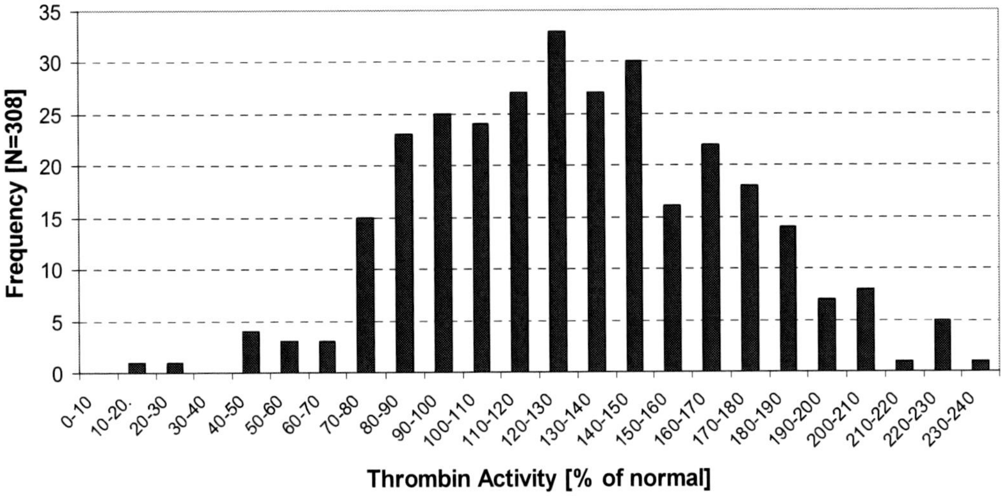

Figure 1. Distribution of thrombin (IIa) activity in unselected EDTA-plasmas of patients. After routine analysis remaining samples of EDTA plasmas of n=308 unselected patients were analyzed for thrombin activity. 50 µl sample was added to 50 µl 2.3 M arginine, pH 8.6, containing 0.77 mM chromogenic thrombin substrate HD-CHG-Ala-Arg-pNA. The ΔA at 405 nm/2h (37°C) was determined and standardized by a 100 % of normal pooled EDTA-plasma.

The correlation between samples stabilized with arginine and samples without arginine stabilization (137 ± 34 % of normal) prior to freezing/thawing was only r=0.835 (Figure 3). Without stabilization, in some of the EDTA-plasmas a higher thrombin activity and in some a lower thrombin activity was generated upon freezing/thawing, i.e. additional plasmatic thrombin activity could be generated or plasmatic thrombin activity could be inactivated, e.g. by protease mediated destruction of the closed α_2-macroglobulin cage.

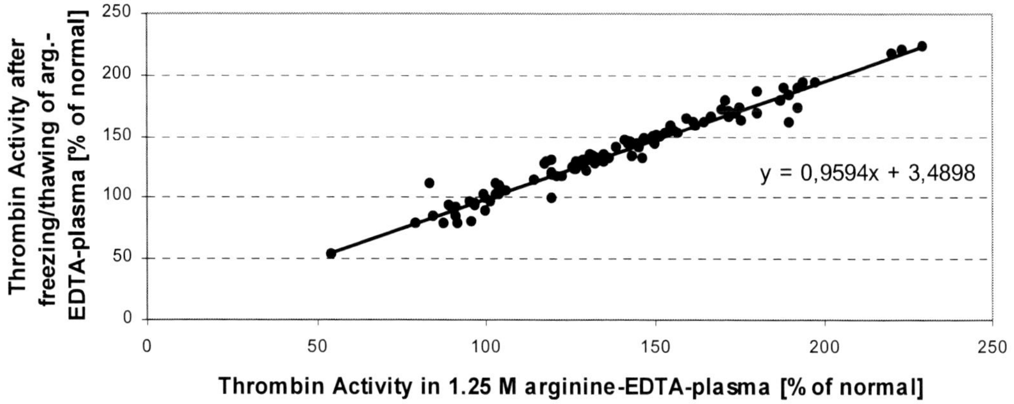

Figure 2. Comparison of unfrozen arginine stabilized EDTA-plasmas with frozen ones. 50 µl of EDTA-plasma of n=94 unselected patients was added to 50 µl 2.5 M arginine, pH 8.6 in a U-well microtiterplate. The plate was analyzed directly for thrombin activity or it was frozen at -24°C, thawed at RT, and analyzed: 10 µl 3.85 mM CS-IIa in H_2O was added and after strongly shaking for 10 s the ΔA at 405 nm was determined.

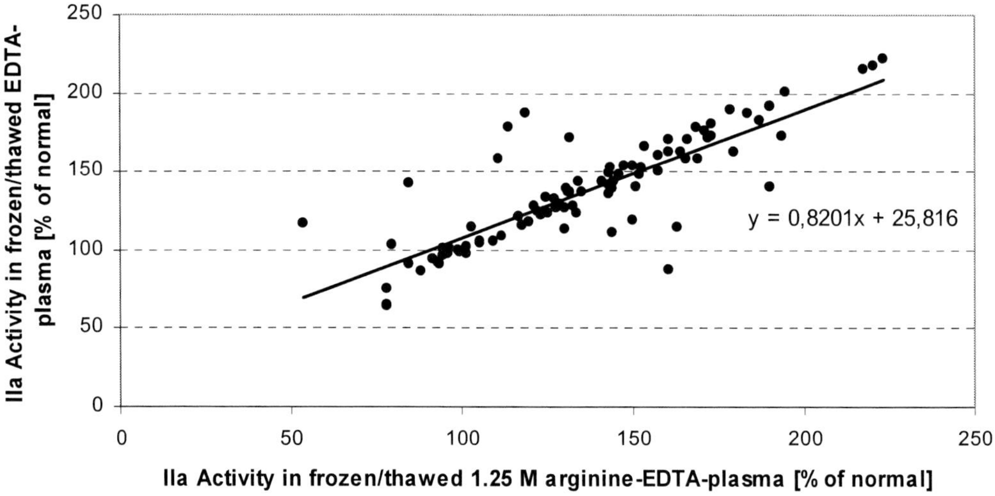

Figure 3. Comparison of frozen arginine stabilized plasmas to frozen unstabilized plasmas. 50 µl of EDTA-plasma of n=94 unselected patients was added to 50 µl 2.5 M arginine, pH 8.6, in a U-well microtiterplate. The plate was frozen at -24°C, thawed at RT and analyzed for thrombin activity. The results were compared to frozen not-arginine supplemented plasma, i.e. the arginine was added after freezing/thawing directly before the 10 µl 3.85 mM CS-IIa. The plate was analyzed directly for thrombin activity or it was frozen at -24°C, thawed at RT, and analyzed: 10 µl 3.85 mM CS-IIa in H_2O was added and after strongly shaking for 10 s the ΔA at 405 nm was determined.

Immediate analysis (138 ± 35 % of normal) correlated to analysis after 2h RT (140 ± 34 % of normal) with r=0.965 (Figure 4).

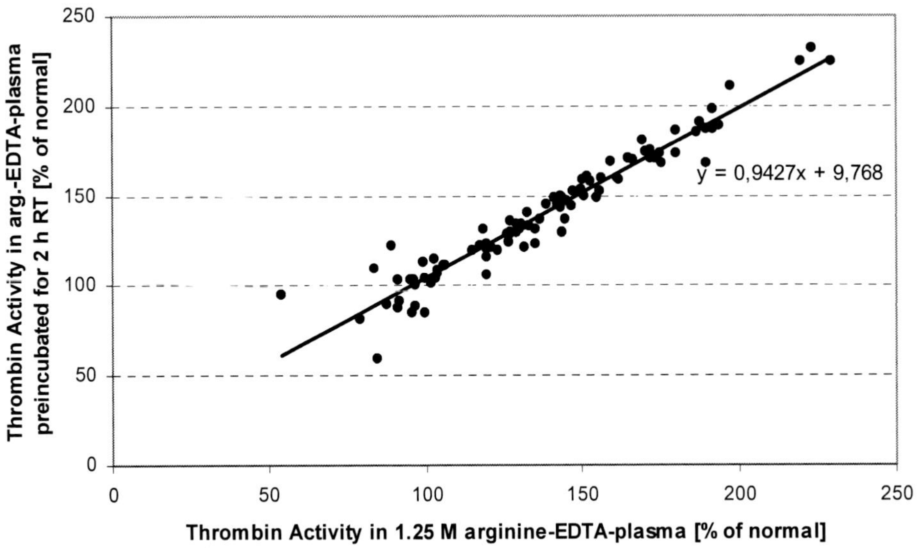

Figure 4. Comparison of arginine stabilized EDTA plasmas without further preincubation to the same ones with a further 2h RT preincubation. 50 µl of EDTA-plasma of n=94 unselected patients was added to 50 µl 2.5 M arginine, pH 8.6, in U-well microtiterplates. The samples were analyzed immediately or after 2h RT for thrombin activity.

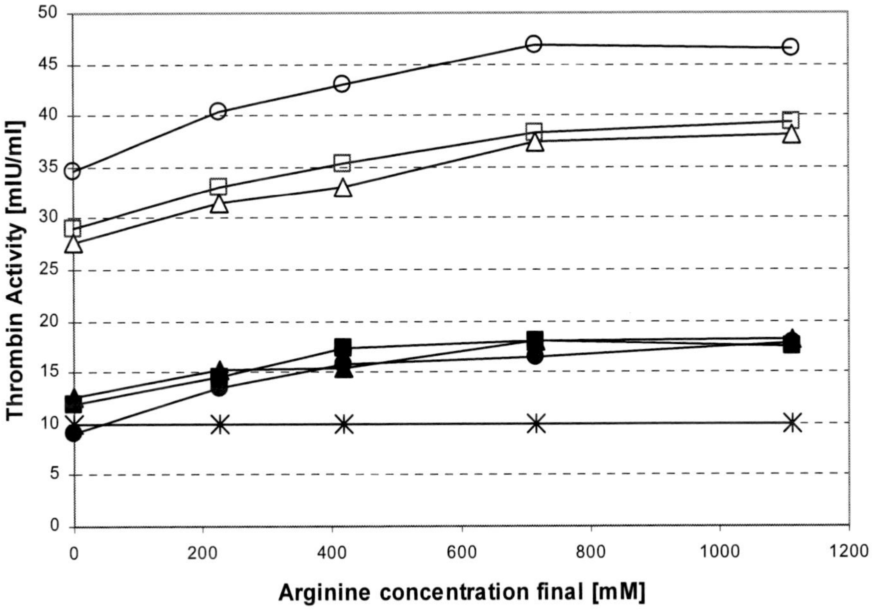

Figure 5. Thrombin activity in plasma in dependence of arginine concentration. Pooled normal citrated plasma, supplemented with 3.2 mg/ml EDTA, was supplemented with 0 mIU/ml IIa (■,▲) or 150 mIU/ml IIa (□,Δ). After 30 min RT 0 IU/ml (squares) or 1 IU/ml heparin (triangles) was added. Pooled normal citrated plasma was also supplemented with 0 mIU/ml (●) or 300 mIU/ml IIa (O). 50 µl samples thereof and of 10 mIU/ml fresh bovine thrombin standard in 7 % human albumin (*) were analyzed for thrombin activity in presence of increasing final concentrations of arginine.

Figure 5 demonstrates that with increasing final arginine concentrations there appeared an up to 50 % increase in basal plasmatic thrombin activity, i.e. more thrombin interacted with its chromogenic substrate by arginine addition (since arginine decreases also the amidolytic activity of thrombin with an approximate 50 % inhibitory concentration of about 0.8 M, the test results are corrected by calibrating against a 10 mIU/ml thrombin standard in 7 % human albumin [2]). This might be due to a reversal of the action of antithrombin I = fibrin, that might have entrapped some thrombin molecules [6,7] or by increased amidolytic thrombin activity within the closed α_2-macroglobulin cage. About 10 % of the added thrombin could still be detected 30 min RT after addition to plasma, i.e. only 90 % of added thrombin was inactivated. Addition of heparin at this time point did not result into a further decrease of thrombin activity. This figure shows also that EDTA plasma is a better stabilizer for thrombin activity than citrate: without arginine addition about 20 % less thrombin activity was detected in citrated plasma without EDTA than in one with EDTA.

If the samples of figure 5 were incubated for 24h at RT and then the thrombin assay was performed, about 50 % of the amidolytic thrombin activity generated by thrombin supplementation was inactivated within 24h at RT (Figure 6). The basal plasmatic amidolytic thrombin activity did not change. Figure 7 shows that an incubation for 24h at 37°C prior to analysis results into an about 75 % decrease of plasmatic amidolytic thrombin activity in the thrombin supplemented samples. The basal thrombin activity decreased only by about 20 % at arginine conc. > 700 mM. Figure 8 shows that a pre-incubation of the samples for 24h at 6°C resulted into an approximately 50 % decrease of thrombin activity in the thrombin supplemented samples, comparable to 24h at RT. The basal thrombin activity did not change. Freezing at -20°C/thawing at RT (Figure 9) resulted into similar results as in 24h RT (Figure 5) or 6°C (Figure 8).

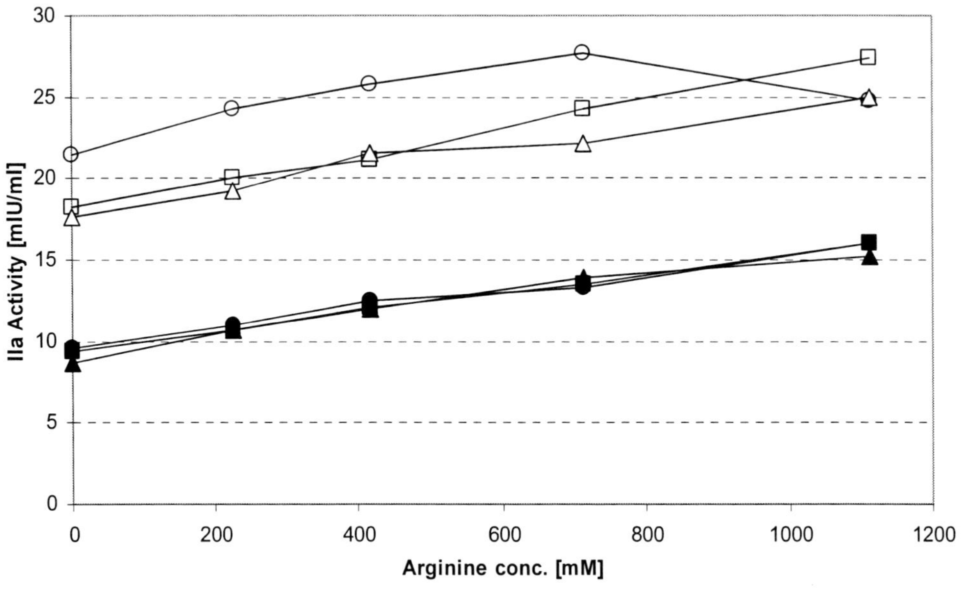

Figure 6. Thrombin activity in plasma after 24h RT in dependence of arginine concentration. The samples of figure 5 were stored for 24h at RT and analyzed for thrombin activity.

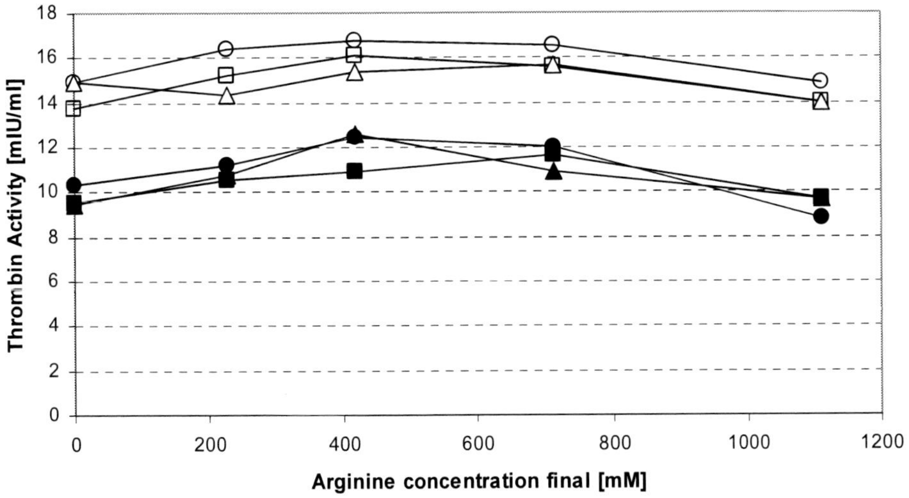

Figure 7. Thrombin activity in plasma after 24h 37°C in dependence of arginine concentration. The samples of figure 5 were stored for 24h at 37°C and analyzed for thrombin activity.

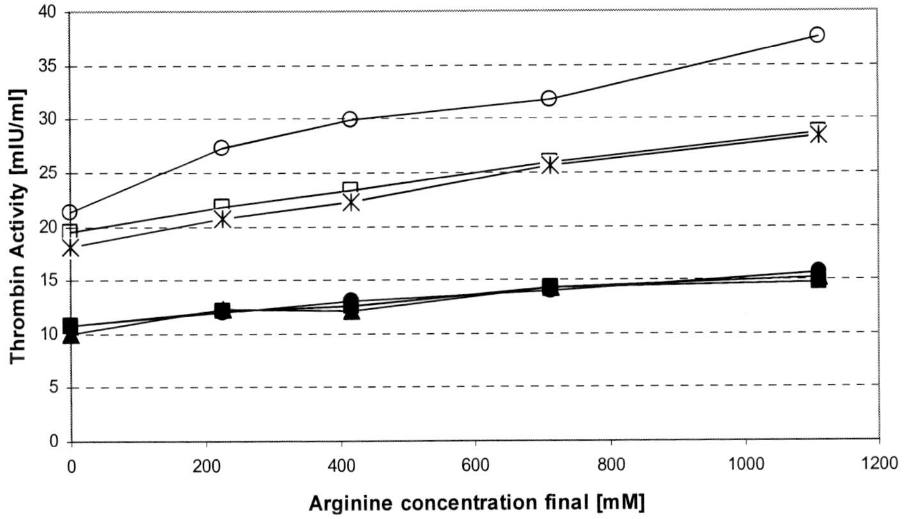

Figure 8. Thrombin activity in plasma after 24h 6°C in dependence of arginine concentration. The samples of figure 5 were stored for 24h at 6°C and analyzed for thrombin activity.

Figures 5-9 demonstrate, that there was nearly no difference between 0 IU or 1 IU heparin/ml, i.e. the assay results were not disturbed by antithrombin III. This lack of influence of heparin on the amidolytic activity of thrombin may indicate that the thrombin was complexed by α_2-macroglobulin and was therefore protected from inhibition by antithrombin. Figure 10 shows that both procedures either pipetting arginine into plasma or pipetting plasma into arginine correlate very well with r=0.982. In the following figures the basal IIa activity refers to the values of 1.1 M arginine stabilized EDTA-plasmas.

Figure 11 shows that an incubation of 24h at RT of the unsupplemented EDTA-samples resulted into a rather low correlation (r=0.824), i.e. EDTA plasma should be stabilized by arginine. Figure 12a demonstrates that the addition of just 52 mM (final plasma concentration) arginine to the EDTA-samples increased substantially their stability, the correlation factor increased from r=0.824 (Figure 11) to r= 0.933. Higher concentrations of arginine improved the stability of thrombin activity up to a correlation factor r=0.962 (Figs. 12b,c,d,e).

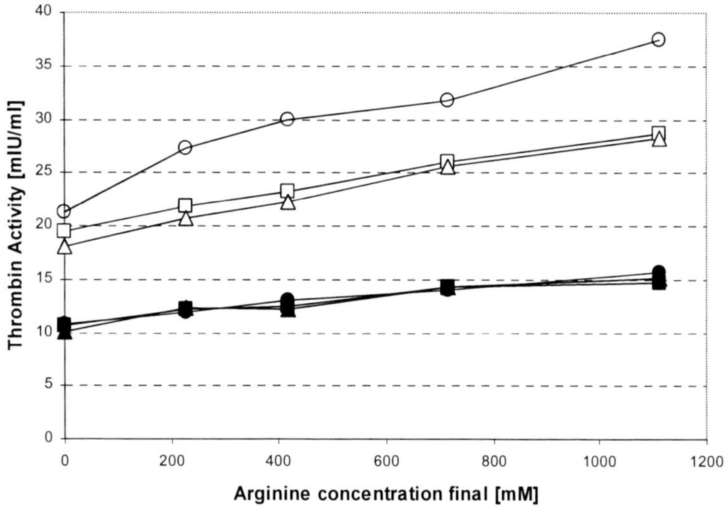

Figure 9. IIa activity in plasma after freezing/thawing in dependence of arginine concentration. The samples of figure 5 were frozen at -24°C, thawed at RT (23°C), and analyzed for thrombin activity.

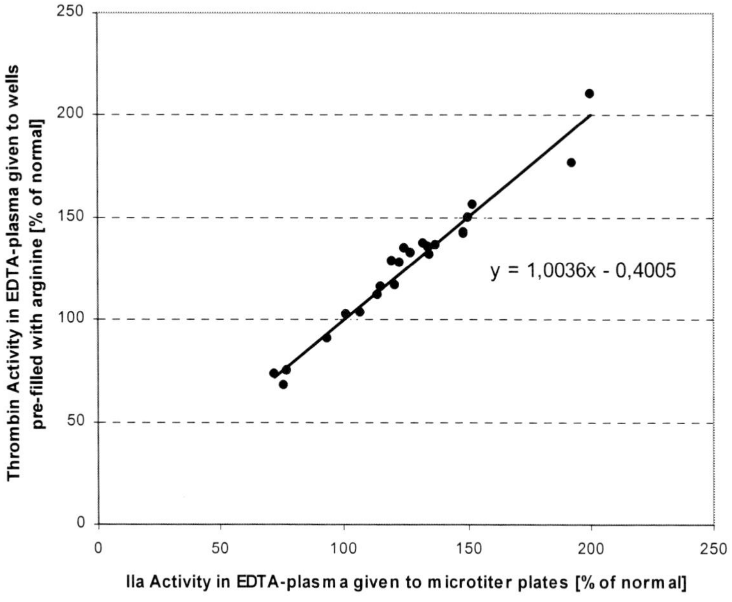

Figure 10. Comparison of plasma+arginine samples to arginine+plasma samples. 50 µl unselected EDTA-plasmas (n=23) were analyzed after 2h RT for plasmatic thrombin activity by addition of 50 µl 0.77 mM CS-IIa in 2.3 M arginine, pH 8.6 and determination of ΔA/3h (37°C). The results were compared with the thrombin activities of the same plasmas stabilized with 1.1 M arginine (50 µl plasma to 50 µl arginine).

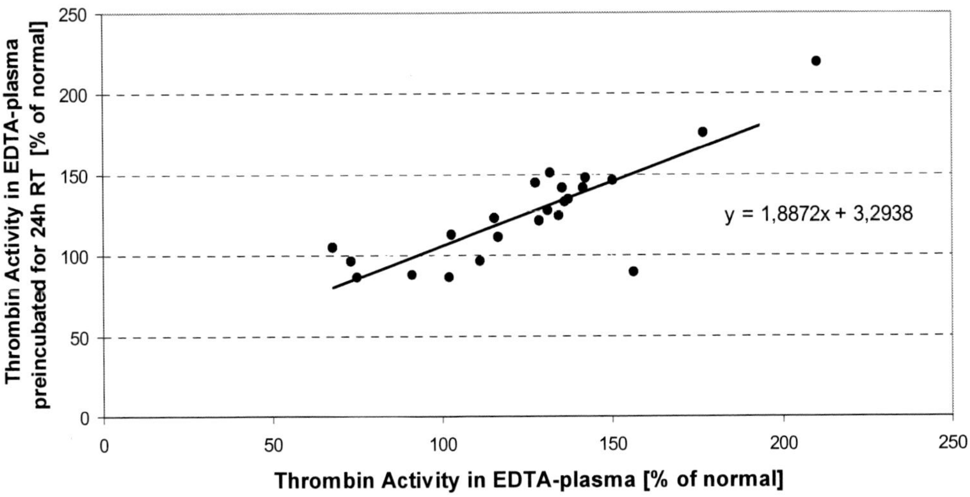

Figure 11. Comparison of thrombin activities after 0h or 24h (RT) plasma storage. 50 μl unselected EDTA-plasmas (n=23; about 3 h RT old) were immediately analyzed for plasmatic thrombin activity by addition of 50 μl 0.77 mM CS-IIa in 2.3 M arginine, pH 8.6 and determination of ΔA/3h (37°C). The results were compared to the thrombin activities of the same plasmas stored for 24h at RT.

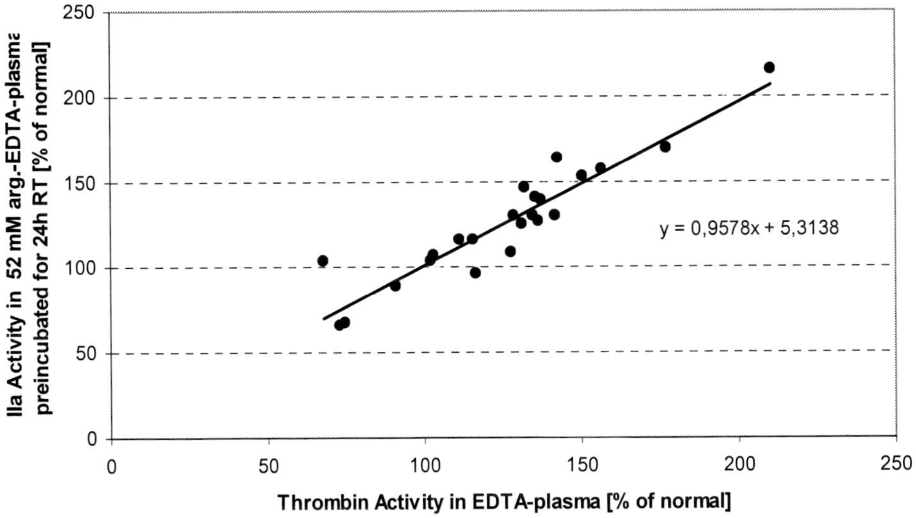

Figure 12a. Comparison of thrombin activities after 0h plasma storage or 24h (RT) storage of 52 mM arginine supplemented plasmas. 50 μl unselected EDTA-plasmas (n=23; about 3 h RT old) were immediately analyzed for plasmatic thrombin activity by addition of 50 μl 0.77 mM CS-IIa in 2.3 M arginine, pH 8.6 and determination of ΔA/3h (37°C). The results were compared to the thrombin activities of the same plasmas supplemented with 52 mM arginine and stored for 24h at RT. The correlation factor was r=0.933.

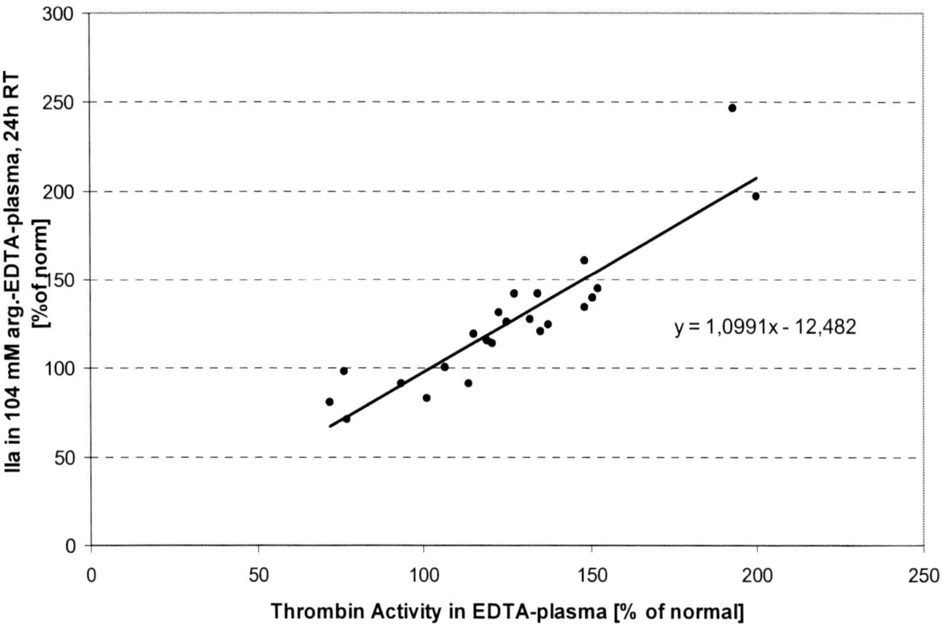

Figure 12b. Comparison of thrombin activities after 0h plasma storage or 24h (RT) storage of 104 mM arginine supplemented plasmas. 50 µl unselected EDTA-plasmas (n=23; about 3 h RT old) were immediately analyzed for plasmatic thrombin activity by addition of 50 µl 0.77 mM CS-IIa in 2.3 M arginine, pH 8.6 and determination of ΔA/3h (37°C). The results were compared to the thrombin activities of the same plasmas supplemented with 104 mM arginine and stored for 24h at RT. The correlation factor was r=0.913.

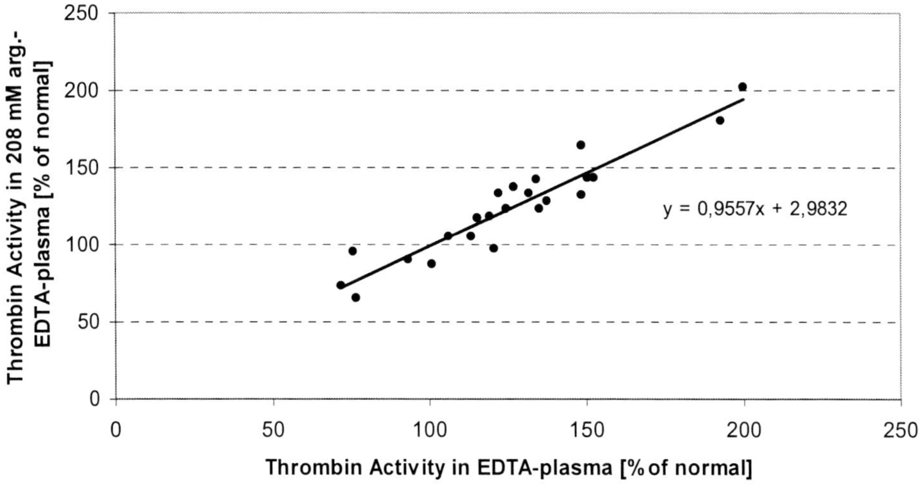

Figure 12c. Comparison of thrombin activities after 0h plasma storage or 24h (RT) storage of 208 mM arginine supplemented plasmas. 50 µl unselected EDTA-plasmas (n=23; about 3 h RT old) were immediately analyzed for plasmatic thrombin activity by addition of 50 µl 0.77 mM CS-IIa in 2.3 M arginine, pH 8.6 and determination of ΔA/3h (37°C). The results were compared to the thrombin activities of the same plasmas supplemented with 208 mM arginine and stored for 24h at RT. The correlation factor was r=0.946.

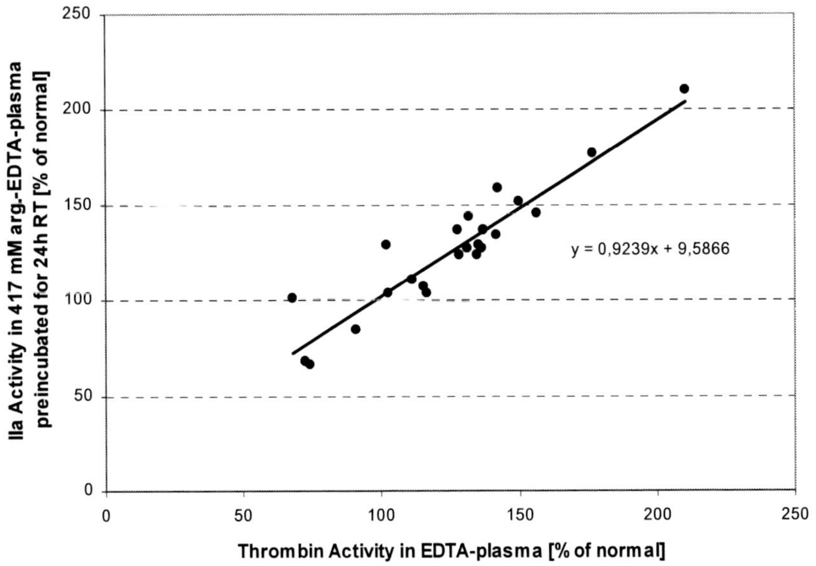

Figure 12d. Comparison of thrombin activities after 0h plasma storage or 24h (RT) storage of 417 mM arginine supplemented plasmas. 50 µl unselected EDTA-plasmas (n=23; about 3 h RT old) were immediately analyzed for plasmatic thrombin activity by addition of 50 µl 0.77 mM CS-IIa in 2.3 M arginine, pH 8.6 and determination of ΔA/3h (37°C). The results were compared to the thrombin activities of the same plasmas supplemented with 417 mM arginine and stored for 24h at RT. The correlation factor was r=0.934.

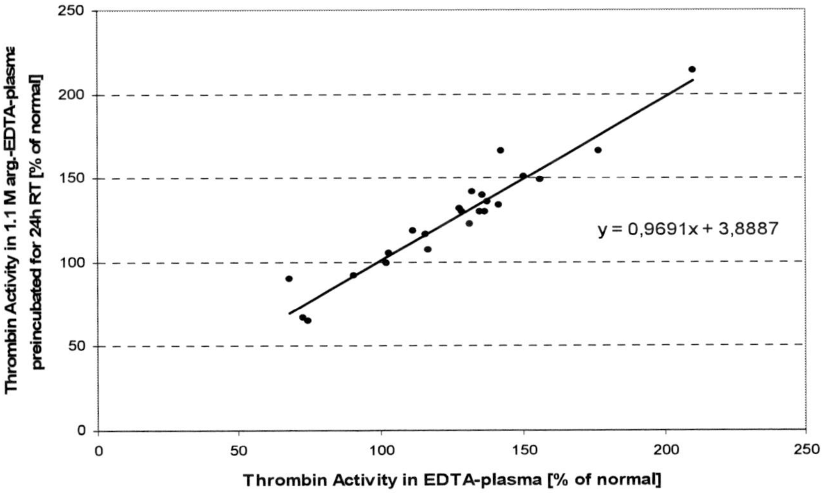

Figure 12e. Comparison of thrombin activities after 0h plasma storage or 24h (RT) storage of 1100 mM arginine supplemented plasmas. 50 µl unselected EDTA-plasmas (n=23; about 3 h RT old) were immediately analyzed for plasmatic thrombin activity by addition of 50 µl 0.77 mM CS-IIa in 2.3 M arginine, pH 8.6 and determination of ΔA/3h (37°C). The results were compared to the thrombin activities of the same plasmas supplemented with 1100 mM arginine and stored for 24h at RT; both procedures correlated with r=0.962.

Conclusions

The present study demonstrates that it is possible to conserve the hemostasis activation state of plasma, even when the sample has to be frozen. The present work contains pre-analytical information, which is derived from thorough studies: plasmatic amidolytic thrombin activity is an interesting new parameter to measure the normal and the pathologic degree of hemostasis activation. Addition of supra-molar concentrations of arginine directly to EDTA-blood might result in severe hemolysis. Therefore, it is suggested, that EDTA-blood is centrifuged as soon as possible and the resulting EDTA-plasma is stabilized by about 1-1.5 M (final conc.) arginine, pH 8.7. The present assay for amidolytic thrombin (IIa Test) is ultra-specific for thrombin (supra-molar concentrations of arginine depolymerize disturbing [6,8] uncrosslinked fibrin and abolish the amidolytic activity of contact enzymes of hemostasis such as kallikrein [9]), ultra-sensitive (lower detection limit = about 0.2 mIU/ml thrombin = 4 % of the 100 % normal plasmatic thrombin activity that is 5.5 mIU/ml), and ultra-precise (intra- and inter-assay CV values < 5 %) [2]. Thrombin activity assays such as IIa generation assays [10,11] or the basal IIa test help to diagnose a pro-thrombotic state in the individual patient [12-15].

References

[1] Stief TW, Weippert M, Kretschmer V, Renz H. Arginine inhibits hemostasis activation. *Thromb Res.* 2001; 104: 265-74

[2] Stief TW. Specific determination of plasmatic thrombin activity. *Clin Appl Thromb Hemost* 2006; 12: 324-9.

[3] Stief TW. Inhibition of thrombin generation in recalcified plasma. *Blood Coagul Fibrinolysis* 2007; 18: 751-60.

[4] Stief TW. Arginine inhibits serpins. *Clin Appl Thromb Hemost* 2007; 13: 213-8.

[5] Stief TW. The fibrinogen antigenic turbidimetric assay (FIATA). The $X^2\bar{x}$ test: the corrected chi-square comparison against the control-mean. *Clin Appl Thromb/Hemost* 2007; 13: 73-100.

[6] Mosesson MW. Update on antithrombin I (fibrin).*Thromb Haemost* 2007; 98: 105-8.

[7] Stief TW, Otto S, Renz H. The intrinsic coagulation activity assay. *Blood Coagul Fibrinolysis* 2006; 17: 369-78.

[8] Rijkers DT, Wielders SJ, Beguin S, Hemker HC. Prevention of the influence of fibrin and alpha2-macroglobulin in the continuous measurement of the thrombin potential: implications for an endpoint determination of the optical density. *Thromb Res.* 1998; 89:161-9.

[9] Stief TW. Kallikrein activates prothrombin. *Clin Appl Thromb Hemost.* 2008; 14: 97-8. .

[10] Al Dieri R, Peyvandi F, Santagostini E, Giansily M, Mannucci PM, Schved JF, Beguin S, Hemker HC. The thrombogram in rare inherited coagulation disorders: its relation to clinical bleeding. *Thromb Haemost.* 2002; 88: 576-82.

[11] Stief TW. Innovative tests of plasmatic hemostasis. Laboratory Medicine 2008; 39: 225-30.

[12] Stief TW. Coagulation Activation by Lipopolysaccharides. *Clin Appl Thromb Hemost.* 2007 Dec 26. [Epub ahead of print]

[13] Stief TW, Ijagha O, Weiste B, Herzum I, Renz H, Max M. Analysis of hemostasis alterations in sepsis. *Blood Coagul Fibrinolysis* 2007; 18: 179-86.

[14] Stief TW, Ulbricht C, Renz H, Max M. Plasmatic fibrinolysis in sepsis. *Hemostasis Laboratory* 2008; 1 (issue 1, in press).

[15] Stief TW. The laboratory diagnosis of the pre-phase of pathologic disseminated intravascular coagulation. *Hemostasis Laboratory* 2008; 1 (issue 1, in press).

INDEX

A

AAA, 192
Abies alba, 187
abnormalities, 18, 104
absorption, 37, 71, 208, 211, 214
abstinence, 172
acetate, 191, 212
acetic acid, 129, 130, 220
acetonitrile, 129, 130
acetylation, 70
acetylcholine, 6, 19, 20, 104
acid, vii, xi, 1, 7, 8, 16, 19, 20, 34, 35, 37, 38, 40, 43, 45, 46, 48, 50, 53, 58, 67, 70, 73, 74, 75, 77, 78, 79, 81, 86, 88, 89, 90, 93, 94, 95, 96, 97, 100, 104, 105, 108, 110, 119, 128, 136, 140, 156, 159, 167, 173, 174, 175, 176, 177, 178, 180, 183, 185, 192, 193, 194, 196, 197, 198, 199, 200, 201, 202, 205, 206, 207, 208, 210, 214, 216
acidic, 6, 150
acidity, 212
ACS, 120
ACTH, 37, 38, 94
actinomycetes, 200
activation, xi, xii, 12, 37, 38, 40, 49, 69, 71, 76, 82, 84, 86, 87, 94, 95, 105, 108, 119, 167, 177, 180, 219, 220, 232
activation state, xii, 220, 232
activators, 96
acupuncture, 98
acute, 27, 29, 110, 113, 120, 122
acute ischemic stroke, 29
acute schizophrenia, 113, 122
acute tubular necrosis, 120
adaptation, 156, 217
administration, viii, ix, 14, 31, 32, 33, 35, 37, 38, 40, 43, 44, 45, 46, 47, 48, 49, 50, 51, 52, 54, 55, 56, 57, 58, 59, 60, 61, 63, 64, 65, 67, 68, 69, 70, 71, 72, 73, 74, 75, 76, 77, 78, 79, 80, 81, 82, 83, 84, 85, 86, 87, 88, 90, 91, 110, 111, 112, 121, 122
adolescents, 19
adsorption, 186
adult, 41, 108, 110
adults, 34
aetiology, 17
AFC, 76, 77, 78
affective disorder, 118
affective reactions, 49
agar, 157, 159
age, 2, 3, 22, 50, 131, 205, 213
ageing, 37
agent, 137
agents, 13, 16, 81, 112, 181
aggregation, 152
aggression, viii, 31, 41, 50, 51, 67, 68, 97
agonist, ix, 39, 103, 105, 107, 110, 112, 113
air, 187
alanine, ix, xi, 103, 108, 111, 118, 119, 178, 183, 184, 185, 192, 206, 207, 209, 210
albumin, 208, 221, 222, 225, 226
alcohol, 140
alfalfa, 213
alkali, 96
alkaline, 150
allosteric, 40
alpha, vii, 1, 34, 35, 72, 73, 94, 96
Alps, 184, 216
ALT, 174
alternative, 50, 112, 140
alters, 104
alveolar macrophage, 180

alveolar macrophages, 180
Alzheimer, 9, 26, 140, 152
amelioration, 112
amendments, 215
amines, 33, 126, 128
aminoaciduria, 120
aminopeptidase, 72
ammonia, x, 106, 107, 117, 171, 173, 175, 176, 178, 179, 180
ammonium, xi, 183, 191, 197, 205, 212, 214, 217
ammonium sulphate, 214
AMPA, 69, 106, 107, 108
amphetamine, 105, 115
amygdala, 70
amyloid, 140
analgesia, 98
analgesic, viii, 31, 43, 51, 67, 69, 90
analysis of variance, 192
aneurysm, 2, 3, 19, 21
angina, 13, 24
angiogenesis, 16
angiography, 3
animal communication, 51
animal models, 112
animal studies, 4
animal tissues, 117
animals, viii, 10, 15, 31, 34, 38, 41, 47, 48, 53, 58, 62, 63, 69, 70, 73, 74, 78, 79, 84, 86, 87, 97
anion, 179
annealing, 141
anorexia, 40
ANOVA, 192
antagonism, 95
antagonist, 32, 38, 40, 68, 99, 108, 111, 115
antagonists, 104, 112, 114
anterior cingulate cortex, 106
anthropogenic, 215
antibacterial, 73
antibiotic, 121
antibiotics, 157
antibody, 71, 74, 75, 76, 77, 78, 86, 87, 93
anticancer, 181
anticonvulsant, 32, 38, 39, 40, 41, 90, 93
antigen, 71, 74, 78, 86, 87, 88, 177, 179
anti-inflammatory, 80, 87, 96
antioxidants, 179
antipsychotic, 109, 110, 112, 121
antipsychotic drugs, 109
antipsychotics, 119, 120, 122
antithrombin, 226, 227, 232
antitumor, 73, 96
antiviral, 72
anxiety, 37, 38, 70, 96, 99
anxiolytic, 37, 38, 41, 87, 90
apoptosis, 37, 89, 91, 93, 179
apoptotic, 37
application, 6, 56, 73, 88, 134, 135, 136, 188
aqueous solution, 131
arachidonic acid, 20, 27
arginine, v, vii, ix, x, xi, 1, 2, 5, 6, 7, 8, 9, 10, 11, 12, 13, 14, 15, 16, 19, 20, 21, 22, 23, 24, 25, 26, 27, 28, 33, 36, 37, 38, 41, 72, 73, 89, 95, 96, 99, 123, 124, 126, 127, 128, 129, 131, 132, 133, 134, 135, 136, 137, 171, 173, 176, 180, 183, 184, 185, 186, 189, 192, 193, 196, 197, 202, 203, 204, 206, 207, 210, 213, 219, 220, 221, 222, 223, 224, 225, 226, 227, 228, 229, 230, 231, 232
arousal, 105
arteries, 3, 4, 6, 9, 10, 11, 14, 15, 27
arteriography, 3
arterioles, 17
artery, 3, 20
aspartate, 33, 36, 116, 119, 185, 209
assessment, 62, 135
assimilation, 202, 208, 211
associations, 217
astrocyte, 178
astrocytes, 106, 107, 108, 111, 116, 117, 180
astroglial, 117
Asymmetric dimethylarginine (ADMA), 17, 134
asymptomatic, 26
ataxia, 108
atherosclerosis, 7, 9, 24, 26
atmospheric deposition, 191, 201
atoms, 6
ATP, 41, 107
attacks, 50, 51, 68
attention, 36, 53, 156
atypical, 112
autocrine, 108
autoimmune, 72
autoimmunity, 98
autonomic, 96
autopsy, 24
availability, vii, xi, 2, 5, 6, 10, 13, 14, 15, 110, 112, 113, 183, 185, 186, 189, 202, 209, 212, 213, 215
avoidance, 45, 47, 48, 67, 90
avoidance behavior, 45, 48, 67
axon, 104

B

B vitamins, 15, 27
bacteria, 140, 149, 152, 156, 166, 167, 168, 211
bacterial, 34, 140, 156, 168, 186
bacterial strains, 156
bacterium, x, 150, 153, 155, 156, 167, 168
barrier, 94, 175, 178
basilar artery, 16, 19, 21, 22, 28
basophils, 97
battery, 121
BCG, 71, 86
beef, 100
behavior, viii, 31, 33, 35, 37, 41, 46, 48, 50, 51, 52, 53, 54, 58, 60, 62, 66, 67, 68, 69, 70, 90, 95, 97, 98, 104, 144
beneficial effect, 13, 15
benzene, 72
benzodiazepine, 32, 38, 39, 70, 93, 95
benzodiazepines, 39, 68
bilirubin, 5, 11, 12, 17, 22, 23
binding, 17, 32, 39, 40, 87, 89, 93, 95, 186
bindings, 40
bioanalytical, 133, 136
bioavailability, 111
biochemical, 14, 36, 176
biochemistry, 176, 216
biodegradation, 214
biological, viii, 13, 15, 24, 25, 31, 33, 41, 67, 68, 76, 90, 95, 124, 128, 133, 134, 135, 136, 200, 212
biological activity, 24, 76, 124, 212
biological systems, 13
biologically, 33, 36
biology, 28, 92, 93, 96, 97, 98, 99, 210, 211
biomarker, 132
biomarkers, 23
biomass, 184, 185, 186, 187, 205, 209, 212, 214
biomedical, 100
biosynthesis, 99, 100, 116, 153, 156, 167
biotechnological, 156, 163, 166
bipolar, 114, 118
bipolar disorder, 114
black, 51, 52, 66, 83, 215
bleeding, 3, 37, 232
blocks, 99
blood, x, 2, 4, 6, 9, 11, 13, 16, 18, 20, 22, 23, 25, 35, 37, 38, 40, 68, 73, 81, 97, 109, 110, 112, 171, 172, 173, 175, 177, 179, 180, 181, 220, 221, 232
blood clot, 4, 11, 13
blood flow, 9, 20, 22
blood plasma, x, 171, 172, 173, 175
blood pressure, 9, 22, 25
blood vessels, 11, 18, 22, 109
blood-brain barrier, 13, 110, 112, 177
body fluid, 34
body weight, vii, 40, 90, 91
bolus, 58
borate, 126, 128, 129, 130, 131
boreal forest, 213
bovine, 23, 27, 39, 40, 95, 137, 178, 180, 221, 222, 225
bradykinin, 20
brain, viii, ix, 2, 5, 9, 10, 12, 18, 19, 27, 31, 32, 33, 34, 35, 36, 37, 39, 40, 41, 67, 68, 70, 86, 87, 89, 90, 92, 93, 94, 95, 96, 97, 98, 99, 103, 104, 106, 107, 108, 109, 110, 111, 113, 114, 115, 116, 117, 118, 119, 120, 137, 175, 178
brain activity, viii, 31, 32, 36, 37, 41, 70, 89, 90, 98
brain development, 104
brain functions, 33, 89
brain stem, 35, 36, 117
brain structure, ix, 32, 36, 68, 70
brainstem, 108
brainstem nuclei, 108
broilers, 38
bubble, 157
budding, 99
buffer, 108, 126, 128, 129, 130, 131, 142, 147, 158, 172
building blocks, vii

C

Ca^{++}, 173, 174
Ca^{2+}, 29, 35
cadmium, 179
calcitonin, 69
calcium, 177, 214
calibration, 221
calorimetry, 167
cancer, 181
candidates, 140
capacity, 33, 35, 71, 156, 190, 212
capillary, 131, 132, 135, 136, 137
CAR, 94
carbohydrate, 168
carbohydrate metabolism, 168
carbohydrates, 157, 205, 213

carbon, vii, x, 1, 11, 106, 139, 141, 142, 145, 146, 150, 151, 157, 159, 167, 184, 190, 196, 209, 210, 211, 213, 214, 215, 217
carbon dioxide, 214, 215
carbon monoxide, 11
carboxylic, 100, 110, 140, 156, 167
carcinoma, 99, 181
cardiovascular, vii, viii, ix, 1, 2, 7, 17, 22, 25, 28, 123, 125, 132, 133
cardiovascular disease, vii, 1, 7, 132
cardiovascular risk, ix, 17, 22, 25, 28, 123, 125
case study, 216
casein, 34, 208
catabolic, 95
catabolism, 24, 25, 33, 34, 35, 72, 178, 179, 180
catalase, 157, 159
catalytic, 7
catecholamine, 68
catecholamines, 38, 89
cation, 128, 129, 130, 172, 190
cations, 105
cats, 40
CBF, 28
CD, 115, 116, 180
CD3, 71
CD3+, 71
CD4, 71, 72
CD8+, 71, 72, 94
CEC, 190
cell, x, 4, 36, 37, 71, 72, 73, 78, 79, 83, 108, 110, 117, 125, 128, 140, 142, 147, 148, 150, 153, 156, 158, 175, 186
cell culture, 72, 125, 128
cell death, 37
cell membranes, 4
cellular homeostasis, 124
central nervous system, 9, 119
cereals, 34
cerebellum, 20, 35, 109, 110, 117
cerebral aneurysm, 2, 17, 18
cerebral arteries, 2, 4, 5, 9, 20, 26, 28
cerebral blood flow, viii, 2, 9, 10, 14, 21, 22, 23, 25, 27, 28, 40
cerebral cortex, ix, 32, 35, 36, 37, 40, 68, 89, 94, 108
cerebral ischemia, 3, 16
cerebrospinal fluid, ix, 4, 16, 17, 19, 21, 23, 25, 28, 103, 106, 115, 119, 127
cerebrovascular, 5, 13, 20
cerebrovascular disease, 13
cervical, 78
channels, 29, 167
chemical, vii, 20, 23, 73, 124, 126, 140, 172, 173, 174, 209
chemical properties, 20
chemical structures, 124, 126
chemicals, 143, 159
chemistry, 25, 95, 98, 100
chemotherapy, 95, 98, 99
chicken, 100
children, 19, 34, 35
chloride, 39, 40, 95
chloroquine, 179
cholesterol, 13, 17
cholesterol-lowering drugs, 13
cholinergic, 68
chromatin, 34, 36, 41, 70, 90, 99, 179
chromatography, 35, 127, 131, 133, 135, 136, 176
chromosome, 109
chronic, 28, 94, 104, 110, 115, 127, 180
chronic pain, 94
chronic renal failure, 28, 127
cigarette smoking, 4
circadian, 38
circulation, 15, 18, 19, 127
classes, 98
classified, 8, 187
cleavage, 177
clinical, 9, 11, 15, 16, 18, 24, 26, 27, 28, 37, 71, 121, 125, 133, 172, 173, 176, 220, 232
clinical diagnosis, 125
clinical trial, 121
clinical trials, 121
cloning, 116, 117
clozapine, 112, 121
CNS, 32, 38, 40, 89, 106, 108, 114, 118, 120
Co, 23, 142, 157, 159
CO2, 208, 212, 214, 215
coagulation, 232
coagulum, 172, 177, 179
coefficient of variation, 220
co-existence, x, 139
cognition, 120, 121
cognitive, 104, 109, 110, 111, 112, 115, 121, 122
cognitive deficits, 112, 113, 121, 122
cognitive disorders, 111
cognitive domains, 121
cognitive dysfunction, 105
cognitive function, 115
cognitive impairment, 104
cohort, 109

collaboration, 26
collagen, 34
colloidal particles, 186
colonization, 205, 209, 211
commercial, 184
communication, 67, 114
communities, xi, 37, 99, 184, 185, 200, 202, 208, 211, 212, 216
community, xi, 184, 187, 197, 200, 201, 202, 208, 210, 213, 214
competition, 185
compliance, 140
complications, 3, 15, 71
components, ix, 23, 48, 50, 58, 67, 69, 99, 108, 123, 150, 179, 180, 213
composition, ix, 32, 34, 36, 172, 176, 177, 178, 179, 190, 191, 196, 205, 209, 212, 214
compounds, 17, 22, 26, 128, 156, 208, 209, 220
computed tomography, 24
computer, 53, 62
concentration, x, xi, 8, 9, 12, 14, 19, 24, 34, 35, 37, 73, 126, 127, 132, 142, 145, 146, 147, 155, 156, 157, 158, 161, 163, 164, 165, 171, 173, 178, 186, 189, 191, 192, 196, 202, 206, 210, 213, 219, 221, 225, 226, 227, 228
conception, 33, 69, 91
concrete, 32, 67, 86, 90, 105
conditioning, 62, 63, 65, 66, 70, 90
conduction, 69
conductive, 4, 9
conformity, 96
congenital heart disease, 19
congestive heart failure, 27, 127, 135
Congress, 3, 92, 94, 96, 97, 98, 99, 100
coniferous, 214
connective tissue, 34
consensus, 121
consolidation, viii, 31
consumption, 97, 201
contractions, 28, 38
control, viii, 9, 10, 15, 16, 24, 31, 36, 38, 39, 42, 43, 44, 45, 46, 47, 48, 49, 50, 51, 52, 53, 54, 55, 56, 57, 58, 59, 60, 61, 62, 63, 64, 65, 66, 71, 74, 76, 77, 78, 79, 81, 82, 84, 86, 87, 89, 94, 96, 98, 106, 109, 110, 119, 127, 134, 177, 217, 221, 222, 232
control condition, 48, 49
control group, 43, 44, 45, 52, 54, 65, 66, 110
controlled, viii, 2, 13, 18, 25, 27, 37, 41, 70, 104, 112, 122, 135
convergence, 114
conversion, v, x, 106, 107, 112, 139, 140, 142, 144, 145, 146, 148, 149, 151, 208
conversion rate, 145, 146
coping, 167
copper, 186
cornea, 73
coronary angioplasty, 13
coronary heart disease, 24
correlation, xi, 25, 37, 71, 120, 219, 220, 223, 227, 229, 230, 231
correlation coefficient, 220
cortex, 34, 68, 106, 110, 119
cortical, 105, 112, 114
cortisol, 38, 89, 96, 99
Corynebacterium, 156, 168
cost-effective, 125
coupling, 9, 125
coverage, 188, 200, 205
covering, 187, 200
creatine, 106, 176, 177, 181
creatinine, x, 171, 173, 176
critical value, 50
CRP, 173, 174
cryopreservation, 178
CSF, ix, 4, 6, 7, 10, 11, 12, 13, 14, 16, 28, 103, 106, 110, 119
C-terminal, 89, 95
cultivation, 37, 153
culture, 12, 17, 37, 93, 95, 142, 212
cycles, 141
cycling, xi, 175, 183, 184, 186, 214, 215
cysteine, vii, ix, 1, 5, 103, 108, 111, 112, 118, 178, 179, 186, 215
cystine, xi, 183, 184, 186, 189, 192, 193, 196, 197, 200, 201, 202, 203, 204, 206, 207, 208, 210, 215
cystinuria, 93
cytokine, 99
cytokines, 89, 94
cytoplasm, x, 7, 140, 142, 150, 186
cytoprotective, 28
cytosol, 127
cytosolic, 7
cytotoxic, 19, 28, 73, 100
cytotoxicity, 27, 72, 117
Czech Republic, 183, 187, 210

D

D-amino acids, 92, 97, 108, 136
D-amino-acid, 117, 118

danger, 3
database, 157
de novo, 156, 178
death, 3, 108, 118
decay, 175
deciduous, 214
decomposition, 185, 209
defects, 94
defense, 67, 68
deficiency, 32, 34, 37, 40, 68, 72, 97, 99, 180
deficits, 105, 111, 112
degradation, 5, 7, 11, 16, 22, 32, 34, 35, 108, 110, 111, 120, 124, 208, 213
degradation mechanism, 5
degradation pathway, 35
degrading, 172
degree, 10, 36, 43, 48, 49, 53, 62, 73, 79, 91, 232
dehydrogenase, x, 35, 93, 171, 175, 178
delta, vii, 1, 21
demand, 9, 36
demineralized, 212
demographics, 21
denaturation, 141
denitrification, 199, 215, 217
denitrifying, 168
density, 4, 39, 142, 147, 153, 185, 187, 189, 205
dentate gyrus, 108
deoxyhemoglobin, 4, 18, 25
deposition, 201, 209, 215
depressed, 70, 88, 99
depression, 38, 70, 96
deprivation, 62
derivatives, x, 111, 126, 127, 128, 129, 130, 131, 132, 134, 135, 171
desensitization, 40, 95
desiccation, 140
desorption, 186
destruction, 223
detection, v, ix, 2, 97, 123, 125, 127, 128, 129, 131, 132, 135, 136, 215, 232
detoxification, 106, 177
developing brain, 110
dexamethasone, 99
diabetes, vii, 1, 7
diabetes mellitus, vii, 1, 7
diabetic, 127
diabetic patients, 127
diagnostic, 94
dialysis, 128, 135
diet, vii, 34, 36, 37, 38, 40, 97, 99
dietary, 34, 37, 127, 134
diets, ix, 32, 33, 34, 91, 97
differentiation, 37, 41, 69, 89, 91, 104, 117
diffusion, 157
dilation, 14
dimethylarginine, 18, 133
dipeptides, 93, 136
direct action, 86
discipline, 96
discovery, 120
diseases, ix, 41, 112, 123, 124, 125, 132
dislocation, 78
disorder, ix, 35, 103, 104, 113
disseminated intravascular coagulation, xi, 219, 220, 233
distal, vii, 1
distribution, 23, 33, 35, 62, 107, 116, 210, 223
disulfide, 177
diversity, 205
DNA, 72, 73, 109, 141, 157, 168
DNA sequencing, 141
dogs, 21
dominance, 185
donor, vii, 2, 8, 15
donors, 7, 8, 14, 25, 28, 72, 221
dopamine, 70, 99, 104, 105
dopaminergic, 115
dorsolateral prefrontal cortex, 109
dosage, 57, 61, 90
double blind study, 37
double-blind trial, 18
drinking, 38
drought, 214, 215
drug treatment, 121
drugs, 73, 112, 113, 114, 120
dry, 196, 208
drying, 140, 209
D-serine, ix, 103, 105, 106, 107, 108, 109, 110, 111, 112, 113, 116, 117, 118, 119, 120, 121, 122
DSM, 160, 165, 167
duration, 53, 54, 55, 56, 57, 58, 59, 60, 61, 62, 63, 64, 66, 72, 73, 74, 76, 79, 153, 168
dysregulation, 20

E

E. coli, x, 139, 140, 141, 142, 144, 145, 146, 147, 148, 149, 150, 151, 152
ecology, 216
ecosystem, 184

ecosystems, xi, 166, 183, 184, 185, 197, 202, 207, 208, 212, 214, 215, 216, 217
education, 113
Egypt, x, 155, 156, 157
ejaculation, 172
elaboration, 33, 41, 63, 64, 65, 66, 90
elderly, 20, 21, 22, 29, 134
electric current, 48, 50, 67, 69
electrical, 37, 40, 49, 67, 74, 90, 91
electrodes, 48, 49
electroencephalogram, 108
electrophoresis, 131, 135, 136, 137, 172
electrostatic, 186
ELISA, 125, 132, 133
email, 171
embryo, 178
emission, 128
emotional, viii, ix, 31, 32, 33, 36, 48, 49, 58, 66, 74, 76, 78, 79, 80, 83, 87, 88, 90, 93
emotional reactions, 90
emotionality, 41, 52, 53, 56, 58, 61, 70
employment, 95, 99
enantiomers, 136
encoding, 7, 117
endocrine, 88, 89
endocrine glands, 89
endocrinology, 99
endogenous, vii, viii, ix, 1, 2, 5, 6, 7, 8, 9, 10, 12, 13, 14, 16, 17, 18, 19, 21, 22, 24, 27, 28, 40, 103, 105, 107, 108, 110, 113, 116, 127, 132, 134, 135, 215
endogenous mechanisms, 14
endoplasmic reticulum, 177
endothelial, 5, 20, 22, 26, 28
endothelial cell, 4, 6, 9, 11, 16, 17, 19, 23, 24, 27
endothelial cells, 4, 9, 11, 16, 17, 19, 23, 24, 27
endothelial dysfunction, 6, 7, 11, 13, 15, 16, 17, 18, 20, 22, 23, 24, 27, 28, 134
endothelin, 16, 19, 26, 28
endothelin-1, 4, 13, 17, 24, 25, 29
endothelium, vii, 1, 6, 9, 15, 18, 19, 20, 21, 22, 24, 26, 29, 127
endotoxemia, 126, 181
end-stage renal disease, 17
energy, 26, 33, 145, 167, 186
engineering, 153
enhancement, 95, 177
environment, 33, 61, 70, 90, 104, 156, 213, 214
environmental, 104
environmental factors, 104
enzymatic, 8, 208
enzyme, ix, 8, 9, 10, 19, 22, 24, 35, 73, 94, 95, 96, 97, 100, 103, 108, 110, 116, 124, 125, 131, 140, 153, 214
enzyme immunoassay, 95
enzyme-linked immunosorbent assay, 125
enzymes, 7, 33, 35, 36, 72, 95, 167, 174, 180, 232
epidemiology, 26
epidermis, 153
epinephrine, 70
erectile dysfunction, vii, 1, 7, 21
erythrocyte, 79, 91
erythrocytes, viii, 31, 74, 78, 79, 86
Escherichia coli, v, x, 139, 140, 141, 152, 154, 156, 167, 168, 177
ESRD, 23
ester, 133
estrangement, 104
estrogen, 12
estrogen, 20
ethanol, 98, 142, 152, 158
ethylene, 220
eukaryotic, 156
euphoria, 105
evapotranspiration, 205
evidence, ix, 6, 16, 18, 36, 70, 71, 76, 84, 92, 103, 104, 109, 114, 119, 213
evoked potential, 92
evolution, 88
evolutionary, 27, 116
excision, 211
excitability, 39, 41
excitation, 69, 128
exclusion, 167
excretion, 8, 17, 25, 29, 37, 38, 99, 124, 125, 127, 153, 186
exogenous, vii, viii, 1, 2, 7, 9, 14, 215
experimental allergic encephalomyelitis, 180
experimental condition, ix, 123, 129
exposure, 42, 43, 44, 45, 46, 47, 48, 53, 67, 74, 78, 153
extracellular, 12, 26, 108, 117, 158, 166, 175
extraction, ix, 93, 123, 127, 128, 129, 130, 136, 157, 158, 191, 199, 212, 215
eye, 73

F

faecal, 37, 38, 53, 56
familial, 35, 94

family, vii, 2, 6, 7, 24, 99, 109, 168, 200
family members, 99
fatality rates, 2, 20
fatty acid, 36, 179
fatty acids, 36, 179
fax, 1, 139, 155, 217
fear, 70
fecal, 99
feedback, 6, 8
feeding, 40, 62, 66
females, 12, 131
fermentation, 140, 152, 153
fertility, 172
fertilization, 178, 181, 208, 217
fiber, 192
fibers, 172
fibrin, 172, 226, 232
fibrinogen, x, 171, 172, 177, 232
fibrinolysis, 177, 233
filters, 192
filtration, 24, 134, 157, 213
Finland, 2, 81
fish, 100
FITC, 132
fixation, 200
flow, 9, 129, 142, 158, 192, 213
flow rate, 129, 142, 158, 192
fluctuations, 202
fluid, 115, 128, 179, 180, 215
fluid extract, 215
fluorescence, ix, 123, 125, 127, 128, 130, 131, 132, 135, 136, 137
fluorogenic, 128, 135, 137
fluorometric, 192, 215
fluvoxamine, 121
focusing, viii, 2
folate, 15
folding, 156
food, viii, 31, 33, 34, 37, 38, 41, 62, 63, 64, 65, 66, 70, 90, 95
food intake, 34, 38
forbs, 185, 187, 212
forebrain, 107, 108, 109
forest ecosystem, 208, 213
Forest Service, 214
forestry, 183, 212, 214
forests, 184, 189, 210, 217
fortification, 99
free radical, vii, 1, 5, 11
free radicals, 11
freezing, xi, 140, 178, 219, 220, 221, 222, 223, 224, 228
frontal cortex, 106, 110
fructose, 159
functional analysis, 116
functional magnetic resonance imaging (MRI), 99
fungal, 208, 209, 214
fungi, 208, 209, 211, 212, 213, 214, 217
fungus, 208, 211
fusion, 175, 176

G

GABA, 32, 35, 38, 39, 40, 68, 70, 89, 90, 95, 96, 104, 106, 114
GABAergic, 93, 114
gamete, 178
Gamma, 177
ganglion, 27
gas, 125, 133, 134, 211
gas chromatograph, 125, 133, 134
Gaussian, 223
gel, 73, 136, 172
gelatin, 157, 208
gels, 172, 176
gender, 3, 12
gene, ix, x, 6, 8, 11, 16, 21, 22, 27, 36, 70, 97, 103, 108, 109, 113, 114, 117, 118, 119, 121, 139, 140, 141, 153, 155, 157
gene expression, 11, 36, 70, 108, 114, 118
generalization, 32
generation, 172, 179, 232
genes, 7, 109, 114, 117, 153, 167
genetic, 4, 16, 18, 35, 104, 109, 118, 119, 163, 166
genetic factors, 4, 109
genetics, 114, 118, 156
genital herpes, 73
genome, 36
genomic, 116
genomics, 215
genotypes, 109
germ cells, 179
Germany, 1, 131, 141, 143, 157, 171, 172, 192, 219, 221
glass, 41, 51, 52, 53, 70, 90, 192
glia, 114, 116
glial, 107, 108, 112, 116, 117, 121
glial cells, 107, 112, 117
globus, 108
glucagon, 35

glucocorticoids, 80, 87
glucose, x, 23, 139, 140, 144, 145, 146, 147, 148, 149, 150, 151, 159, 167
glucosuria, 120
glutamate, x, 33, 36, 37, 69, 89, 104, 105, 106, 107, 112, 114, 115, 116, 117, 119, 145, 153, 168, 171, 175, 178, 185, 209, 213
glutamatergic, ix, 103, 104, 109, 114, 115, 120
glutamic acid, x, xi, 114, 117, 171, 173, 175, 183, 185, 192, 196, 206
glutamine, 106, 107, 115, 116, 117, 127, 173, 175, 178, 180, 185, 209
glutathione, 175, 177
glycine, ix, x, xi, 103, 105, 106, 110, 111, 112, 113, 116, 120, 122, 171, 173, 175, 178, 183, 185, 192, 200, 215, 216
gold, 3
gold standard, 3
gonadotropin, 93
government, 217
grading, 26
gram-negative, 156, 159
grants, 113
granulocyte, 88, 91
graph, 5
grass, 202
grasses, 185, 189, 209, 212, 216
grassland, 185, 212, 214, 217
grasslands, 212, 213, 216, 217
grazing, 184
grey matter, 68, 107
groups, vii, 41, 43, 45, 46, 47, 52, 53, 54, 56, 58, 59, 62, 70, 71, 74, 75, 76, 86, 87, 192
growth, x, 34, 37, 95, 143, 155, 156, 157, 167, 189, 209, 211, 212, 213
growth inhibition, 156
growth rate, 212
growth temperature, 143
gyrus, 100

H

haemoglobin, 19
half-life, 5, 191
hallucinations, 104
haloperidol, 105, 112, 121
halophilic, x, 140, 155
handling, 37, 38, 41
haplotypes, 118
harm, 104, 215
harmony, 215
harvest, 165
head, 48, 49, 176
headache, 3
health, vii, viii, 1, 2, 98, 133
hearing, 104
heart, 16, 28, 128, 133
heat shock protein, 125
heating, 42, 140
height, 42, 51, 53
hematoma, 26
heme, vii, 1, 4, 5
hemisphere, 3
hemodialysis, 128
hemoglobin, 4, 18, 19, 23, 28
hemorrhage, 14
hemostasis, xi, xii, 219, 220, 232, 233
hemostatic, 176
hepatocyte, 72
herbal, 98, 132
herbs, 185
herpes, 73, 92, 95, 98
herpes simplex, 92, 95, 98
herpes simplex virus type 1, 98
heterogeneity, 40, 186, 190, 203, 217
heterogeneous, 25
heterotrophic, 213, 214
high-performance liquid chromatography, v, ix, 17, 123, 125, 133, 134, 135, 136
hippocampal, 116
hippocampus, 36, 40, 70, 106, 108, 110, 119
histamine, 70
histidine, xi, 173, 178, 183, 185, 192
histochemical, 36, 72
histological, 73
histone, 8, 19, 34, 36, 97
HIV, 98
homeostasis, 22, 40, 106
homocysteine, 8, 15, 26, 28
homogeneous, 172
homology, 22
horizon, xi, 183, 189, 190, 191, 193, 194, 195, 196, 197, 198, 199, 200, 201, 202, 203, 204, 205, 206, 207, 208, 209, 210, 217
hormone, 25, 37, 92, 93
hormones, 33, 92, 98
hospital, 10
host, 208, 209, 212
hostility, 105

HPLC, ix, 123, 125, 126, 128, 130, 131, 132, 137, 142, 158
HSV-1, 72, 73
human, 8, 11, 15, 16, 17, 18, 19, 21, 22, 23, 26, 32, 33, 35, 81, 91, 93, 94, 95, 97, 98, 99, 100, 106, 112, 115, 117, 118, 119, 120, 121, 124, 125, 127, 128, 131, 132, 133, 134, 135, 137, 172, 176, 177, 178, 179, 180, 181, 221, 222, 225, 226
human brain, 106, 118, 119, 120
human cognition, 115
human neutrophils, 100
humans, 4, 9, 10, 14, 22, 34, 37, 70, 104, 115, 117, 133
humic substances, 186, 212
humus, 186, 209, 216
hydration, 167
hydro, 21, 133
hydrogen, 73, 111, 157, 179, 186
hydrogen peroxide, 73, 111, 157, 179
hydrogen sulfide, 186
hydrolases, 8
hydrolysis, vii, 2, 5, 8, 10, 14, 15, 25, 117, 128, 135, 157
hydrolyzed, 34, 159
hydrophilic, 133
hydroxyl, 127, 140
hygiene, 94
hyperactivity, 109, 178
hyperalgesia, 69, 95, 96, 100
hypercholesterolemia, vii, ix, 1, 7, 11, 14, 17, 23, 26, 123, 125, 127, 134
hyperhomocysteinemia, ix, 123, 125, 127
hyperinsulinism, 178
hypersensitivity, viii, 31, 32, 33, 69, 74, 78, 79, 80, 81, 87, 88, 90, 91
hypertension, vii, ix, 1, 7, 19, 23, 24, 27, 123, 125, 126
hypothalamic, 40, 89, 99
hypothalamus, 38, 40, 68, 70, 108
hypothesis, 4, 7, 14, 25, 104, 110, 113, 114, 115, 119, 120, 189
hypoxic, 6

I

IASP, 92, 94, 96, 97, 98, 99, 100
ibuprofen, 96
id, 13
identification, 19, 21, 40, 97, 158, 167, 168
IgG, 173, 174
ileum, 38, 68
imaging, 94
immobilization, xi, 183, 191, 196, 197, 198, 199, 200, 201, 203, 204, 205, 209
immune cells, 87, 89
immune reaction, viii, 32, 87
immune response, viii, 31, 32, 33, 71, 74, 75, 76, 77, 78, 79, 80, 84, 86, 87, 88, 90, 91, 92, 93, 96, 175, 180
immune system, 32, 33, 71, 73, 74, 86, 88, 89, 90, 91, 96, 99
immunity, viii, 31, 180
immunization, 74, 86
immunodeficiency, ix, 32
immunohistochemical, 95
immunological, viii, 2, 71, 74, 94, 180
immunomodulation, 94
immunoreactivity, 6, 25, 108
immunosuppression, 78
impairments, 105
in situ, 17, 108, 216
in situ hybridization, 108
in vitro, 6, 9, 11, 13, 38, 68, 72, 97, 98, 100, 109, 135, 177, 178, 179, 180
in vivo, ix, 5, 8, 9, 12, 13, 25, 99, 108, 111, 113, 120, 123, 125, 135, 180
incidence, 2, 3, 6, 12, 26
incubation, x, 81, 139, 141, 142, 143, 144, 145, 146, 148, 149, 150, 151, 158, 159, 160, 166, 189, 191, 199, 221, 222, 226, 227
incubation period, 199
incubation time, 144, 145, 146, 150, 160, 221, 222
independence, 69
indices, 43, 45, 51, 52, 58, 62, 66, 76, 79, 80, 81, 82, 83, 84, 85, 87, 88, 90, 91
indigenous, 212
induction, 16, 72, 140, 175
industry, 34
infarction, 26
infection, 95, 98
infectious, viii, 2
infertility, 178
inflammation, 23, 68, 69, 71, 78, 79, 80, 81, 87, 96
inflammatory, 81, 87, 96
informed consent, 221
infusions, 25
inherited, 232
inhibition, vii, 2, 5, 8, 9, 10, 11, 13, 20, 39, 70, 87, 110, 111, 112, 132, 179, 227

inhibitor, vii, 1, 2, 5, 7, 8, 9, 18, 19, 20, 21, 24, 27, 28, 36, 69, 96, 110, 111, 112, 120, 122, 134, 135, 177, 181
inhibitors, vii, 1, 6, 16, 19, 24, 111, 112, 113, 120, 121, 124
inhibitory, 35, 51, 68, 69, 72, 88, 97, 106, 179, 226
inhibitory effect, 72
injection, viii, 19, 31, 35, 37, 40, 49, 69, 78, 82, 87, 91, 98, 127, 192
injections, 57, 61, 79
injury, 71, 87
innervation, 89, 97
innovation, 113
inorganic, 179, 185, 208, 211
insertion, 189
instability, 128, 134
intensity, viii, 31, 37, 48, 50, 51, 53, 67, 68, 71, 76, 78, 79, 84, 87, 88, 90, 187, 209, 212
interaction, viii, 31, 32, 38, 74, 86, 90, 91, 96, 109, 133, 167, 179
interactions, 22, 41, 67, 68, 73, 88, 89, 99
interface, 214, 216
interference, 15, 128
interferon, 71
interleukin-2, 72
international, 26, 220
interneurons, 114
interpretation, 189
interrelationships, 89
interstitial, 99
interval, 42, 48, 50, 62, 74, 106, 197
intima, 6
intracranial, viii, 2, 16, 18, 20, 24, 25, 26, 27
intracranial aneurysm, viii, 2, 16, 20, 25, 26, 27
intracranial arteries, 18
intraperitoneal, viii, 31, 38, 40, 50, 74, 90
intravenous, 10, 13, 24, 35
intravenously, 9
intrinsic, 22, 232
inventories, 211
Ion-exchange, 176
ions, 35, 40, 153, 192, 205
iron, vii, 1, 5, 186
irritation, 51, 67, 74
ischemia, 14, 16
ischemic, 3, 26, 27, 127
ischemic stroke, 3, 127
ISO, 192, 212
isoforms, 7, 8, 11, 119, 125
isoleucine, xi, 34, 183, 192
isotope, 133

K

K^+, 35, 173, 174
K-12, 167
ketamine, 104, 105, 106, 115
kidney, vii, 1, 7, 24, 34, 108, 110, 127, 128, 131, 136
kidney failure, vii, 1, 7
kidneys, 72, 125
kinase, 17, 112
kinetic parameters, 217
kinetics, 40, 216
knockout, 108, 113, 117, 118
KOH, 142, 157

L

labeling, 95, 135, 185, 202
lamella, 48
land, 184
land use, 184
larvae, 180
laser, 92, 131, 132, 136, 137
latency, 42, 62
latex, 81
lattice, 74
law, 68
LDH, 173, 174
LDL, 11, 17
leaching, 185, 189, 191, 201
lead, 10, 11, 14, 16, 105, 185, 205
leakage, 186
learning, viii, 31, 33, 36, 62, 63, 64, 65, 66, 90, 92
learning process, 33, 63
legume, 216
leucine, xi, 34, 173, 183, 185, 192
leukemia, 23
leukemic, 72, 95
leukemic cells, 72
leukocyte, 99
leukocytes, 82, 86
LIF, 131
lifetime, 104
ligand, 40
likelihood, 158, 168
limitation, 112, 211
linear, 129, 192, 221, 222
lipid, 179, 181

lipid peroxidation, 179, 181
liquefaction, 171, 172, 174, 175, 177
liquid chromatography, 133, 135, 136, 158
listeria monocytogenes, 152
literature, 89, 180
lithium, 73
liver, 25, 34, 35, 36, 37, 72, 94, 97, 108, 110, 125, 178
livestock, 187
LMW, 175
localization, 7, 27, 36, 68, 97, 116, 117
location, 53
locomotion, 58, 61
locus, 109, 118
long-term, x, 70, 112, 180, 183, 186
long-term memory, 70
loss of consciousness, 3
losses, 201
low molecular weight, 174
low-density, 11
low-density lipoprotein, 11
LPS, 181
LSD, 192
Lycopersicon esculentum, 185
lymph node, 72
lymphocyte, 180
lymphocytes, 71, 72, 74, 75, 76, 77, 78, 176, 180
lymphoid, 37, 72, 89, 93, 95
lymphoid cells, 95
lymphoid follicles, 72
lymphoid organs, 89
lymphoid tissue, 93
lysine, viii, ix, x, xi, 31, 32, 33, 34, 35, 36, 37, 38, 39, 40, 41, 42, 43, 44, 45, 46, 47, 48, 49, 50, 51, 52, 53, 54, 55, 56, 57, 58, 59, 60, 61, 62, 63, 64, 65, 66, 67, 68, 69, 70, 71, 72, 73, 74, 75, 76, 77, 78, 79, 80, 81, 82, 83, 84, 85, 86, 87, 88, 89, 90, 91, 92, 93, 94, 95, 96, 97, 98, 99, 100, 171, 173, 183, 184, 185, 186, 189, 192, 193, 196, 197, 202, 203, 204, 206, 207, 210
lysis, 186

M

M.O., 212
macrophage, 19, 20, 179, 180
macrophages, 71, 84, 96, 180
magnesium, 105
magnetic, 9, 68, 106, 116
magnetic resonance, 9, 106, 116
magnetic resonance imaging, 9
magnetic resonance spectroscopy, 106, 116
maintenance, vii, 33, 40, 46, 62, 65, 66, 70, 87
males, 74, 78, 81, 131
mammal, 33
mammalian brain, 96, 107, 116, 117
mammalian cell, 27
mammalian cells, 27
mammals, 18, 34, 36, 93, 99, 107, 108
management, 21, 24, 28, 110, 184, 185, 187, 208, 210, 213
management practices, 208
manganese, 186
manipulation, 110
mapping, 21
maritime, 139, 155
mass spectrometry, 125, 133, 134
matrix, 168
maturation, 37, 177, 178
MCC, 81
mean arterial pressure, 9
meanings, 45
measurement, ix, 123, 158, 165, 176, 215, 232
measures, 13, 48
mechanical, 13, 69
media, 11
medial prefrontal cortex, 106, 116
mediation, 27, 119
medication, 105
medicine, 92, 93, 96, 97, 98
Medline, 100
medulla, 108
membranes, 39, 40, 95
memory, 66, 70, 72, 93, 95, 96
men, 21, 25, 104, 176, 177
Mendel, 183, 212, 214
menstrual, 37
Merck, 131
mercury, 179
mesencephalon, 70
messengers, 5, 89
meta-analysis, 212
metabolic, ix, 9, 16, 20, 23, 70, 93, 95, 100, 103, 110, 113, 119, 134, 186
metabolic pathways, ix, 103, 113
metabolism, 8, 9, 15, 20, 22, 26, 32, 33, 93, 94, 97, 100, 111, 115, 116, 117, 124, 177, 178, 180, 216
metabolite, 6, 35, 40
metabolites, 10, 24, 28, 34, 35, 36, 93, 127, 133, 134, 135, 217

metabolizing, 72
methamphetamine, 121
methionine, xi, 8, 15, 28, 34, 97, 173, 183, 192, 203
methyl group, vii, 2, 5, 8, 15, 36
methyl groups, 36
methylation, vii, 2, 7, 10, 13, 15, 17, 25, 70, 124, 128, 133, 135
$MgSO_4$, 142, 157
mice, viii, 6, 15, 31, 32, 37, 38, 71, 74, 78, 79, 80, 81, 93, 96, 99, 108, 109, 110, 111, 112, 113, 118, 119, 121, 122
microbes, 93, 95, 200, 212, 214
microbial, 22, 166, 167, 186, 191, 196, 197, 198, 199, 200, 201, 203, 205, 207, 208, 209, 210, 211, 212, 213, 214, 216, 217
microbial cells, 186
microbial communities, 208
microbial community, 186
microclimate, 205
microdialysis, 135
microflora, 217
microorganism, 141
microorganisms, 152, 156, 184, 185, 186, 201, 202, 205, 214, 217
microsomes, 97
midbrain, 110
middle-aged, 2
migration, 76, 110, 132
mineralization, 184, 196, 199, 200, 212
mineralized, 205
minerals, 213
Ministry of Education, 210
mitochondria, 35, 36
mitochondrial, 175, 179
mitogen, 72
MK-80, 106, 122
models, 14, 41, 70, 90, 114
modulation, 20, 71, 97, 98, 100, 113, 120
moisture, 190, 199
moisture content, 199
molecular mass, 73
molecular mechanisms, 39
molecular structure, 39
molecular weight, 172, 174, 175
molecules, 70, 89, 134, 226
monkeys, 10, 11, 18, 20, 27
monoclonal, 22, 100
monoclonal antibodies, 100
monoclonal antibody, 22
monocytes, 176
monograph, 32
monolithic, 131, 136
mood, 104, 115
mood disorder, 115
morbidity, 3
morphological, 32, 73
mortality, 2, 3, 9, 21, 209
motor activity, 56, 58
mountains, 184
mouse, 39, 97, 108, 109, 113, 118, 119, 125, 128, 178, 179, 180
movement, 54, 55, 56, 57, 58
mRNA, 6, 11, 14, 20, 107, 110, 117, 118
MRS, 106, 116
multiple sclerosis, 25, 128
muscle, 23
muscle cells, 23
muscles, vii
mutant, 119
mutation, 140
mycelium, 209, 211
mycorrhiza, 200, 210, 211, 212
myelin, 8, 125
myelin basic protein, 8, 125
myoglobin, 174
myo-inositol, 106
myosin, 34

N

Na^+, 35, 108, 117, 118, 174
NAA, 106
N-acety, 106, 179
NaCl, ix, x, 74, 139, 142, 143, 144, 147, 149, 150, 155, 157, 158, 159, 160, 161, 162, 163, 164, 165, 166, 167, 172, 221, 222
naphthalene, 137
National Institute of Neurological Disorders and Stroke, 1
National Institutes of Health, 1
natural, x, 4, 20, 34, 96, 171, 180, 200, 209, 217
natural killer cell, 180
natural laws, 96
needles, 206
neocortex, 108
nephrotoxic, 111
nephrotoxicity, 111, 120
nerve, 27, 37, 89
nerves, 20
nervous system, 86, 89, 104, 115

neural tissue, 136
neuralgia, 73
neurochemistry, 33, 89, 96
neuroendocrine, 38, 95, 96, 115
neuroglia, 36
neurokinin, 69
neurological deficit, 3, 26, 27
neurological disorder, 35
neuromodulator, 35, 40, 116
neuronal apoptosis, 41
neuronal excitability, 40, 108
neuronal migration, 104, 110, 119
neuronal survival, 104
neurons, ix, 32, 36, 40, 69, 103, 106, 107, 111, 114, 116, 117, 180
neuropathology, 114
neuropeptide, 89
neuropeptides, 89
neuropsychiatric disorders, 114
neuroscience, 96
neurotoxicity, 116, 152, 178
neurotransmission, ix, 94, 103, 104, 105, 111, 112, 116, 117, 119, 120
neurotransmitter, 33, 35, 37, 38, 41, 68, 69, 70, 86, 89, 90, 91, 104, 115, 121
neurotransmitters, 89, 106, 135
neutrophil, viii, 31, 32, 33, 71, 74, 81, 82, 83, 84, 85, 86, 87, 88, 91, 93, 180
neutrophils, 81, 82, 84, 88, 179
Nevada, 217
nimodipine, 18
nitrate, 10, 22, 26, 127, 159, 172, 189, 193, 202, 212, 214, 217
nitrates, xi, 183, 205
nitric oxide(NO), vii, ix, , vii, 1, 4, 5, 16, 18, 19, 20, 21, 22, 23, 24, 25, 26, 27, 28, 123, 124, 132, 134, 135
nitric oxide synthase, vii, 1, 18, 19, 20, 21, 22, 24, 25, 27, 28, 124, 132
nitrite, 18
nitrogen, vii, xi, 1, 6, 34, 106, 176, 183, 184, 187, 190, 193, 194, 195, 196, 197, 198, 199, 201, 202, 206, 210, 211, 212, 213, 214, 215, 216, 217
nitrosative stress, 25, 175
NMDA, ix, 69, 103, 104, 105, 106, 107, 108, 109, 110, 111, 112, 113, 114, 115, 116, 120
NMDA receptors, ix, 103, 104, 105, 106, 107, 109, 110, 111, 113, 116, 120
N-methyl-D-aspartate, ix, 103, 104, 116, 117, 119
N-mineralization, 210
NMR, 115, 142, 152, 158, 166, 168
NO synthase, 134
nociception, 69, 98
nociceptive, 42, 67, 68, 69, 98
noise, 53
non-invasive, 9
nonparametric, 95, 192
norepinephrine, 38, 68, 89, 99
normal, x, xi, 11, 14, 18, 20, 23, 25, 37, 38, 99, 106, 108, 109, 119, 144, 171, 172, 175, 176, 178, 219, 221, 222, 223, 224, 225, 232
normal distribution, 119
normalization, 38
Norway spruce, xi, 183, 184, 186, 187, 189, 206, 209, 210, 212, 214, 215, 216
NOS, vii, viii, 1, 2, 5, 6, 7, 8, 9, 10, 13, 14, 15, 24, 124, 125
novelty, 70
N-terminal, 8, 36, 97
nuclear, 7, 19, 25, 34, 36, 68, 70, 73, 77, 86, 87, 89, 125
nuclei, 25, 32, 35, 36, 68, 89, 108
nucleotide sequence, 158
nucleotides, 160
nucleus, 38, 68, 108
nucleus accumbens, 108
nutrient, 40, 99, 142, 157, 158, 159, 162, 164, 165, 185, 211, 213
nutrient cycling, 185
nutrients, 185, 211, 213
nutrition, xi, 183, 184, 185, 197, 200, 201, 202, 208, 209, 210, 211, 214
nylon, 192

O

object recognition, 122
obligate, 153, 168
observations, 6, 10, 14, 25, 104, 177, 206
ODS, 130
oligomerization, 27
omission, 108
online, 100, 135
on-line, 127, 130, 136
oocyte, 175, 176
oocytes, 178
open field test, 58
opioid, 76, 87
optical, 81, 232
optical density, 81, 232

optimization, 166
oral, 14, 25, 26, 28, 34, 73, 110, 111
organ, 74, 75, 76, 77, 78
organic, xi, 153, 156, 179, 184, 185, 186, 191, 192, 194, 196, 200, 202, 203, 205, 206, 209, 210, 211, 213, 214, 215, 216, 217
organic compounds, 200, 213
organic matter, 185, 200, 205, 211
organism, 33, 73, 88
organization, 67, 96, 99, 116, 179
orientation, 187
ornithine, 127, 131
osmotic, v, x, 140, 153, 155, 156, 158, 160, 161, 162, 163, 164, 165, 166, 167, 168, 169
osmotic challenges, 167
osmotic pressure, 156
overgrazing, 184
oviduct, 178
oxidants, 176
oxidation, 5, 11, 12, 23, 35, 97, 108, 186
oxidation products, 5, 11, 12, 23
oxidative, 11, 27, 72, 73, 111, 175, 179
oxidative damage, 179
oxidative stress, 11, 27, 73, 111
oxide, 5, 18, 19, 20, 21, 23, 24, 25, 27, 28, 132
oxides, 23
oxygen, viii, 9, 32, 33, 74, 81, 82, 85, 86, 87, 88, 90, 91, 100, 175, 179, 199
oxygen consumption, 9, 179
oxygenation, 40, 69
oxyhemoglobin, 4, 25

P

pain, viii, 31, 32, 33, 42, 43, 44, 46, 47, 48, 49, 50, 51, 67, 68, 69, 90, 92, 94, 96, 98, 100
paper, 3, 141, 192
parabolic, 48, 67, 90
paracrine, 108
paralysis, 3
parameter, 216, 220, 232
parenteral, 33, 37
particles, 81, 200
particulate matter, 185
pastures, 184
pathogenesis, ix, 12, 70, 96, 103
pathology, 179
pathophysiological, vii, viii, 1, 2, 5, 17, 108, 113, 114
pathophysiology, ix, 2, 17, 103, 104, 105, 113, 114, 124
pathways, 5, 14, 34, 69, 100, 140
patients, ix, xi, 2, 3, 4, 7, 9, 10, 12, 13, 14, 15, 16, 17, 19, 20, 21, 22, 23, 24, 25, 26, 27, 28, 29, 38, 71, 73, 103, 104, 106, 109, 110, 112, 115, 116, 119, 120, 126, 127, 128, 132, 134, 135, 219, 221, 222, 223, 224, 225
PBMC, 99
PCP, 104, 105, 106, 110, 112
PCR, 141, 157, 160
peat, 208
pentylenetetrazol, 38, 39, 93
peptide, 29, 99, 128, 136, 172
peptide bonds, 172
peptides, 70, 89, 93, 96, 100, 136, 185, 200, 208, 211
percentile, 131
perception, 50, 51, 67
perceptions, 67, 104
performance, 62, 135, 136, 137, 158
perfusion, 5, 9, 22
peripheral blood, 72
peritoneal, 128, 180
permeability, 39, 175
permit, 46
peroxisomes, 35
PET, 28
pH, x, 96, 126, 128, 129, 130, 131, 140, 141, 142, 144, 147, 149, 150, 151, 157, 158, 159, 172, 190, 191, 209, 212, 214, 217, 221, 222, 223, 224, 225, 228, 229, 230, 231, 232
pH values, 150, 151, 212
phagocytic, viii, ix, 31, 32, 33, 71, 74, 81, 82, 83, 84, 85, 86, 87, 88, 90, 91
phagocytosis, 81
pharmaceutical, 13
pharmacokinetics, 111
pharmacological, viii, 2, 16, 86, 96, 110, 113, 120
pharmacology, 95, 98, 118
pharmacotherapies, 114
phencyclidine, 104, 106, 114
phenol, 167
phenolic acids, 214
phenotypic, 159, 168
phenylalanine, xi, 183, 185, 192, 196, 205, 207, 208
phosphate, 129, 142, 158, 173, 174, 177, 215
phospholipids, 179
phosphorus, 187, 206
phosphorylation, 7

phototrophic, 156, 167
phylogenetic, 157, 160, 168
phylogenetic tree, 168
physical properties, 212
physiological, vii, ix, 1, 2, 5, 7, 9, 18, 19, 33, 34, 35, 36, 86, 91, 93, 95, 96, 108, 113, 118, 123, 157, 159
physiology, 97, 156, 179, 212
pig, 38, 92
pigments, 16, 33
pigs, 9, 38, 68, 73, 98, 99, 134
pituitary, 37, 89, 97
placebo, 13, 18, 25, 27, 37, 104, 112, 122
plants, xi, 34, 35, 184, 185, 189, 197, 200, 202, 205, 207, 209, 210, 211, 212, 213, 214, 215, 216, 217
plasma, ix, x, xi, xii, 4, 8, 9, 13, 14, 15, 16, 17, 18, 19, 20, 23, 24, 25, 26, 27, 28, 29, 35, 38, 40, 71, 92, 97, 99, 111, 115, 120, 123, 124, 125, 126, 127, 128, 129, 130, 131, 132, 133, 134, 135, 136, 137, 171, 172, 173, 174, 175, 176, 177, 219, 220, 221, 222, 223, 224, 225, 226, 227, 228, 229, 230, 231, 232
plasma levels, 13, 14, 15, 20, 23, 25, 27, 127
plasma proteins, 175
plasmids, 141
plasminogen, 181
plastic, 189
plasticity, 98
platelet, 178
platelets, 4
play, 9, 11, 15, 35, 68, 89, 110, 185, 200
PMI, 106
polyethylene, 192
polymorphism, 6
polymorphisms, 22, 121
pools, 128, 186, 201
poor, 37, 145, 150
population, 2, 72, 94, 104, 109, 118, 176
positron emission tomography, 115
postmenopausal women, 12, 25, 27
postmortem, 106, 110, 119
postsynaptic, 39, 106, 107
post-translational, vii, 2, 7, 15
potassium, 18, 41, 142, 158, 180, 187, 220
potassium channels, 18, 41
power, 53
PPI, 111
precipitation, 128, 187
preclinical, 9
predictors, 4
preeclampsia, vii, 1, 7, 18, 20, 23, 25, 26
preference, 212
prefrontal cortex, 70
pregnancy, 18, 20, 126
pregnant, 26, 126
pregnant women, 26
premature infant, 29
preparation, 38, 40, 111, 120, 132, 148, 149, 151, 190
pressure, 184, 199
presynaptic, 40, 106, 108
prevention, viii, 2, 25, 27
primate, 6, 10, 14, 16, 20, 21, 25
primates, 6, 10, 13, 14, 23, 25, 27
privatization, 184
probe, 142
procedures, 189, 191, 199, 227, 231
production, vii, 2, 5, 6, 8, 9, 10, 13, 14, 15, 17, 18, 22, 71, 72, 93, 108, 113, 116, 124, 134, 140, 153, 156, 157, 159, 166, 167, 185, 186, 187, 205, 209, 213, 215, 216
productivity, 166, 185
prognosis, 71
prognostic value, 12
program, 53, 62, 129, 185, 192
progressive, 50, 168
prokaryotic, 156
proliferation, 37, 41, 71, 72, 89, 91
prolyl endopeptidase, 136
promote, 86, 107, 112, 178
promoter, 119, 121
promoter region, 119, 121
property, 72
prophylaxis, 95
propionic acid, 106
proteases, 172, 209, 220
protection, 96, 140
protein, vii, 2, 5, 6, 7, 8, 10, 11, 12, 14, 16, 17, 19, 21, 23, 24, 27, 32, 40, 70, 92, 100, 108, 109, 110, 112, 117, 119, 128, 135, 156, 174, 175, 176, 177, 179, 208
protein function, 7, 156
protein hydrolysates, 135
protein kinases, 6
proteins, vii, ix, 7, 14, 17, 32, 33, 34, 36, 37, 40, 41, 70, 72, 73, 89, 90, 94, 97, 112, 124, 133, 136, 140, 156, 172, 174, 176, 185, 200, 205, 208, 209, 210, 211, 214
proteinuria, 120
proteolysis, 140, 152

proteolytic enzyme, 172, 208
prothrombin, 232
pro-thrombotic, 232
protocol, 41
pseudomonas, 153
psychiatric disorder, 121
psychiatric disorders, 121
psychopharmacological, 105
psychopharmacology, 97
psychoses, 115
psychosis, 104, 113, 114, 115, 121
psychosocial, 95
psychosocial stress, 95
psychotic, 120
public, 37
pulmonary hypertension, 19, 25
pulse, 9
purification, 157
putrescine, 127
PVC, 189, 199
pyramidal, 32, 36
pyramidal cells, 32
pyridoxal, 215
pyrimidine, 140, 156, 167

Q

quality control, 177
quantum yields, 128
quarantine, 41

R

racial groups, 2
radical, 20
radioactive isotopes, 185
radiological, 18
random, 191
range, x, xi, 3, 8, 9, 33, 37, 78, 86, 90, 104, 125, 155, 157, 159, 205, 207, 208, 209, 219, 221, 222
RAS, 94
rat, 9, 16, 17, 21, 23, 24, 27, 35, 36, 37, 40, 41, 43, 45, 48, 50, 51, 52, 53, 54, 56, 58, 59, 62, 63, 65, 71, 75, 76, 77, 78, 79, 81, 86, 87, 90, 94, 97, 99, 107, 108, 110, 112, 116, 117, 120, 125, 127, 130, 131, 132, 135, 136, 137, 180
rats, viii, 6, 9, 13, 28, 31, 32, 34, 36, 37, 38, 40, 41, 42, 43, 44, 46, 47, 48, 49, 50, 51, 53, 56, 57, 58, 59, 60, 61, 62, 63, 65, 66, 67, 68, 70, 74, 75, 76, 77, 81, 82, 83, 86, 87, 88, 90, 91, 92, 93, 95, 97, 98, 99, 111, 112, 122, 127, 180, 181
reaction time, 62, 63, 65, 66
reactive nitrogen, 23, 25
reactive oxygen, 11, 176, 179
reactive oxygen species (ROS), 11, 176, 179
reactivity, ix, 32, 49, 71, 73, 74, 90, 128
reading, 104
reagent, ix, 123, 126, 127, 128, 131, 135, 136, 143, 159, 221
reagents, 126, 128
recall, 105
reception, 33
receptor agonist, 38, 39, 40, 68, 121
receptors, ix, 38, 39, 40, 68, 69, 70, 87, 89, 94, 97, 98, 99, 100, 103, 104, 105, 106, 107, 110, 111, 113, 117, 121
recognition, 40
recovery, 128
red blood cell, 4, 12
red blood cells, 4, 12
redox, 175, 177
reduction, 6, 11, 15, 18, 21, 48, 49, 50, 81, 159, 217
refractory, 210
regional, 22, 116, 117
regular, 184
regulation, 7, 10, 14, 17, 20, 23, 27, 35, 72, 116, 124, 133
regulators, 33
reinforcement, viii, 31
relationship, 3, 97, 180, 208
relationships, 93, 119
relaxation, 6, 11, 19, 21
relevance, 28, 121
reliability, 132
renal, 8, 9, 22, 23, 97, 125, 132, 133
renal disease, 97
renal failure, 9, 132
repair, vii
repression, 97
reproduction, 63, 70, 72, 81, 87, 98
research, viii, xi, 2, 3, 73, 76, 86, 87, 89, 183, 184, 187, 189
researchers, 73
reserves, 81
residues, 7, 11, 14, 15, 17, 36, 70, 72, 97, 124, 185
resin, 127
resistance, 9, 83, 88, 90
resolution, 3, 10, 127, 221
respiration, 156, 185, 209, 213, 214, 216, 217

respiratory, 167
restoration, 214
retardation, 34, 35, 104
Rhizobium, 156
rhizosphere, 214, 216
ribonucleic acid, 97, 168
ribosomal, 168
ribosomes, 36
rice, 211
rings, 9
risk, 2, 12, 17, 18, 24, 25, 27, 118, 127, 134
risk factors, 24, 118, 134
RNA, 36
rodent, 106, 118
rodents, 92, 96
room temperature, xi, 126, 130, 219, 220, 221
ROS, 176
rotations, 49, 81
Royal Society, 130, 131
RP-HPLC, 134

S

Saccharomyces cerevisiae, 97
saline, viii, 31, 41, 43, 48, 53, 62, 74, 78, 79, 81
salinity, 140, 142, 143, 149
saliva, 6
salmonella, 167
salt, 156, 166, 168, 220, 221
salts, 156
sample, ix, 107, 119, 123, 126, 130, 172, 175, 191, 192, 220, 221, 223, 232
sampling, 189, 191
schizophrenia, v, ix, 103, 104, 105, 106, 109, 110, 111, 112, 113, 114, 115, 116, 117, 118, 119, 120, 121, 122
schizophrenic patients, 115, 116, 119
Schmid, 115
science, 96, 97, 98
scientific, 89
Scots pine, 211, 213
SDS, 172, 175, 176
secretion, 37, 38, 68, 89, 176
seedlings, 189, 209, 211, 212, 213, 214, 217
seizure, 38, 40
seizures, 38, 40, 93, 108, 118
selectivity, 118
selenium, 179
semen, 172, 173, 174, 176, 177, 178, 179
seminal vesicle, 176
semi-natural, 213
senescence, 97
sensitivity, viii, 31, 33, 40, 42, 43, 44, 45, 48, 49, 67, 84, 88, 90, 93, 98, 129, 130, 132, 168
sensitization, 69
separation, 127, 128, 129, 130, 131, 136, 193, 214
sepsis, 233
sequencing, x, 155, 157
series, 38, 43, 72, 74, 111
serine, ix, x, xi, 103, 106, 107, 108, 109, 110, 111, 112, 113, 116, 117, 118, 119, 120, 171, 173, 183, 185, 192, 200, 215, 220
serotonergic, 96
serotonin, 16, 32, 38, 68, 69, 70, 87, 99, 104, 121
serum, ix, 10, 13, 16, 18, 28, 71, 81, 103, 109, 119, 123, 125
services, iv
shear, 11, 14, 24
sheep, 180
shock, 41, 50, 51, 68, 98, 153, 158
shocks, 37, 74
shoot, 205
short period, 207
shortage, 184
short-term, 70, 178, 209, 214, 215, 216
shrubs, 189, 209
shuttles, 175
sickle cell, 28
side effects, 112
sign, 10, 42, 44, 50, 93
signal transduction, 114
signaling, 20, 89, 116, 132, 178
signals, 175
signal-to-noise ratio, 127, 129
signs, 48
silica, 131, 136
silver, 172, 175, 176
similarity, 160, 161
simulation, 41
single nucleotide polymorphism, 109, 118
sites, ix, xi, 39, 40, 48, 73, 103, 108, 110, 112, 113, 120, 183
skeletal muscle, 36
skeleton, 128
skin, 37, 41, 67, 69, 73, 98, 99, 100, 156
skin conductance, 37
small vessel disease, 21
smoking, 4
smooth muscle, 6, 11, 19, 29
smooth muscle cells, 11

social, 112
socioeconomic, 184
sodium, 6, 129, 131, 136, 192
software, 157, 168
soil, x, 156, 183, 184, 185, 186, 187, 189, 190, 191, 194, 196, 199, 200, 201, 202, 203, 205, 207, 208, 209, 210, 212, 213, 214, 215, 216, 217
soil particles, 186
soils, xi, 183, 185, 186, 207, 208, 209, 211, 213, 214, 215
solar, x, 155, 156, 157
solid phase, 127, 129, 130, 136
solutions, 215, 221
solvent, 192
somatostatin, 69, 89, 99
soy, 97
spastic, 6, 15, 20, 21, 27
species, 9, 25, 32, 68, 94, 157, 159, 160, 165, 166, 168, 179, 184, 185, 200, 205, 207, 209, 213, 215, 217
specificity, 8, 19, 24, 27
spectra, 142, 166
spectrophotometry, 18, 192
spectroscopy, 158, 166
spectrum, vii, 1, 5, 73, 76, 166
speech, 35, 37, 104
sperm, x, 171, 172, 173, 175, 176, 177, 178, 179, 180, 181
sperm function, 177
spermatogenesis, 181
spermatozoa, 172, 175, 176, 178, 179
spermidine, 127
spermine, 127
spinal cord, 34, 69, 90, 108, 176
spleen, 37, 74, 75, 76, 77, 78, 86, 132
stability, xi, 128, 130, 132, 152, 219, 228
stabilization, xii, 140, 209, 220, 223
stabilize, 175
stages, 37, 211
standard deviation, xi, 172, 219, 220, 222, 223
standards, 131, 175, 200
statin, 24, 27
statins, 13
sterile, 74
stiffness, 9, 22
stimulus, viii, 31, 67
storage, xi, 178, 192, 219, 222, 229, 230, 231
strain, x, 70, 139, 140, 141, 143, 144, 145, 146, 150, 153, 155, 156, 158, 159, 160, 161, 162, 163, 164, 165, 166
strains, 156, 157, 159, 160, 165, 168
strategies, ix, 32, 186
stratification, 205
streptomyces, x, 139, 141, 153
stress, viii, 11, 14, 24, 31, 32, 33, 37, 38, 67, 70, 74, 76, 77, 78, 79, 80, 81, 82, 83, 84, 85, 86, 87, 88, 89, 90, 91, 93, 94, 96, 99, 140, 156, 166, 167
stress factors, 140
striatum, 68, 108
stroke, 3, 7, 9, 26, 28, 29, 127, 134
subarachnoid hemorrhage, v, viii, 1, 2, 16, 17, 18, 19, 20, 21, 22, 23, 24, 25, 26, 27, 28, 29
subcutaneous injection, 68
substances, vii, 4, 11, 96, 98, 105, 106, 210
substitution, 160
substrates, 8, 216
sucrose, 145, 150
sugars, 145, 159, 213, 214
sulfate, 136
sulfur, 156, 186, 210
summer, xi, 72, 184, 202
supernatant, 125, 158, 166
superoxide, 21, 179
superoxide dismutase, 21
supplemental, 38
supply, xi, 140, 150, 184, 189, 200, 213
suppression, 72, 79, 87, 93
suppressor, 96, 180
suppressors, 76, 87
surgery, 21, 29
survival, 108, 152, 153, 168, 175, 179
susceptibility, 22
swelling, 175, 178
switching, 130, 131, 136
symbiosis, 200, 212, 216, 217
symbiotic, 208
symbols, 147, 162
sympathetic, 37, 89
symptoms, ix, 3, 73, 103, 104, 109, 110, 111, 112, 120
synapses, 107
syndrome, 35, 178
synthesis, ix, 5, 7, 9, 14, 15, 16, 17, 18, 23, 25, 28, 33, 34, 36, 38, 72, 73, 89, 106, 108, 117, 123, 135, 140, 143, 144, 153, 156, 164, 165, 166, 167, 178
synthetic, ix, 103, 110, 113, 140, 143, 150
Syria, 99
systematic, 20, 25, 28
systematic review, 20, 25, 28

systemic circulation, 110
systems, viii, 31, 32, 33, 36, 41, 68, 69, 70, 73, 86, 89, 90, 91, 98, 104, 105, 133, 184

T

T cell, 179, 180
T cells, 179
T lymphocytes, 94
taiga, 214
tail-flick, viii, 31
tandem mass spectrometry, 133, 136
taxa, 160
T-cells, 72
teens, 104
telephone, 217
temperature, 41, 42, 142, 143, 144, 185, 187, 190, 192, 213
temporal, 179
tendons, vii
tension, 48, 49, 50
terminals, 89, 108
tetracycline, 157, 159
tetrahydrofuran, 126
thalamus, 40, 97, 106, 108
thawing, xi, 219, 220, 221, 222, 223, 224, 226, 228
theoretical, 33
therapeutic, ix, 13, 15, 19, 103, 110, 111, 112, 113, 114, 120
therapeutic agents, 120
therapeutic approaches, 13, 15
therapeutic targets, ix, 103, 110, 120
therapeutics, 140
therapy, xii, 12, 16, 25, 26, 27, 73, 94, 95, 220
thermal, 42, 67, 68, 69, 90
thermostability, 152, 168
thiobarbituric acid, 177
thioredoxin, 167
threatening, 68, 220
threonine, x, xi, 97, 171, 173, 183, 185, 192, 203, 207
threshold, 42, 43, 47, 48, 49, 50, 51, 63, 69
thresholds, 45, 46, 47, 48, 49, 50, 51
thrombin, x, xi, 171, 172, 177, 219, 220, 221, 222, 223, 224, 225, 226, 227, 228, 229, 230, 231, 232
thrombosis, 177
thrombotic, 19
thymidine, 72
thymus, 7, 92
timber, 185
tissue, 15, 22, 32, 36, 37, 71, 72, 95, 97, 128, 132, 172, 176, 177, 181
T-lymphocytes, 72, 73
T-maze, viii, 31, 41, 62, 66, 90
Tokyo, 22, 23, 26, 123, 130, 142, 143, 157, 158, 159
tolerance, 93, 140, 168
tonic, 38
topsoil, 203, 210
toxic, 72, 175
toxicity, 180
toxicology, 95
toxin, 22
trade, 185
training, 96
trait anxiety, 37, 95
transcription, 17, 70, 141
transcription factors, 70
transfer, 215
transferrin, 173
transformation, 192, 212
transformations, 215
transfusion, 178
transgenic, 15, 140, 153
transgenic mice, 15
transglutaminase, 175
transient ischemic attack, 9
transition, 175
translocation, 36, 208, 211
transmission, 40, 69, 109
transplantation, 71
transport, xi, 35, 36, 72, 94, 108, 177, 219, 220
transportation, 38, 99, 124
trauma, 71, 100
treatable, 3
trees, xi, 158, 168, 183, 184, 185, 187, 189, 191, 206, 207, 208, 210, 213
trial, 19, 27, 37, 43, 45, 46, 48, 50, 62, 105, 112
Trichoderma viride, 96
tritium, 72
Trojan horse, 178
tuberculosis, 94
tumor, 73, 100, 181
tumor cells, 100, 181
tumorigenic, viii, 2
tumours, 73
tundra, 184, 212, 214
turgor, 167
turnover, 11, 199, 216
type 2 diabetes, ix, 16, 24, 123, 125, 134
type 2 diabetes mellitus, 16

tyrosine, x, xi, 171, 173, 183, 185, 192, 196, 205, 208

U

ubiquitin, 70
ultrastructure, 213
uniform, 187
urea, 14, 38, 99, 176
uric acid, 176
urinary, 14, 127, 133
urine, vii, ix, 2, 21, 92, 123, 127, 133, 134, 180
USDA, 187, 189, 214, 217
UV, 156, 158
UV radiation, 156

V

vaccination, 94
vacuum, 18
valine, xi, 34, 183, 185, 192
values, 35, 38, 42, 43, 44, 45, 46, 47, 48, 49, 50, 51, 52, 53, 54, 55, 56, 57, 58, 59, 60, 61, 63, 64, 65, 66, 71, 76, 77, 78, 79, 81, 82, 84, 85, 86, 87, 88, 128, 129, 143, 144, 145, 146, 147, 148, 149, 151, 152, 160, 161, 162, 163, 164, 165, 172, 227, 232
variability, 215
variable, 7
variance, 190, 203
variation, 163, 221, 222
vascular, 4, 9, 11, 17, 19, 20, 22, 23, 24, 29, 212
vascular system, 20
vasculature, 23
vasoactive intestinal peptide, 89, 94
vasoconstriction, 4, 11, 16
vasoconstrictor, 4, 29
vasodilatation, 27
vasodilation, 6, 24, 127
vasodilator, 4, 5, 9, 22, 41
vasoneuronal coupling, 28
vasopressin, 37
vasospasm, viii, 1, 2, 3, 4, 5, 6, 7, 9, 10, 11, 12, 13, 14, 15, 16, 17, 18, 19, 20, 21, 22, 23, 24, 25, 26, 27, 28
vector, 141
vegetation, xi, 183, 184, 189, 193, 196, 197, 198, 200, 201, 202, 203, 204, 205, 209, 210
vein, 11, 20
velocity, 177
vessels, 2, 6, 9, 11, 17, 19, 109
virus, 72, 73, 92, 99
virus replication, 72
vision, 165
visual, 51, 67
vitamin B1, 8, 15
vitamin B12, 8, 15
vitamins, 97
vocalizations, 49
vocational, 112
vulnerability, 109

W

water, vii, 34, 41, 73, 158, 161, 162, 165, 167, 184, 185, 190, 192, 206, 212
wavelengths, 128
wells, 221
wet, 132, 217
wetting, 209
wheat, 34, 37
white blood cells, 4, 180
white matter, 108
windows, 40
winter, 72, 189
Wistar rats, viii, 31, 41, 74
withdrawal, 104
women, 23, 25, 37, 104
wood, 215
working memory, 105

X

xenobiotics, 72

Y

yeast, 117, 142, 157
yield, v, x, 140, 143, 144, 150, 153, 155, 156, 160, 161, 162, 163, 164, 165, 169
young adults, 20

Z

Zea mays, 185, 214
zinc, 73, 98, 105, 180
zinc oxide, 73
zygotes, 178

B

β-amyloid, 152